MODEL ENGINEERING

A FOUNDATION COURSE

Peter Wright

SPECIAL INTEREST MODEL BOOKS

SPECIAL INTEREST MODEL BOOKS
Special Interest Model Books Ltd.
P.O. Box 327
Poole
Dorset
BH15 2RG

First published by Nexus Special Interests 1997

Reprinted 2003

This edition published by Special Interest Model Books 2005

Reprinted 2007, 2009, 2010, 2013

ISBN 978-1-85486-152-8

Typesetting by Kate Williams, London
Printed and bound in Malta by Melita Press

Contents

Preface

Judging by the comments and questions one overhears at exhibitions and the like, or indeed the questions which one is personally asked at club open days, there are many who would dearly like to participate in the hobby of model engineering but who do not in the least understand how to start.

Many of us who are now practising model engineers were fortunate in our younger days to have had the opportunity to follow what would nowadays be termed a craft apprenticeship, or perhaps, as in my case to have followed a sandwich course which included a worthwhile introduction to the processes of mechanical manufacture at a time when there was still great reliance on the manual skills of individual craftsmen.

This seems not to be the case these days, as more and more the craft apprenticeship is becoming a short-term education programme which seems incapable of producing engineers with a wide range of skills who somehow know how the system of making things has to work in order for it to be successful.

This is obviously not true of all industries since the assembly of aircraft, for example, is still largely a manual trade, but the introduction of numerically controlled machine tools has revolutionised the manufacture of the component parts and in the field of small batch production, or the realm of development engineering which is my background, operators of these computer-controlled machines are gradually taking over from the time-served men who were their forerunners. Consequently, people with the wide range of skills which operators in development workshops used to have are becoming a thing of the past.

These wide-ranging skills are unfortunately just those that are needed by the amateur worker, who must of necessity be a Jack of all Trades, since he will need to carry out most of the processes of manufacture himself (but usually not all of them) given only the modest workshop facilities which his hard-earned money will buy. It is to provide the essential general introduction to these processes that this book has been written. It provides an introduction to the techniques which may be used in an amateur workshop to perform tasks for which more complex, specialised (and more expensive) equipment would be employed in industry.

As a substitute for this more expensive machinery, the amateur usually develops manual skills, and not a little ingenuity, to enable him to create in the home environment a miniature locomotive or a full-size long-case clock, or whatever, and it is hoped that this book will encourage the development of such skills.

Many of the workshop operations described and photographed are actual 'work in progress'. This means that the workshop is not always tidy and the photographic backgrounds are sometimes more cluttered than is desirable. I have also learned as I have progressed, and I hope the photographs (in particular) show an improvement, although it may not be obvious to you since the illustrations do not appear in chronological order.

My own interests and facilities naturally limit what can be shown, but I have been fortunate to have access to other people's workshops and tools, and my thanks go to them for putting up with the endless questions, "Have you got…?" and "Can I come over and photograph it?".

My thanks go especially to my father-in-law, Peter Feast, not only for his extensive collection of hand tools and long experience of their use, but also for being available when experienced hands are needed for the photographs. My grateful thanks go also to his daughter for her support and encouragement, for her willingness to help, particularly as photographer's assistant, holding up backgrounds or reflectors, or firing the shutter when my hands are holding the tools. She has also learned uncomplainingly, over the years, the true meanings of the phrases "I'll just go and process a few words" and "I'm going to have ten minutes in the workshop".

Before the book itself, a word of encouragement. Do not think, that because you have not previously been involved in this fascinating hobby you will not be successful. Do not think, having visited a local or national handicrafts exhibition and admired those superb examples of the craft, that you must desist because you will not be able to produce such fine models. Very few of us can! Certainly not at the first attempt. All that matters is that you should enjoy the work which you do.

If, at the end of the day, you do not feel that your results should be displayed for others to see, there is no compulsion for them to be displayed. In any event, you are much more likely to feel reasonably proud even of your first efforts, for to complete a large project is itself a pleasure and a triumph, and what does it matter if, at the end you feel that you could have done it better? It will be better next time, and that's when even greater pleasure will come.

If the book encourages some of you to make a start when you have previously only dreamed of owning some hand-made mechanism, the efforts of its preparation will be amply rewarded.

Peter Wright
March 1997

This book is dedicated to the memory of JAB

Although not immediately interested in the subject matter,
he was always interested in the products of the workshop
and full of enthusiasm for the completion of this book

1

Introduction

What is model engineering?

Before commencing, it would be as well to define what is meant by 'model engineering', for a definition is not easily found. Looking through the list of competition classes at the International Model Show provides no real guide since the classes adopted for these competitions reflect the wide range of interests which modellers have, rather than assisting in forming a definition of what model engineering actually is. Thus, one sees on display large-scale model steam locomotives and road vehicles, internal combustion engines, model boats and aircraft, alongside which are to be found scenic model railways, farm carts, model soldiers, model cars and radio-controlled everything. So a true definition of model engineering is not to be found at the site of our annual pilgrimage.

The term itself is so ingrained into the language that one hesitates to suggest a change, but if that opportunity did exist I should be inclined to propose 'mechanical engineering in the home workshop'. But no-one would prefer to use that, even though it suits the purpose of this book admirably.

Some such definition is required, however, for we cannot possibly regard the construction of a clock as model engineering unless the clock produced is indeed a model of an actual mechanism, but this is seldom the case. Clock-makers prefer to make a dining-room size clock which has some practical value rather than make a model, yet their activities are catered for by the International Model Show organisers and by the supporting trade stands, just as the builder of model steam locomotives is supported. So we clearly need some alternative definition.

The difficulty in defining the term arises from the very diverse nature of the items which are produced, and from the fact that some branches of the hobby are so well supported by the trade that bolt-together kits may be purchased. For some models, kits are sufficiently low in cost to allow volume sales to be achieved, and in some areas of interest almost all activity centres around commercially produced components. This applies particularly to electrically powered model car racing, for example, for which virtually all components can be purchased from specialist retailers.

There are, however, other branches of the hobby in which the models produced comprise so many parts that commercial production is incredibly costly, and although bolt-together kits are available, they sell in relatively small quantities, and most models are amateur-made. This applies particularly to large-scale model steam locomotives.

Poor support from the trade is also the case when certain minority interests are considered. These include the design and manufacture of hot-air engines and the building of scale model cars for static display. These aspects are not yet sufficiently popular to encourage trade support although hot-air engine design details are published from time to time in the model press, encouraged by the annual competition for these engines at the International Model Show.

The models in the competition classes at the exhibition therefore embrace the whole range of modelling skills, from the complete design and manufacture of every part of a large-scale steam locomotive (including the production of patterns and castings) to assembly and detailing of a kit, either of the bolt-together type, or the type in which basic parts are provided but the builder must shape the parts, assemble and finish the model, such as in model boat or model aircraft building.

The common theme which links together many of the models seen at the exhibition is that extensive use of a machine tool has been a requirement in their production. This is not true of all of the items, for many fine models are constructed without the use of relatively expensive machinery, but for many the term model engineering implies the use of such a machine and for a large part, the use of machine tools forms the subject of this book.

As far as possible, this book has been filled, not with pictures of models or a great deal of theory, but with practical information related to the making of things. To some extent, the information provided attempts to cover subjects with which I myself found some difficulty in the early days of establishing my modelling workshop, having only a rather hazy knowledge of some of the techniques employed.

The making of things requires skill. Unfortunately, skills cannot be learned from a book, only by experience. It is therefore important that you have plenty of practice, but this must be based on sound techniques, and I hope that the basics have been covered here and in sufficient detail for you to approach the practice with confidence. I also hope that there is indeed sufficient information within this book to practise with confidence, even if, at first, the practice may not be perfect.

In order to build a model you do not need to have the capability to design it. Indeed, designing models is equally as difficult as designing the prototype and is best left until you at least have some experience as a builder. Building is therefore started by the purchase of the relevant drawings, followed by the materials and castings, if these are required. Once these basics are available, building may start. Sources of information and drawings, and advice on the purchase of an essential stock of material, is given in Chapter 2, while Chapter 3 provides an introduction to the materials used in the manufacture of models, and to the subject of heat treatment.

Chapter 4 describes the requirements for the workshop itself, and its possible locations. The ways in which a workshop might be fitted out are described, together with the requirements for lighting and the need for magnifying viewers. The general types of small hand tools are described, and the equipment needed for soldering and brazing.

Once manufacture commences, you will need measuring and hand tools for marking out, cutting and filing, the capability to drill and size holes and the machinery to carry out the two essential operations of turning and milling. The basic hand skills are covered in Chapters 6 and

7, while Chapter 5 describes the measuring equipment which is normally utilised.

Chapters 8 and 9 describe methods of joining metals, considering separately what can be described as mechanical methods (attachment by screws, riveting and pressing parts together) and other methods in which materials such as solders and adhesives are introduced into the joint.

The later chapters describe the essential machinery which is required, and introduce the basic machining processes of hole production, threading, turning and milling. Turning is carried out on a lathe, the essential features of which are described in Chapter 12. Its use, and adaptation to other processes are described in Chapters 13 to 16 and Chapter 17 contains a description of the major considerations to be borne in mind when purchasing a lathe and contains advice on how to inspect a second-hand machine and set it up.

A lathe is absolutely essential for the sort of modelling envisaged. Although it is designed for turning, it can readily be adapted for milling, provided that the work is not too large, and with the addition of a few accessories can become an extremely versatile machine. The amateur is frequently equipped only with a lathe, but this is usually supplemented by a drilling machine since much drilling is necessary, and although holes can be drilled on the lathe, it is not readily adapted for drilling holes in large plates, and this is frequently required. A brief description of the main features of a drilling machine is given at the beginning of Chapter 10. The subject of making holes is quite diverse and this occupies most of Chapter 10.

Chapter 11 contains an introduction to the more commonly used screw threads, and to methods of cutting threads using hand tools. Thus, what might be described as the basics of handwork are described in Chapters 5 to 9, while that essential machine tool, the lathe, and its use, are covered by Chapters 10 to 16.

Don't be tempted to skip the basics. If you take the trouble to acquire the basic skills, there is much pleasure to be had even from the simplest task of sawing off a piece of material. Performed carefully and skilfully, sawing squarely and close to the line, the item you are building will progress rapidly, with little frustration. If you approach sawing as a necessary evil, racing to finish it, using an unsuitable saw, you will almost certainly not saw close to the line, will break more blades and will have more filing or machining to do in the end. And the acquisition of the basic skills, which only needs to be done once, saves time over and over again.

There is never enough modelling time, so saving time is as vital as saving money. In some ways, time and money are interchangeable commodities, and money invested in a tool, or machine tool attachment, for example, may save valuable time.

Branches of the hobby

Steam locomotives

For many, model engineering simply means messing about with live steam locomotives. In lots of ways, live steam models are very satisfying things to be involved with since one gets the

Figure 1.1 My 5-inch gauge pannier tank to LBSC's *Pansy* design.

3

Figure 1.2 Steam raising during a visitors' day at the Bracknell Track.

Figure 1.3 Vic Newport and his great grandson admiring a visiting locomotive at their local track.

Figure 1.4 Cliff Irving and his outside-cylindered *Maid of Kent.*

Figure 1.5 A finely detailed backhead on a 5-inch gauge GWR *Dukedog.*

pleasure of the workshop time, which satisfies the urge to make things which most of us seem to have, but it also allows the builder to take his model to the local track and turn what were once just lumps of metal in a store into a living, breathing thing.

This branch of the hobby developed out of the early commercial model railways which were generally built to a much larger scale than is currently the case. The locomotives were also normally steam powered, and to be interested in model railways in the early part of the century meant being immersed in the technicalities of steam, spirit and coal firing and suchlike topics.

At that time, much model steam locomotive design and construction was most definitely in the toy train category. This meant that simple mechanisms and simple techniques were the order of the day and when repairs became necessary, quite simple tools and techniques were all that were required to effect a repair. Much that was needed for a model could, therefore, be hand made and it was largely for wheels and cylinders that a lathe was required. As the availability of small lathes increased, there came into the hobby of model engineering many of the model railway enthusiasts. Naturally, the lathe made the manufacture of domes, chimneys, buffers and so on that much easier and with quite simple facilities the amateur could develop his own locomotives or improve various of the commercial offerings, many of which appear to have had in-built shortcomings.

In those days, 2½in. and 3½in. gauge models were quite common commercial sizes and it was natural, therefore, that enthusiasts should be drawn to the idea of making sufficiently powerful models capable of hauling passengers. Through the efforts of writers in *Model Engineer*, notably LBSC who started his 'Live Steam' column in that magazine in 1924, basic techniques were developed for the building of boilers, motion work, cylinders and fittings for workable passenger-hauling models and by the late 1930s the hobby was well established.

As hobbyists have become more affluent over the years, particularly since 1945, models have tended to become larger. The support offered by the trade in the form of drawings, castings and general materials has increased and there is nowadays a marked trend towards larger models, made on larger machines, costing relatively more than was once the case.

Currently, the most popular scales for passenger-hauling locomotives are ¾in. per foot (19.05mm per foot) running on 3½in. gauge track, and 1⅟₁₆in. (27mm) per foot, running on 5in. gauge track. The elevated track is still the most common arrangement for passenger hauling and most modellers join a local club in order to have access to a track. Although 3½in. and 5in. gauges are the most popular, there are still club tracks that can accommodate 2½in. gauge models.

Many clubs with more spacious grounds available have established ground-level tracks which provide for 5in. and 7¼in. gauge models, these latter being built to 1½in. (38mm) per foot scale. Models designed for 7¼in. gauge are naturally more powerful than their smaller counterparts. Being larger, they are also capable of being made with almost all details to scale and yet can still be driven by 'over-scale' drivers, whereas this is not always possible in the smaller gauges if reliable steam fittings, lubricators and so on are to be made.

A further current trend is also evident in the increasing numbers of narrow-gauge steam models which are being made, influenced by the preserved lines. The advantage of a narrow-gauge locomotive is the increased size (and passenger-hauling capability) one gets for a given gauge. Thus, a narrow-gauge model designed to operate on 3½in. gauge track might represent a 2ft gauge locomotive and is therefore built to a scale of 1¾in. (44.5mm) to the foot. It is consequently more than twice the scale of a regular 3½in. gauge model and can have larger cylinders, boiler and fittings. It is easier to handle on the track, more powerful and more robust.

There is, too, a charm about narrow gauge which makes an attractive model and there is the further consideration that, particularly on 7¼in. gauge, ground-level track, one easily gets the impression that one is running a real locomotive.

At the other extreme, the last few years have seen a revival in interest in 'O' and '1' gauge live steam models. Again, this has arisen from the availability of commercial models and track in these gauges, many of the suppliers concentrating on the production of 'freelance' and narrow-gauge models having simple mechanisms and boilers and selling at reasonable prices. The availability of these models has stimulated amateurs to improve on the commercial offerings and to develop more true-to-scale models and much interesting work has been done by the members of societies such as the Gauge "O" Guild and the Gauge "1" Model Railway Association.

Figure 1.7 A double-ended, battery-powered locomotive from John Brotherton's stable.

Battery-powered locomotives

There has recently been a rapid growth of interest in battery-powered locomotives. The considerable adhesive weight provided by a 12-volt car battery, or perhaps two of them, ensures adequate haulage capacity, and if the type of 'boxy' outline provided by a diesel shunter is adopted, there is plenty of room to house a large-capacity battery, which ensures that a full

Figure 1.6 David Elen driving the Easthampstead Sewage Works' *Lady Muck.*

afternoon's passenger hauling can be undertaken without the necessity to have a spare battery available.

Some of the simplest (and cheapest) arrangements utilise a simple chain drive, power being provided by a second-hand car generator energised through a home-made electronic controller. If you are into electronics and like tinkering, this is a useful field for experimentation. However, some mechanical knowledge is also called for since skill is required in arranging a drive which allows some rise and fall of the driven axle without changing the centre-to-centre distance between the driven and driving shafts. In this respect, there is some merit in using a belt drive, which allows a little give, an internally toothed belt perhaps being ideal.

One of the advantages of the battery-powered model is the general simplicity of the design. The number of components which must be made is reduced in comparison with even the simplest model steam locomotive and there is no boiler to construct, thus removing the need for the equipment which is necessary for boiler building.

There are some advantages on track days too. There is no need for the lighting up, and afterwards the cleaning down of the dirtier parts of a steam model. A quick oil-round and connection of the battery usually suffices at the start of the session, and an equally quick wipe round with an oily rag to remove the general grime picked up from the track, together with disconnection of the battery, is all that's needed at the end of the day.

A further advantage of this type of model is the ready availability of kits of parts and ready-assembled models. The kits require finishing and painting but costly machine tools are not necessary and the simple ability to drill fixing holes, clean up, bolt together and paint is all that may be required. Provided that funds are available, the purchase and assembly of a kit can be a quick way into the hobby and certainly gets you down to the track, participating, in a short time.

Traction engines

Model traction engines are deservedly popular. They are normally coal-fired, steam models (unless built for the showcase or sideboard) and therefore provide that nostalgic coal-oil-and-steam smell which is so characteristic of the

Figure 1.9   An award-winning showman's tractor.

steam era. They possess the decided advantage of not needing a track on which to run and yet are eminently suitable for doing some real work in the form of passenger hauling. The steam side of things needs managing correctly if

Figure 1.8   An example of the *Minnie* traction engine seen at a modelling exhibition.

Figure 1.10   Fine detail on the tender of a large-scale model traction engine.

the engine is to perform satisfactorily and the exposed motion work gives much visual satisfaction unlike that on inside-cylindered locomotives which is normally all but invisible.

Strictly, what are colloquially termed 'traction engines' comprise several distinctly different vehicle types. The most common is the general-purpose engine which is effectively a self-propelled prime power source. It is normally capable of towing a trailer-mounted load but its prime purpose is to act as a local power source for agricultural and light industrial use. It can, therefore, drive threshing machinery, circular saws and other light-to-medium duty machines.

A self-propelled machine known as a tractor or road locomotive is used for other than local haulage. This provides weather protection for the crew in the form of a short awning above the footplate. Tractors may be physically somewhat larger than the general-purpose machines but are otherwise broadly similar.

The Rolls-Royce of the tractor world is what is usually called the 'showman's road locomotive'. This type has an all-over canopy supported on polished brass 'Olivers' and is usually more ornately decorated in comparison with other engines. This type is generally the largest of the tractors, its length extended by the usual front-mounted platform for the generator which allowed the machine to provide the power for the electrically operated rides at the fairgrounds, and for lighting at the circuses, which the showmen ran. Models of showmen's locomotives are enormously popular and drawings and castings are available from several commercial sources.

Also on the massive scale were the agricultural engines which were used to perform heavy-duty tasks on the land such as ploughing, dredging and the cutting of land drains. These machines make interesting models, but for some, there is the disadvantage that they were frequently used in pairs thus requiring twice the work to model authentically. While this is true, it will not necessarily take twice as long to make a pair since machine setting-up time does not increase, but an alternative is for two modellers to make one each and then to combine their efforts for the making of accessories such as plough, cultivator mole drainer or whatever seems to be appropriate.

The popular scales for model traction engines (to use the colloquial term) range from 1-inch scale to 4-inch scale (25 to 100mm per foot) or larger. In the smallest of these scales, a model engine is a relatively small model, about 12in. high (300mm) and 18in. long (460mm). Such a model is capable of pulling the owner along, but it is a hard struggle on grass, and there has been a move towards the larger scales over the last few years. A popular size is the 1½in. to the foot scale (37mm), giving a model that is capable of real work and yet not too heavy to move about nor too large to store conveniently.

However, the trend is towards "bigger and better" and 2-, 3- and 4-inch scale models are now quite common and well catered for commercially. Before embarking on a model, your capacity to turn the hind wheels should be considered. If, say, the prototype's wheels were 6ft in diameter (and some were larger) you will have to be able to cope with a 24in. (600mm) diameter to machine the hind wheel rims in 4-inch scale and this will therefore limit the scale of model that you can consider. Large boilers may present some problems too, unless you are a capable welder and competent enough in the making of pressure vessels to consider steel for the boiler material, and this aspect is also worthy of consideration before a start is made.

Stationary and marine engines

Stationary and marine steam engines have been popular modelling subjects since the early days

Figure 1.11 A nicely displayed Stuart Turner beam engine.

Figure 1.12 A horizontal steam engine to the popular *Perseus* design.

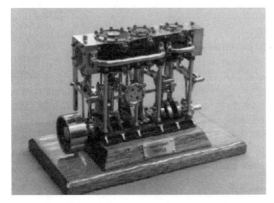

Figure 1.13 A Stuart Turner triple expansion engine.

Figure 1.14 Testing an example of Tubal Cain's *Kitten* before final polishing and assembly to its base.

of the hobby and the range of kits produced under the name of Stuart Turner includes designs which date from before the First World War. Kits and drawings are also available from many other suppliers for engines as simple as a single, oscillating-cylinder engine which requires no valve gear as such, to the more complex winding, roller-mill and beam engines.

The small, oscillating-cylinder types are intended for incorporation into simple model boat power units, when married to a suitable boiler, or may be incorporated into the types of toy steam plant which are always popular. Some simple designs are available also for small model cranes using this simple type of non-reversible steam engine.

Some of the larger stationary models may be non-reversing but they are usually provided with a slide valve and single eccentric arranged for running in one direction but without the second eccentric and some kind of link arrangement to allow running in either direction at will. These engines provide a useful introduction to the steam engine generally, if one is needed, but do not require the manufacture of large numbers of components and can therefore be completed in reasonable time. A variety of materials will be encountered during the manufacture (cast iron, brass or gunmetal, mild steel) which provides valuable modelling experience and only a small boiler is required to generate steam if a working model is wanted rather than a sideboard or showcase model.

Model boat enthusiasts may well wish to build a small, reversing steam engine for use in a model boat or launch hull. This provides a valuable extension of the skills of model boat building which have perhaps already been acquired and takes the modeller into the quiet realms of slow-speed engines and a leisurely progress of the vessel.

Internal combustion engines

The building of internal combustion (IC) engines utilises quite different techniques from those used in the manufacture of steam engines. In comparison with an IC engine, the steam engine is relatively crude. It is slow revving and its reciprocating parts are not normally balanced. Its motion work is not usually enclosed, and its lubrication is therefore most frequently by simple oil can. Also, pressures and temperatures existing in the steam engine cylinder are quite low, and simple (and rather crude) means

Figure 1.16   A four cylinder, water-cooled petrol engine.

are usually employed in providing the seal between piston and cylinder bore (at least in models), graphited yarn still being the most common material for this duty.

IC engines usually utilise finely lapped bores having lightweight pistons which are fitted with cast iron rings, although pistons are sometimes used without rings, especially in the smaller sizes. However, pistons are usually closely lapped to the cylinder bore. Connecting rods are normally polished, and are matched for mass (weight) in a multi-cylinder engine and a rudimentary balancing exercise is

Figure 1.15   A 15cc *Kiwi* petrol engine.

Figure 1.17   Two air-cooled IC engines.

routinely undertaken to provide a smooth-running, high-speed engine. A high-pressure lubrication system may also be incorporated, but even if simple 'splash' lubrication is used, this is obviously more sophisticated than the locoman's occasional squirt from the oil can.

Drawings and castings are available from the trade for a very wide range of IC engines, embracing designs for petrol, glowplug and diesel types, both air and water cooled. Those strictly intended as model power range from less than 1cc capacity up to 30cc or so, providing power for a wide range of model boats and aircraft and adaptable to model cars or locomotives should your interests lie in that direction. Complementary designs for carburettors, contact breakers and magnetos are also available and several suppliers of miniature ignition equipment and sparking plugs can satisfy these essential needs.

If you do intend to use the engine in an aircraft, car or boat, be sure to enquire about the type and capacity which are allowed by the rules of the various competition classes, before starting, otherwise the model may not be usable or saleable.

In common with other branches of the hobby, there is increasing interest in scale models of particular IC engines, rather than designs intended specifically as power sources for models, and beautiful (working) models of rotary aero engines are now quite common at club exhibitions. Working models of the Rolls-Royce Merlin engine have also been exhibited, at least one of which is capable of sustained running through the use of scale magnetos. There is thus much scope for fine workmanship in this field and a wealth of simpler designs on which to cut one's teeth before progressing to the fine-scale end of the range.

Clocks and scientific instruments

At one time, principally between the wars and immediately after World War 2, there was much amateur interest in the making of scientific or semi-scientific instruments which were either not readily available, or impossibly expensive for ordinary mortals to acquire. There was also a modest interest in the home manufacture of various common devices such as electric clocks and harmonographs and designs for these devices were published by *Model Engineer*. Drawings for some of these devices are still available and they may still be made, together with other devices such as a microscope or slide projector, although these latter do require that you have available, or can obtain, the necessary lenses and other specialised components.

Figure 1.18 A lantern clock.

Figure 1.19 A well-polished skeleton clock and monogrammed key.

Figure 1.20 A medal-winning cased clock.

Today, such items can readily be purchased, but there remains an underlying amateur interest in various types of scientific devices which were once produced, mostly by hand, to suit the scientific interests of the past. Amateurs having these interests tend to work alone, producing whatever they fancy, based on the originals and the methods of manufacture employed to make them. Very little appears to be published in the model press but quite delightful and well-made examples appear regularly at various shows and exhibitions.

Clockwork devices figure greatly among the items exhibited, and clockmaking generally appears to go from strength to strength. Much is published concerning the design and manu-facture of clocks, and drawing sets and books are readily available to assist the beginner to make a start.

Clock work is generally quite different from most other branches of model engineering. Since a clock mechanism is naturally mostly wheels and pinions, equipment is required for accurate circular division and specially prepared cutters are needed to cut the various tooth forms which are required, so some initial investment is required in equipment for cutting wheels.

Apart from the cutting of wheels and pinions, much of the work required in making a clock, is, or may be, carried out by hand. This includes the cutting and truing of plates, the crossing out of wheels, burnishing of arbors, and so on, and there is much personal satisfaction to be drawn from this kind of work since the builder feels that he (or she) has made a significant contribution by his or her own efforts.

If you have a leaning towards woodwork as well as metal work, some sort of cased clock should satisfy your needs. If not, skeleton clocks or metal-cased types provide an alternative and a transparent dome or cover can be obtained to protect the mechanism, once it is complete. One additional pleasure which may be derived from the making of a clock is that of engraving one's name or initials into the mechanism, a practice which is so well established that it no longer seems pompous or affected to do this. If your engraving is not up to scratch (sorry!) there is always the alternative of cutting the key in the form of your initials.

If you are content to assemble a clock in the first instance, without cutting plates and wheels yourself, there are some kits available which require the assembler to contribute by cleaning up and polishing commercially made components and then to assemble the clock. These kits are not inordinately expensive and can in any event sometimes be purchased in stages, thus spreading the cost. These kits provide an extremely useful introduction to the art of clock assembly and are a valuable means for a potential clockmaker to gain some experience without the necessity to establish a full workshop at the outset.

Workshop tools and accessories

There are two views concerning the making of tools and accessories, one which regards their building as being complementary to, but an interruption of, the real business of model-making, and the second which maintains that a well-made tool is a joy to use and care should

Figure 1.21 A small drilling machine, machine vice and accessories.

Figure 1.22 A *Quorn* tool and cutter grinder.

Figure 1.23 A non-geared rotary table to the G H Thomas design.

therefore be exercised in its manufacture. Both views are equally supportable, and no-one would totally disagree with the second. What the opponents of this view claim is not that their tools are poorly made, but that they are essentially simple and not therefore consumers of large amounts of workshop time in their making.

Whatever your view, it is undoubtedly true that much of what is required in the workshop may be manufactured once a lathe has been purchased. Many of the available tool designs date from the days when most amateurs could not afford to purchase accessories and machine tools, beyond the basic lathe, since such were not manufactured for the amateur market and commercial (industrial) prices were simply too high. Commercial machinery was also often too big to be housed in the amateur's workshop.

There was also a general lack of availability of such lathe accessories as collet chucks (at least for the size of lathe in which the amateur was interested, or could afford) and vertical slides. Publication of design details for these items, suitable for manufacture on the amateur's limited machinery, was therefore commonplace and drawings, castings and those hard-to-get materials were readily available. In spite of the fact that there is today a much

wider availability of some of these items, information and materials are if anything more readily available and there is significant choice if you need to reduce the cost, by making items yourself.

Among the major items which may be manufactured on a small lathe are bench grinders, drilling machines, sawing attachments or circular saw tables for the lathe, and motorised hacksaw machines. It is common practice to make ready-finished components available whenever their manufacture presents difficulty to the amateur equipped with only limited machinery, and this allows relatively large items such as vertical milling machines to be completed on a 3½ inch lathe. Various modellers' suppliers specialise in the supply of materials and information for tools and other accessories and those retail outlets which carry a wide range of items for the model engineer usually have some items for sale in this category.

Model boats

Model boats embrace such a wide range of different classes that one must be careful whether they are covered by the general definition of model engineering adopted for the purposes of

Figure 1.24 The *African Queen.*

Figure 1.25 A more formal steam launch with two saloons.

this book. Unless one wants to adopt a metal-plate construction for a model ship (and there are some who do) a hull constructed in wood or glass-reinforced plastic is likely to be used. Its construction is more likely to be allied to woodworking and other manual skills than to model engineering. Boat modelling therefore attracts those model engineers also having skills, or interests, in woodworking, or allied handicrafts.

Given these prerequisites, there is much scope for experimentation and development work. If your interest lies in the building of high-speed power boats, together with the production of their power plants and drive assemblies, there can be a real engineering involvement and much room for private experiment, not only with the mechanics but with hull design, and the transformation of engine power into speed over the water.

Such development work is encouraged by a strong network of local clubs and societies and by organisations such as the Model Power Boat Association (MPBA), in the UK, and the internationally recognised Naviga which has established classes which define groups and subgroups of models so that competition craft conform to certain basic formulae.

The recognised classes number some 28, 23 of which are for working boats, as distinct from non-working, scale models. These encompass straight running and steering classes, radio-controlled models, with both IC and electrical power plants, and tethered (round-the-pole) hydroplanes in both airscrew- and waterscrew-driven types. There is a very wide range of choice should you be interested in the competitive side of the hobby, but it naturally pays to seek out a club whose activities correspond with your own interests since not all classes may be vigorously supported within all clubs.

If your interest lies outside the area of competition boating, radio control, nowadays readily available even to those without knowledge of electronics, makes possible the sailing of power or sail boats (model yachts) as a relaxing leisure activity. There has consequently been an increase in all forms of pleasure boating, more than adequately supported by the kits, fittings and radio-control equipment available through the model trade.

Electrically powered scale model boats are straightforward to build, either from kits, or scratchbuilt using commercially available plans. They are clean and quiet in operation and within reason can be operated on almost any stretch of water to which the public has access. Some mechanical design knowledge is required in arranging the drive, and the different movements which are controlled by radio, but these can often be arranged using commercial components and significant workshop facilities are not necessarily essential.

At the other extreme, if steam is your scene, plans for hulls suitable for power plants as simple as a single-cylinder oscillating engine, or as complex as a triple-expansion power plant, are readily available. There is thus, as in other branches of the hobby, a wide range of activities available for those who are interested.

Radio control of live steam models is well established in the smaller model railway scales (principally gauge 'O' and gauge '1') and this form of control is ideal also for steam-powered boats. Those afflicted with the three model 'diseases' of live steam, model boats and radio control naturally find much of interest, and scope is available for trying out your own ideas, or just doing your own thing.

Most sailing model boats are of the yacht type, having what is known as a fore-and-aft rig, as distinct from a square rig. The most popular models are of the yacht type, having jib and main sails, and sometimes even a spinnaker for use when sailing downwind. Again, there are internationally recognised standard classes, formulated to allow competitions to be staged, and many of the club activities revolve around the racing of class boats.

Radio control is naturally widely used, and deservedly popular, but racing of vane-controlled models is still widely organised and if you are interested in sailing a model influenced only by the wind, once released, without the encumbrance of radio control, a local club is almost sure to exist to satisfy the need for facilities and to organise competitions.

Radio-controlled sailing of a model yacht requires control of the sail settings and also of the rudder. Two servo-controlled winches and a tiller control are therefore required and a good basic knowledge of sailing generally is needed to obtain the best out of a particular model under the prevailing wind and weather conditions. Again, various recognised classes are established to allow competitive sailing to be organised, radio-control model racing being organised and run pretty much as is the full-size sport.

If the competitive element does not attract you, a more true-to-scale type of model may be more appropriate (racing models tend to be devoid of any non-functional parts). Once again, within reason, this type of leisure sailing can be undertaken on almost any publicly available water.

Model aircraft

Like model boating, the flying of model aircraft attracts model engineers with an interest in the building of power units, and the production of engines of appropriate size and having the desired power:weight ratio can be a challenging field for experimentation.

Competitive flying is still the basis of club activities and models are classified into three major groups: free-flight, control-line and radio-controlled models. Control-line models are naturally all powered, but even so, several sub-groups are established including stunt and speed flying, team racing and combat. There is also an interest in the control-line flying of scale models but most of these appear in the free-flight group. Free-flight also embraces rubber-powered models, gliders and radio-controlled gliders, while powered models provide opportunities for aerobatic competitions and pylon racing.

The confirmation of the feasibility of making and flying model helicopters has also provided a field for amateur development of anything from the mechanics to the complete machine, although most modellers still prefer to purchase a kit or ready-assembled unit. However, the possibility exists, provided that you have the requisite theoretical knowledge and the capability to design and manufacture the required parts.

Model cars

Two basic types of model car are produced by model engineers: the strictly scale model, and the functional, sporting or racing car, normally radio controlled and used simply for fun, or for competitive club racing. These racing models are most commonly electrically powered since this allows clean, quiet models to be produced which are suitable for racing both indoors and

Figure 1.26 A fine model Bugatti.

out, thus permitting all-year-round racing.

Club racing is, however, well organised for IC-engined cars, the models being made to ⅛ scale and powered by 3.5cc glowplug engines. Full radio control is fitted, as it is for electrically powered racing, modern radio equipment allowing several cars to race together, simply by using different crystals. Due to the noise and smoke, and the possibility of fuel spillages, IC-engined models are normally raced only on outdoor circuits.

Electrically powered racing has developed since the introduction of fast-charge nickel-cadmium batteries. A battery pack of this type allows perhaps 7 or 8 minutes racing before exhaustion, but may then be fully recharged in 20 or 30 minutes, allowing further racing. Clubs usually arrange a full racing programme, individual events lasting 5 minutes plus one lap, the day's racing culminating in various group finals according to performance during the preliminary events.

Due to the popularity of electrically powered racing there is significant support from the trade and it is not normally necessary to be a fully-equipped model engineer in order to participate. Extremely comprehensive kits are available and a range of screwdrivers and spanners is all that is required in order to assemble a commercial kit. Assembly instructions are usually well written and a mechanically minded 12

year-old can readily carry out the assembly, given a modicum of adult supervision.

This is not to say that the models are simple. Four-wheel drive through differentials, together with independent suspension and Ackerman steering, are normal, together with oil-damped shock absorbers. Variations in treads and wall rigidity in the tyres, and the type of shock absorber, allow experiments in relation to the suspension and road holding. There is also the never-ending search for the 'best' motor (within the limits allowed by the class rules) and the selection of gear ratios to provide what one considers to be the best compromise in terms of top speed, maximum acceleration and battery duration.

To contain these experiments within a formal framework, there are recognised racing classes, relating broadly to motor type, and operating voltage, and determining whether motor modification is permitted. Naturally, if you wish to join the club racing scene, you must adopt one of the recognised classes of vehicle. The choice lies between IC engine or electric motor drive, the choice for this latter type being either a straight racer or a buggy.

Unlike model car racing, scale model car building is not well supported by the trade and is therefore a scratchbuilder's activity, usually based on a detailed study of particular prototypes. Such models may be built in any materials with which you are familiar, but the very best are made with the 'correct' materials as far as possible and may also be internally correct if built to a sufficiently large scale and level of detail. For this type of model, a lathe and a capability to carry out milling are essential.

Modelling and making

Modelling, in relation to model engineering, can simply be interpreted as 'making miniatures'. Modelling is thus distinctly different from making real things which are not miniatures, and what will do for de-burring a hole in a bracket for hanging a shelf on will not do for de-burring a hole in the frames of a model locomotive.

If you are building a model which is one-tenth the size of the prototype, a .010in. chamfer on the edge of a hole is equivalent to 0.1in. on the prototype. This may be acceptable in the case of a hole, but surface imperfections of as little as .005in. are almost sure to show, and in any event represents .050in. on the prototype, at one-tenth scale.

Surface blemishes must be considered for each part of the model. Many of the materials which are bought have surface imperfections of one sort of another. Bright-drawn steel rods and bars frequently have inclusions in the surface which can well render them unsuitable for use if the need is to model a polished component and it is often prudent to buy the next largest size and machine the item all over, rather than take the easier route of removing just the minimum of material and trying to polish out the blemishes on unmachined surfaces.

When sheet materials are required, they are usually produced by shearing off the required size from a large sheet using a guillotine. Guillotined plates are often more seriously flawed than rods and bars. They frequently show considerable distortion along a cut edge since the cutting process tends to turn the edge over, as well as cutting it, and the result is sometimes very distorted. Considerable roughness can also be evident on the edge, and the very minimum which needs to be done is to file the edges smooth. This means that the plates need to be purchased a little larger than the required final size, and in view of this it is well worthwhile getting the sheets cut well oversize, afterwards cutting and filing the edges to the correct size and profile.

In practice, many of the models which are built are smaller than one-tenth scale. A 3½-inch gauge locomotive is one-sixteenth full size, and even a 5-inch gauge model is a little smaller than one eleventh of its full-size counterpart. This is an extremely important point to bear in mind if the aim is the production of a fine model. Everything must be in scale.

On a perfect model, this means that the heads of bolts and screws are the correct type and size, that all fastenings are present, or appear to be so, and the model appears to be constructed as was the prototype. This means, for example, that the fastenings will be hexagon-head bolts and scale-size rivets, and the more convenient slotted-head screws should almost never be used as fasteners on a model. It means that there is a need for quite small fastenings. A 1-inch bolt scales at .088in. for 1 1/16 inch to the foot scale (a 5-inch gauge locomotive) and this means using an 8BA bolt on the model. If the model is built to ¾ inch to the foot scale (3½-inch gauge) a 1-inch bolt scales at 1/16in. which means something between 10 and 11BA. Similar considerations apply to rivets, and once again, small sizes which are in proportion with the scale of the model are essential to achieve correct appearance.

Of course, this doesn't mean that all of the rivets need to be present. They certainly do not all need to be holding the parts together. Provided that there is sufficient security, some of the rivets can be just dummies, and it is possible to solder parts together and insert rivets, simply for appearance sake. The tanks of a model tender can often be better treated in this way, and it is usually simpler to put the dummy rivets in position in a plate, countersinking them on the underside, and leaving just one or two here and there for actual attachment, and then soldering the tank together.

Care needs to be paid to the fit of parts. If you are making a plant stand or a firescreen in steel scroll work, a 1/16in. gap between parts which nominally touch one another is unfortunate, but not usually a disaster. On a 5-inch gauge locomotive (1$\frac{1}{16}$ inch to the foot scale), such a gap represents over $\frac{5}{8}$in. and if it isn't supposed to be there, its presence will severely mar the model and spoil its scale appearance.

Achieving scale appearance also means paying particular attention to surface finish, and to the thicknesses of parts, particularly the plate work, in situations in which the edges of the plates are visible. This means using thin plates, since a $\frac{1}{4}$-inch plate on a prototype locomotive must be represented by .022in. on a 5-inch gauge model, and by .0156in. on its 3$\frac{1}{2}$-inch gauge counterpart. Thus, quite flimsy structures can be produced and this means handling the complete, or half-completed model with suitable care.

It is arguable that such a model is too fragile to take out on the track and this then raises the question of whether there are different standards of models which are made for different purposes – practical working models, and those made for exhibition or for the domestic sideboard. The working model may be more robustly made in certain parts, and may not strictly be a scale model. In this case, it still needs to look like the prototype it represents, and should aim to capture the spirit of the original rather than be a faithful miniature copy. It may be more simply made, have fewer rivets, be modified to make it easier for a 'non-scale' driver to handle, have parts omitted for access, and so on. Nevertheless, if it gives the impression of being a miniature version of the original, the designer and the builder have collaborated well in producing a good model.

In truth therefore, absolute scale is not necessarily essential, but correct proportion is, and it is vital that this sense of proportion is carried through in every aspect of the model. It pays to have a clear idea of the intention right at the outset since it is disappointing if the finished model doesn't quite come up to expectations at the end.

Since large models naturally take a long time to make, and there are many parts, your expectations and ability at the end will be different from those with which you set out at the beginning. It is, therefore, important to choose something easily achievable for your first model so that your skills and ambitions don't have too long to change during its building. Provided that the aims don't change during the building of a model, you should feel at the end that the initial objectives have been achieved.

By keeping your mind firmly on the initial objective, which ought to be 'build a simple steam engine', 'build a simple clock' or whatever, you shouldn't divert yourself by becoming convinced that parts made at the beginning need to be remade. You should not keep remaking parts, and you should avoid abandoning a project on the basis that it doesn't meet your current aims. The current aims ought properly to be applied to the next project.

Workshop machinery

The machinery with which the workshop is equipped is dependent on many factors. Your own inclinations and interests naturally play a part, but opportunities for acquiring machinery also vary from one individual to another and if you work in a situation in which secondhand items occasionally become available, you may be fortunate enough to obtain a wider range of basic machines and accessories than other modellers.

The operations which need to be performed in the workshop are essentially two-fold: cutting material from the work so that it assumes a circular cross-section, and the alternative of cutting material away so that a flat surface is produced. Cylindrical shapes, or those with circular cross-sections, such as cones, are produced by rotating the work and bringing a cutting tool into contact with it.

This process is illustrated in Figure 1.27 which shows a steel billet which is rotating, while a cutting tool is used to reduce its diameter by being pushed progressively along its length. This process is known as turning (describing the motion of the work) and is the operation for which a lathe is designed.

Although rectangular shapes can be formed on a lathe, flat surfaces are usually formed on a machine in which the cutting tool rotates and the work is fed into the path of the cutter in such a way that the desired shape is produced. One simple form of this operation is shown in Figure 1.28. A tool having cutting edges on its outside diameter is being used to profile the edge of a thick steel plate.

This process is known as milling and the machine on which the process is carried out is a milling machine. These machines are produced in two basic types, known as vertical and horizontal milling machines, depending upon whether the cutter rotates about a vertical or a horizontal axis. The two types of machine differ fundamentally in the manner of their construction, and a horizontal mill (to use the colloquial

name) is not simply a vertical mill turned on its side. The uses of the two types of mill differ, and while a vertical mill will not do all of the things that a horizontal mill is capable of doing, the vertical mill is really the one to buy, since a horizontal mill is not suitable for much of the work which the modeller will wish to perform.

Manufacturers of milling machines overcome the problem posed by the different capabilities of the two types of machine by making a vertical head which effectively converts a horizontal mill into a vertical mill. Such machines thus provide the major capabilities of both types of mill.

Most real milling machines are large, floorstanding structures provided with a large table which can be moved by calibrated feedscrews in three direction: along and across horizontally, and up and down, vertically. The work is mounted on the table and can be positioned in all three directions for a cut to be applied.

In recent years, small bench-mounted vertical mills have been introduced onto the market. Due to the bench-mounted arrangement, the table cannot be moved in the up-and-down direction. Instead, the spindle is provided with a down feed operated either by a lever or a calibrated handwheel. The lever-operated down

Figure 1.27 A straightforward turning operation – reducing the diameter of a circular steel billet.

Figure 1.28 Milling the profile of a steel plate.

feed arrangement is precisely like that on a drilling machine and a mill of this type is described as a mill/drill to distinguish it from the more usual type of industrial machine.

Most of the operations for which a vertical milling machine is used industrially can be performed by adapting a lathe, using a rotating cutter in the position normally occupied by the work, and cutting material mounted in the position usually reserved for the tool. The lathe must be provided with additional accessories in order that this adaptation can be successfully applied, but this is far less costly than buying a milling machine or a mill/drill. It also takes up virtually no space, and is therefore frequently the way in which milling is performed in the amateur's workshop.

One way to obtain both a lathe and a milling machine is to purchase a combination. Several manufacturers produce lathes for which supplementary vertical milling attachments are available. Since the lathe is provided with a table which has calibrated and controlled movement in two directions, it is possible to perform milling by fitting a column-and-spindle assembly to the rear of the lathe. This saves space and money, both valuable commodities, but brings the disadvantage that the workholding table on the lathe may be smaller than that on the equivalent mill/drill.

Some manufacturers provide a solution to this problem by producing both a mill/drill and a lathe in modular form, so that the column assembly of the mill/drill can be fitted to the lathe by using an adaptor. The column assembly naturally also fits the sturdy base and large co-ordinate table of the mill/drill, so the two machines can be bought progressively, first the lathe, then the vertical head, and afterwards the co-ordinate table.

Because the simplest adaptation of the lathe for milling, (without a supplementary vertical head) has its shortcomings, principally a restriction on the size of the workpiece, the amateur production of rectangular work is sometimes done on a shaping machine. In this type of machine, the work is bolted to a table, and a tool not unlike a lathe tool is passed over it, acting much in the way that a plane does for wood. Indeed, there is little difference in principle between a planer and a shaper, as far as metal cutting is concerned.

The shaper seems to have fallen out of favour in recent years, but it was at one time readily available in the form of small bench-mounted models suitable for amateur use. Many of these were operated by hand, called hand shapers, but were often also available in motorised form. They are useful in allowing larger rectangular items to be shaped, and this is an operation which can be difficult or slow if only a small lathe is available. A shaper is less costly than a milling machine, takes up less space and represents a useful addition to the small workshop if space and money do not allow a mill/drill to be obtained.

A third operation, which complements the production of circular and rectangular work, is that of drilling holes in items made by the other processes. If you are lucky enough to be able to obtain a mill/drill at the outset, this may well be used for much of the drilling which is required. It is not absolutely ideal for this work since in a mill the drill cannot pass through the table as it usually can on a small drilling machine, and this means that packing must be put below the work to allow penetration, unless use can be made of the clearance provided by the workholding slots. Even so, this doesn't allow a drill or reamer to be put right through the work, and a drilling machine is certainly not a luxury even if a vertical mill or mill/drill is available.

The requirement for a small drilling machine is considered at the beginning of Chapter 10.

If workshop space and money are readily available, there is much choice of machinery available. For most people, a start is usually made with a lathe and a drilling machine, but some basic accessories are required to complement these two items. The lathe requires workholding chucks and a faceplate, together with a drill chuck for the tailstock and the drilling machine (drill) requires a suitable machine vice. Tool bits for the lathe, and drill bits for both machines will naturally be required and there must be some means available for sharpening these. A small, general-purpose grinder thus complements the two basic machines.

Once these machines are purchased and set up, the basics are available.

If your needs are very specialised, and you habitually undertake much heavy milling, or need to cut gears and pinions frequently, it may be too time-consuming to utilise an adaptation of the lathe, or in the case of heavy milling, may unnecessarily strain the machine, or be beyond its capability. In these instances, you will need to consider the purchase of specialised machinery to suit your particular needs. Possible additional items will suggest themselves as your modelling progresses.

As noted elsewhere, time and money are usually in short supply, but in some ways they are interchangeable, since money spent on a specialised machine (and the space to house it) frequently leads to a saving in time, as it is not necessary to install (and later remove) an adaptation to another device. A specialised machine might, in any event, be more appropriate and thus inherently more productive.

Safety

General

Someone once said that there are only three types of tool: those that cut you, those that hit you, and those that fall on the floor and roll under the bench out of sight! Note that two-thirds of all tools are therefore dangerous. Many of the processes which are employed in the amateur's workshop are governed by the *Health and Safety at Work Act* or by other legislation, when used in industry. Industrially, machines too are governed by legal requirements which demand that guards be correctly fitted, and be used at all times, in order to try to ensure that the more obvious dangers posed by rotating machinery and work do not become sources of personal injury.

Employees in industry are expected to contribute to their own safety by adopting a style of dress that avoids such obvious dangers as flapping clothes or flimsy shoes, or to avoid possible dangers caused by too-long a hairstyle, and it is obligatory for many operators to wear protective spectacles or goggles as a matter of course. For some occupations, the wearing of protective footware is obligatory and for amateurs, a reasonably stout shoe or boot is a sensible precaution: even a lathe chuck key dropped on to a toe inside a flimsy shoe can cause painful injury.

We should take note of this. There are just as many potential hazards in the amateur's workshop. In fact, there may well be more, since our machines are, for the most part, manually operated and we are therefore that much closer to the actual cutting processes than some of our industrial counterparts. In the home workshop there are also likely to be more processes carried out than in the average industrial 'shop' since we normally carry out machining, heat treatment, chemical processes such as pickling and etching, spray painting and even the casting of metal. Greater, rather than less, care is therefore needed at home.

Machining

Perhaps the greatest potential hazard to hands and eyes are the machining operations which are performed on lathes, drilling and milling machines. Either the workpiece or the cutter is rotating and there are consequently immediate dangers from the rotating item. So keep hands and fingers well out of the way.

There is often a temptation to 'feel' the finish on the work while either it or the cutter is still rotating. Don't do it. Under some circumstances it may be less dangerous than under others, but if the work or cutter is asymmetrical (has a 'lump' going round well off the axis of rotation) as is often the case, the 'lump on the side' will catch you a nasty bang and part of a finger will be lost, or even worse. So, once again, don't do it. If you must remove swarf while drilling or milling, then use a brush.

Swarf comprises the curls of metal that are removed from the workpiece by the cutting tool. This comes off in different forms depending upon the tool, the cutting process and the material, but all swarf is potentially dangerous. It may be hot enough to burn, sharp enough to cut and flying about all over the place, more or less all at the same time. Some of the old-time machinists used to say that you hadn't become a professional until you had burned a few hairs off your chest (ladies excepted, of course). This is amusing at first sight, but it is also symptomatic of the kind of bravado that used to override matters of safety at one time, especially for a generation of machine shop men who were brought up in the days of machines driven by overhead line shafting and totally unguarded belts.

It may be bearable if a piece of very hot swarf sticks to the forearm and produces a small burn, or goes down the front of an open-necked shirt to stick somewhere on the neck or chest. Unfortunately, such burns always come at an unexpected moment and what you might do to shake off the offending chip causes as much danger as the hot chip itself. It is not an affectation to wear a warehouse type of coat in the workshop. Its tapering sleeves, narrowing at the cuff, protect the arms from the abrasive

and burning effects of the swarf which is produced and, being tapered, are not likely to present a hazard by becoming caught up in the works.

The danger to the eyes is obvious. Not only may the swarf be very hot, but the individual chips can be quite large under some circumstances. They also have sharp and very ragged edges and will do much damage if they come into contact with the eyeball. All-enveloping safety goggles will protect the eyes completely from these dangerous particles of swarf, and will also provide complete protection should the unexpected happen, such as the shattering of a cutter.

Safety spectacles may be more comfortable to wear, but they do provide less comprehensive protection than safety goggles, and it is vital that you keep your head as far as possible out of the way of flying swarf, particularly if the machining operation is producing its swarf as the result of an interrupted cut. In these cases the shortish cuts produce distinct chips which simply fly into orbit once they are removed from the workpiece.

If the cut is continuous, swarf tends to come off in relatively long strands due to being still attached to the workpiece at the point at which the cut is being made. Even when detached, the longer lengths of swarf, having more mass, are not expelled from the work (or tool) with the same velocity and are therefore less dangerous on this account. The greater mass also tends to allow heat to dissipate more rapidly along the chip and, in any case, some parts of a long chip have been removed from the work for a relatively long time so they are cooler.

The machining operations posing the greatest hazard are therefore when fly cutting or when facing the end of a rectangular bar in the lathe. Both of these operations produce short pieces of swarf which are completely separated from the work, the short curls of swarf produced when cutting across the corners of a square bar being much hotter than when cutting continuously near the centre of the bar. They are also much more likely to fly about due to their complete detachment from the job. Protective goggles are readily available from local DIY stores and should be used whenever there is a likelihood that swarf will be flying about.

Drilling machines also produce their fair share of swarf which naturally presents the same hazard as that produced when milling or turning. The drilling of steel frequently produces long curls of swarf that wind up the drill flutes and go round with the drill. These often become quite long and project from the top of the drill shank. Since they are rotating with the drill, they pose a threat to the hands, which are usually in the immediate vicinity of the work, and also to the eyes, especially if the drill is bench mounted as is often the case.

Curls of swarf produced by a large drill are shown in Figure 1.29.

The way to prevent the occurrence of these long curls is to ease the pressure on the drill marginally from time to time during the drilling process. This reduces the rate of cut at the drill point and therefore reduces the thickness of the curl, weakening it and causing it to break off. Reducing cutting pressure completely stops the cutting action and the curl becomes detached from the work, so the long curls can be prevented from forming by varying the cutting pressure. If the drill flutes do get jammed

Figure 1.29 Curls of swarf.

up with swarf, stop the machine and clear them before any damage or injury is caused.

The machining of some forms of brass produces a quite different swarf from that produced by steel. This comprises fine, needle-like chips, some of which are visible but many of which are not. The smallest chips are akin to cut hair, and being small in diameter they are extremely sharp and readily penetrate the skin and can be very painful, and very difficult to extract, if they lodge in the fingers. Once again, the danger to the eyes is obvious, but easily forgotten, and if you are turning brass rod (particularly) in the lathe, take great care not to rub the eyes unless you are absolutely sure that no brass swarf is on the hands or fingers.

Soldering and brazing (silver soldering)

There are always some dangers posed by soldering and silver soldering, even the making of a simple electrical joint using soft solder. Materials are being heated to around 200°C and a large, high-temperature heat source is in use. There is thus the potential danger of fire. It is therefore important that these operations are carried out in a safe environment, with adequate facilities to prevent accidents, and even the ordinary, electrically powered soldering iron must be provided with a protective stand if accidents are to be avoided.

When brazing or silver soldering, the dangers are greater. Much higher temperatures are required – up to 750°C. A workpiece such as a locomotive boiler can be very large and a large source of heat is required to raise it to the correct temperature. Heat is commonly provided by a propane gas torch having a large diameter and projecting its flame 12in. (300mm) or more from the burner head. The dangers are obvious. The workpiece becomes very hot and there is much radiated heat. Very hot air surrounds the burner flame and this can readily ignite any flammable material in the vicinity, even though this is not in the direct flame. A clean, dust-free location is essential if silver soldering is to be carried out as is some dedicated area containing a steel-framed hearth.

There are also subsidiary potential dangers. Some silver solders can decompose and give off poisonous gas if maintained at high temperatures for some time in the liquid state. Therefore it is vital that adequate ventilation is available so that any gaseous emissions are diluted by a good air supply.

The dangers posed by leaks of the bottled propane gas are also obvious and valves, hoses and equipment generally must be kept in good condition and be inspected regularly to ensure that no accidental leaks occur. Hoses must be connected to the gas bottle through a hose failure valve (as a minimum) which is designed to cut off the gas should excessive flow rates be detected. A regulator is the alternative connection between hose and bottle and will normally also incorporate a safety device.

It must also be borne in mind that many of the fluxes used during soldering are mildly poisonous, and care must be exercised during storage, mixing and use, that they are not accessible to children or pets and cannot contaminate food. It goes without saying that the hands should always be washed thoroughly after handling any chemicals.

One further danger posed by brazing is the use of an acid 'pickle' for cleaning the work and removing flux residues. The pickle used is dilute sulphuric acid. Even in its dilute form, this acid is highly corrosive. For such activities as model boilermaking it is needed in relatively large quantities and safe methods of storage and use are absolutely essential.

However, the greatest danger comes when preparing the dilute acid. It is supplied in concentrated form, which is even more corrosive, and it must be mixed with water to provide the working solution. **ON NO ACCOUNT MUST WATER BE ADDED TO THE CONCENTRATED ACID.** The first drops of water to touch the acid cause a chemical reaction not unlike an explosion, thus spreading acid in all directions. **ALWAYS ADD ACID TO WATER.**

Further information relating to all of these hazards is given in Chapter 9.

Painting

Ordinary brush painting poses no particular hazards other than those posed by the flammability of the solvents and thinners used, also not forgetting that prolonged contact with any chemicals can lead to skin problems in particular cases.

However, much model painting is today undertaken by the use of miniature spray guns and air brushes, when paint and the medium which carries it (the thinner) is atomised into fine droplets and sprayed onto the work. Rapid evaporation of the medium then occurs, as the paint dries, thus 'loading' the atmosphere with further amounts of chemical.

Several inherent dangers are present. The very fine droplets of paint and medium can be inhaled readily and will contaminate nasal and bronchial passages. Also, the air in the immediate vicinity of the operator can become so loaded with solvents that unconsciousness can occur, and the results can be fatal. To avoid these dangers it is vital to ensure that there is adequate ventilation available to prevent the build-up of chemicals in the atmosphere. It is helpful if the work is contained within a box and enclosed on all sides except the front. This allows excess paint and overspray to be contained within the box, and also helps to contain the airborne solvent medium.

To prevent ingestion of paint droplets, a simple fabric- or paper-based mask is generally adequate. Such masks are available from DIY stores, but it should be borne in mind that they are frequently intended to prevent the ingestion of dust during cutting and grinding operations and therefore do not provide a barrier to the inhalation of solvents so their use does not remove the need for adequate ventilation.

To prevent excessive build up of solvent vapour in the atmosphere, it is helpful to carry out spray painting in short sessions, allowing plenty of time for the air to return to normal before continuing. If a great deal of spray painting is likely to be required, a small booth ventilated by an electrically driven fan should be considered as essential.

Metric or imperial?

Going metric presents a particular problem for the modeller, principally because many of the items one wishes to make will be based on designs which were prepared when the only conceivable measuring system to use in the UK was the imperial one.

This means that the whole device will have been designed to use imperial materials. All dimensions will be in imperial units and it follows therefore that imperial measuring equipment will be the most convenient to use. Unfortunately, in the UK we are now in a non-imperial, but at the same time a non-metric, environment. The preferred British Standard drill sizes are now metric; some, but only some, of the materials which we purchase are now specified in metric dimensions; many of the drawings which are used are totally imperial and many of us only have imperial measuring equipment.

A further problem has also arisen, in that it is likely that many readers of this book will be totally unfamiliar with the imperial system, having passed through their school years at a time when the government was totally committed to the conversion of industry to metric measurements. Consequently, their schooling was entirely metric and for them, the mere mention of sixteenths and sixty-fourths of an inch, or the sight of an imperial rule, makes the whole business of making measurements seem about as complicated as degree mathematics.

There are thus likely to be problems for those unfamiliar with the imperial measuring system, since very few sets of drawings are available in metric scales. One basic problem which arises when using an unfamiliar measuring system is that it is not always possible to visualise a size simply by knowing what it is. Thus workers in imperial units may have no idea how big a 10mm-wide strip of metal is, and metric workers may not be able to visualise how big a ¼in. diameter hole is. To avoid problems which might be caused by the adoption of one system rather than the other for use in this book, where dimensions are given, they are given in both systems, imperial first, followed by the metric equivalent in parentheses.

This policy has, however, caused some inconsistencies to appear since some metric dimensions given are rounded equivalents of the imperial size, whereas others are the commercial equivalent size of the imperial. For example, the metric equivalent size for a ⅜ in. square high-speed tool bit is 10mm and this is referred to as ⅜in. (10mm). On the other hand, the rounded metric equivalent of ¹⁄₁₆in. is 1.6mm, making it appear that ⅜in. is the same as 9.6mm. It is hoped that the context will provide sufficient clues for you to decide the intention in each case.

It should be particularly noted that there are no accepted metric equivalents of the larger model railway gauges, from 2½-inch gauge upwards. If you build in one of these gauges you must at least ensure that your wheel standards (flange width, tread width, back-to-back measurement etc.) comply with the accepted standards or you may not be able to run your model on the available tracks.

2

Information and supplies

Background

Some research and background reading is a necessary part of taking up any new hobby. The most obvious sources of initial (general) information are the bi-monthly, monthly or fortnightly magazines which are now so readily available from newsagents. The periodical which most readily comes to mind is *Model Engineer* (published by Nexus Special Interests) due to the fact that it has been published since the very beginning of the 'home workshop experimentation' era and carries the name by which the hobby is now known.

The editorial staff of *Model Engineer* have always tried to represent as wide a range of interests as possible in the magazine, although they are obviously dictated by readers' interests in that they publish only the articles which their contributors write. Over the years, a tremendous amount of information has been published in *Model Engineer* and much of this is still available in the form of sets of drawings. There is also a healthy trade in secondhand copies of the magazine and bound volumes are quite as collectable as are other types of secondhand books.

In the period prior to and immediately following the 1939–1945 war, *Model Engineer* reflected the gradual widening of interests in modelling activities and at that time published articles covering model power boats and their power plants, model yachts, aeromodelling, stationary steam engines, locomotives and traction engines, together with electrical and optical equipment such as microscopes and the then relatively rare slide projectors.

The hobby was perhaps less well supported by the trade in those days and in any event, modellers probably had relatively less disposable income that could be devoted to a hobby, and much that would nowadays be available as castings had to be fabricated. Treadle-operated lathes were also still quite common until the 1940s and this influenced the way that lathe work was approached since one tends to notice pretty quickly that the tool is blunt if the power for cutting is provided by the leg muscles. Reading some of these older issues of the magazine is quite instructive, and it is surprising how many ingenious dodges were developed at that time to overcome the limitations of the available machine tools.

Drawings for many of the major projects designed during the 1930s, and since then, are still available and access to these and to the more recent designs is described under the heading "Sources of drawings and designs" below.

Currently, *Model Engineer* provides articles covering large-scale steam locomotives, traction engines, agricultural machinery, tools and workshop items for home construction together with reviews of equipment, notes on the prototype and so on. The increased interest in small-scale live steam for garden railways is also reflected in occasional articles covering this subject.

A newer magazine which provides information on a similar range of topics is *Engineering in Miniature*. This magazine commenced publication in 1978 and it is now well established as a monthly periodical which complements *Model Engineer* in many ways rather than just being a competitor by covering the same range of topics.

As the various branches of the modelling hobby have become more specialised, new magazines have appeared on the bookstalls in order to allow the editorial content to expand. *Aeromodeller* has been published for many years now, but more recent titles are *Model Boats*, *RCM&E*, *R/C Model Cars* and *Radio Modeller*. Younger still, is *Model Engineer's Workshop*. This was originally introduced as a quarterly magazine, but is now issued every two months. It covers all aspects of workshop practice and is an invaluable source of information for metal workers, of whatever persuasion. (*Aeromodeller*, *Model Boats*, *RCM&E*, *R/C Model Cars*, *Radio Modeller* and *Model Engineer's Workshop* are all published by Nexus Special Interests.)

If your local newsagent does not have additional copies of these magazines for display on his shelves for casual sales, he will be quite willing to place an order for you for delivery with your daily paper or for collection, if you do not have a daily delivery. Another way to receive a magazine regularly is to take out a subscription and have it posted to you by the publishers.

Moving away from strictly modelling activities, clockmakers are catered for by the monthly *Clocks* magazine (also published by Nexus Special Interests).

Becoming a regular reader of a modelling periodical brings one into the ranks of the armchair modeller. This is a desirable step forward since it means only a modest financial outlay yet it brings you into the learning phase. It should allow you to become aware of the existence of local clubs, their open days and exhibitions, meeting dates and so on. Periodicals will also provide the opportunity to see what others are making, and the sort of techniques which are used, and will generally allow you to become familiar with model engineering drawings and, through published photographs, allow comparison between drawings and the items which are produced from them.

Through the advertisements you will become aware of the retailers that can supply drawings, castings, materials, tools and general supplies and as your ideas take shape you will be able to obtain a few catalogues which are relevant to your interests. From these you should be able to judge the sort of financial commitment that will be involved so that you can plan the transition from armchair to workshop. Some advice about the choice of a first project is given below.

Sources of drawings and designs

It is perfectly possible to prepare your own drawings for construction of a model, clock or scientific instrument, but there are pitfalls, even for the experienced, and as a beginner it is probably best to work to a published design.

Fortunately, there is no shortage of drawings for items covering the whole range of activities which go under the heading of model engineering.

Without doubt, the most important collection of drawings is that associated with *Model Engineer* magazine. This periodical has specialised over many years in presenting descriptions of the making of many different items, accompanied by reduced-size drawings inset into the descriptive text. Most of the drawings utilised over the years are still available. Those likely to be of interest to model engineers are catalogued in *Plans Handbook No. 2* (published by Nexus Special Interests) which lists the sets of model engineering and model boat drawings which can be purchased from the publishers.

This Plans Handbook is effectively a catalogue of all of the drawing sets which are available, covering model locomotives, traction engines, stationary and marine steam engines, boilers, steam-powered toys, clocks, workshop tools and accessories, in addition to model boats of all sorts.

These original drawings are normally presented at full size for the model so that a large item such as a passenger-hauling, live steam locomotive occupies several large sheets of drawings. Prints of these are available from the Nexus Plans Service (Nexus House, Boundary Way, Hemel Hempstead, Herts HP2 7ST, tel: 01442 66551) and will generally be found to provide all of the details necessary for the building of the item concerned.

A useful feature of *Plans Handbook No. 2* is that it lists the content of each of the individual sheets that go to make up the full set for a large model. Normally, one will need the full set of drawings in order to construct the model as designed, but if only the tender of a locomotive is of interest, for example, then the sheets can be purchased individually and there is therefore no unnecessary expense. This Plans Handbook is also useful in listing suppliers from whom

castings are available, should the item need them, and this provides a valuable indication of the support offered by the trade in respect of particular designs.

A further good feature is that the list contains references to the *Model Engineer* volumes in which the design was originally described, assuming that it did originally appear in the magazine, thus providing a ready reference to what is usually described as the 'words and music'.

Many retailers are agents for the Nexus Plans Service and so provide an alternative local source of supply from this important collection of information. Some suppliers have developed their own designs over the years and are able to provide both the drawings and the castings that are required for these models or workshop items. The advertisements in the model press will provide details of these as will the individual suppliers' catalogues.

Choosing the first project

Your first project must be chosen with care. It must be relevant to your interests, and interesting and challenging to make, but above all, it must be a project which has an achievable conclusion. It is a mistake to commence your modelling with a too-detailed model as it may well take so long that you will become discouraged and the project will founder due to lack of progress. Better to choose a simple glow-plug single-cylinder engine than go immediately for the 9-cylinder radial that you really have your eye on. There is only disappointment to be found if a project founders due to lack of experience.

There is no need for this to happen. First of all, it must be appreciated that a complicated or extremely detailed item naturally takes longer to complete than a simple, straightforward one. Secondly, as a beginner or inexperienced modeller, there will inevitably be techniques to learn which will slow down the 'production rate', making completion that bit further away. The learning process is naturally essential to your progress as a modeller, and it must therefore take place, but it is best that it takes place on a simple model and that the project progresses to a conclusion.

It may not be true for every aspect of the hobby, but it certainly is for model locomotives, and may well be for other fields, that the older the model design, the simpler it is likely to be. There is a current tendency for models to be designed more as scale models and less as broad look-alikes for the prototype which they represent. Designers are tending to attempt a much improved scale appearance for their models and they are consequently becoming more detailed. This means that there are more parts to make, more rivets to insert, smaller fittings to assemble and increased problems of making and assembling the parts. Consequently there are more drawings to purchase, and, naturally, an increase in cost. There may also be an increase in the cost of the special castings, and many more of them.

On the other hand, locomotive designs by LBSC, and others, who popularised the hobby of model locomotive building, are intended to produce models without excessive detail. The drawings are simple and uncluttered, and although they may be lacking in detail concerning some actual dimensions and materials to be used, can generally be relied upon to produce workable models. They are ideal as a source of information for your first attempts. You will learn much from having to make your own decisions regarding materials, hole sizes and so on, and in addition you will be working with a set of drawings which were designed to produce a straightforward working model, without too many frills, which can be built by a beginner in a reasonable time.

Conventions used on drawings

Projection drawing

It is conventional to present the external outlines of an object by drawing elevations and a plan view, as illustrated in Figure 2.1. The illustration shown is presented in what is known as 'third angle projection'. In this method of presentation, the views of the object are presented in such a way that each shows what would be seen by looking on the near side of the adjacent view. Thus, the plan is drawn immediately above the roof and is the view which would be seen by looking down on it. The view of the right-hand end of the house is placed at the right-hand side of the front view, or front elevation, and shows what would be seen by standing at the right-hand side of the house and looking towards it.

An alternative method of showing the plan and elevation is also in use. This presents the plan and elevation of what can be called the 'garage' side of the house in the juxtapositions

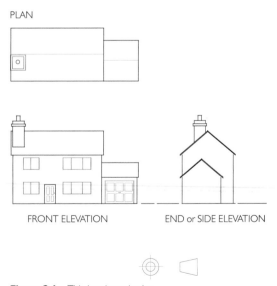

PLAN

FRONT ELEVATION   END or SIDE ELEVATION

Figure 2.1 Third angle projection.

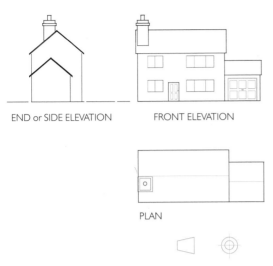

Figure 2.2 First angle projection.

The example of a full set of views in third angle projection which is shown in Figure 2.3 will serve as an example of drawing methods.

Views of a rectangular block are presented which show all six faces. The plan has the two side elevations above and below it. Together, they show that the top surface has a circular recess in it, since the two side elevations show its outline in dotted form, indicating that it is hidden.

The two end elevations appear on each side of the lower of the two side elevations. They also have the dotted outline of the circular recess indicated and show that a deep groove having a shallow groove within it is machined down the length of the block. The hidden edges of these grooves are shown dotted on the plan view.

shown in Figure 2.2. Therefore, although the views presented are the same, the use of first angle projection causes the end elevation and plan to appear on the opposite sides of the front drawing elevation from those in which they appear in third angle projection.

Modern drawing practice requires the projection which has been adopted to be shown in the margins of a drawing so that there can be no doubt of the draughtsman's intention. A small symbol should accompany the statement, as shown in the figures.

This convention is not generally followed on drawings for model engineers, so modellers must be aware that the two projections are in current use. Most model drawings known to me seem to adopt first angle projection, which was at one time the preferred projection in the UK. This is a pity since I was brought up on third angle projection, and as a consequence, at least one item in the workshop was made the wrong way round! The lesson to note is to be sure to use all of the information available on the complete drawing set to determine which way things are to be made. There is not usually a front door and roof to provide orientation!

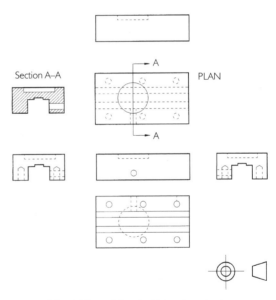

Figure 2.3 A full set of views in third angle projection.

The side elevations also indicate the presence of drilled holes in the downward extensions of the block, the dotted lines again indicating hidden features.

One side elevation and the view of the underside show that there are seven holes: six,

31

described as 'blind' since they do not penetrate the block, drilled in the bottom face, and the seventh drilled through from the side, into the central groove.

In a practical case, the drawing would carry dimensions to specify the sizes of all of the features, details of the holes and so on. Dimensions cannot always be shown on a simple outline drawing and there is frequently the need to provide a cross-section to show details which would otherwise not be evident.

The recess in the top surface of the block can be used as an example. This is revealed by drawing a section of the block, effectively slicing it along the line A–A and presenting the view which would be seen by looking in the direction of the arrows A. The portion of the block which has been 'cut through' by the sectioning process is hatched, whereas those parts not cut are shown without hatching, thus revealing the recess and the drilled hole which lie on the line A–A.

The sectional view is used whenever internal or hidden features must be shown in detail, or when dimensions need to be specified in some detail. Clearly, this is not always the case, since it is possible to specify the dimensions of the recess in the top surface of the block as ¾in. diameter × ³⁄₁₆in. deep (19mm diameter × 4.75mm deep) without the necessity to draw the section.

Sectional views are much used on model engineers' drawings, most commonly as what might be described as 'sectioned assembly' views, an example of which is shown in Figure 2.4. This shows an assembled water gauge for a model locomotive. The drawing has been sectioned to reveal the internal features so that it is possible to see the internal passages, the fit of the gauge glass, the seals at top and bottom of the glass and details of the blowdown valve in the bottom fitting.

On an actual drawing, the threads and sizes of the passages are specified, together with the

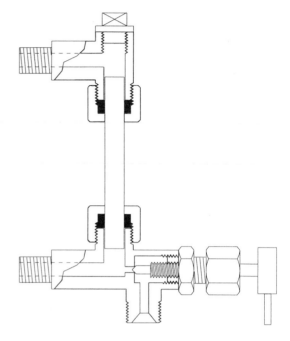

Figure 2.4 A locomotive water gauge shown as a sectioned assembly.

sizes of the hexagonal nuts and the general external dimensions of the parts, and the size of the glass tube.

Detailed dimensions are not usually provided for every last feature and it is sometimes necessary to measure the dimensions on the drawing to determine the sizes which are unspecified. To assist this process, the items are drawn at a suitable size to allow easy measurement. This is called the 'scale' of the drawing.

The most usual scale is full size for the item shown, but if the features are very small, as is frequently the case, items are drawn to a larger scale, typically twice full size and will be identified as such. Dimensions shown on such a detail are, however, given as the actual dimension required, and a dimension shown as ¼in. (6mm) is drawn at twice this size, a fact which can be confirmed by measuring the drawing with a rule. It is thus necessary to work to the stated dimensions, but if a dimension is unspecified and needs to be measured, the size required is naturally one half of that indicated by the rule.

Castings

If items having complex shapes are required, they are frequently produced as castings which embody the essential basic shape. A casting generally needs to be machined to produce the flat, circular and cylindrical sections which are its working or attachment surfaces, but it does not necessarily need to be machined all over.

If some surfaces can be left 'as cast', the drawing may indicate this by identifying those which must be machined. Several different symbols have been used to indicate the machined surfaces and two may still be encountered on model engineers' drawings.

The first method is illustrated in Figure 2.5 which shows a section through a cylindrical sleeve or bearing. It is made from a casting, and only the bore and the registers at each end need to be machined. This is indicated by using a tick to show each machined surface. Where there is

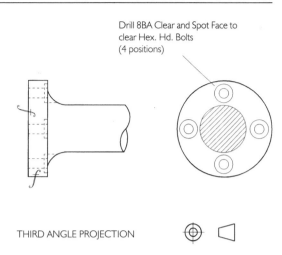

Drill 8BA Clear and Spot Face to clear Hex. Hd. Bolts (4 positions)

THIRD ANGLE PROJECTION

Figure 2.6 Indication of surfaces to be machined on a casting.

room, the tick is placed on the surface which is to be machined, but in cases where this is not possible, the tick touches a dimension line, or a similar line drawn especially for the purpose.

An alternative method of indicating machined surfaces is shown in Figure 2.6, which shows a cast shaft with a flange at one end. Only the end face and outside diameter of the flange need to be machined and these two surfaces are identified by a letter 'f' which is drawn through the surface to which it relates.

One further point about the use of castings is also shown in Figure 2.6. The flange of the shaft needs to be drilled through in four positions for attaching bolts and the unmachined cast surface of the flange needs to be machined locally so that the bolt heads, or nuts, bed down on to a squarely machined surface. This operation is called 'spot facing'. It is performed by a flat-ended cutter which is generally known as a counterbore but which might equally be called a spot facer on account of this method of use. Counterbores are described in Chapter 10.

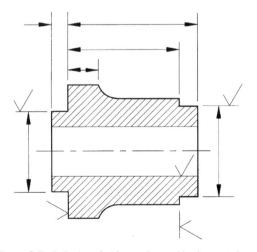

Figure 2.5 Indication of surfaces to be machined on a casting.

Dimensions and tolerances

General practice

If you are familiar with the sort of conventions adopted on the formal drawings used in industry, you will find model engineering drawings surprisingly lacking in detail. First of all, very few tolerances will be found on the details, these only being shown where a press (interference) fit is to be used or a defined clearance is to be provided to allow *Loctite* or one of the other modern adhesives to be used to secure two items. The situation which immediately comes to mind is the fitting of locomotive wheels to their axles, or the fits required when building up a crankshaft.

For running fits there is also seldom any indication of the sort of clearances which are envisaged, and it is usual to find simple instructions such as 'ream ½in.' on the detail drawing of the bearing and an equally simple '½in. diam.' on the shaft which fits the bearing. There is thus a need to understand what one is trying to achieve when making the different parts. This subject is described in more detail in Chapter 10.

It must also be appreciated that in many instances actual sizes do not matter and the amateur builder is expected to be reasonably competent in assessing the sort of fits and clearances that are required and to know which dimensions are critical and which are not. This is particularly true for running fits, where what matters above all is that there should be 'nice' clearances to suit the duty envisaged.

In clockmaking, the builder approaches the provision of some bearings in exactly this way, the bearing holes in clock plates being polished by broaching until the shaft 'feels' right when mounted between the plates. Model locomotive builders naturally need to provide good bearings without significant play when considering the valve gear components, but for some components of the running gear (axles particularly) sloppy fits are beneficial, within reason, and the model will ride the better because of them.

The general lack of tolerances arises because our models and other items are one-offs for the most part, since not many of us will want to build a 'production batch'. Therefore there is no need for interchangeability between two parts which are nominally identical since it is sufficient that each individual part fits those with which it immediately mates. This does not mean that there is necessarily scope for sloppiness in manufacture, only that the parts which I make do not have to be interchangeable with yours. Equally, if a 2-cylinder engine is being built, it is not absolutely vital that both bores are exactly the same size. It is sufficient that each piston fits the bore for which it is intended.

For this reason also, many minor or unimportant dimensions are sometimes omitted from the drawings, and the builder is left to determine his own actual sizes. In model loco-

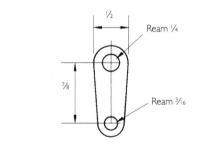

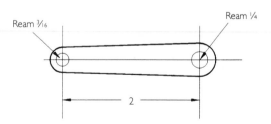

Figure 2.7 Incompletely dimensioned levers.

motive work, this seems to occur particularly with small links and pull rods which have rounded ends. The main dimensions are given, together with hole sizes, but the end radii are not always specified. Blend radii are similarly not identified. Figure 2.7 shows typical links taken from one of LBSC's designs of the 1950s. For these, and similar, links an end radius equal to the diameter of the hole will be found satisfactory and this two-to-one ratio in diameters may be used for all such links unless space or the shape of the prototype preclude this.

'Full' and 'bare' dimensions

Figure 2.8 shows the reversing lever and its stand for a model locomotive. The arrangement is typical. A long lever is pivoted near the bottom of the stand and is fitted with a latch and trigger which allow the lever to be locked at various positions throughout its full range of movement. The lever is connected to a reach rod which connects to the valve gear in some way, allowing full forward or full backward gear to be selected, or positions in between.

The reach rod is ⅛in. (3mm) thick and is linked to the handle by a shouldered screw which allows the necessary rotation between the two parts as the handle rotates about its pivot. To provide the clearance for this to occur, the shouldered screw is specified as being ⅛ FULL below the head, meaning that this dimension must be slightly greater than the thickness of the reach rod.

On the other hand, the fitted bolts holding the stand to the frame must not be longer in the unthreaded portion than the total thickness through which this part of the bolt passes. Since this is ⁷⁄₁₆in. (11mm) in this particular case, the dimension on the bolt is specified as '⁷⁄₁₆ BARE', i.e. barely (less than) ⁷⁄₁₆in.

Possible errors

It is unfortunately true that errors and inconsistencies do sometimes remain in a set of drawings when they go on sale to modellers. Model engineers seem not to be worried by the fact that drawings sometimes contain errors. Indeed, it is surprising that errors seem to persist for many years without them being reported to the publisher of the plans. There is doubtless much swearing when errors are discovered (and this is frequently after parts have been fully made) but after a couple of days one is usually reconciled to the fact that something must be remade, or perhaps a modification has already been schemed out that allows the incorrect parts to be used. Thus, what commences as a disaster is usually turned into a triumph and this somehow increases the pleasure of the hobby.

But it obviously does pay to talk to other modellers who have built the item concerned as they may remember if there were any particular

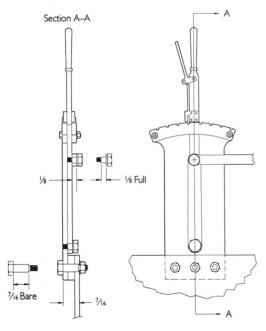

Figure 2.8 'Full' and 'bare' drawing conventions.

difficulties which could be corrected by amendments to the drawings.

Problems don't always become well known so it pays to look out for errors before it is too late. A certain amount of time must be allocated to just looking at the drawings, learning your way around them and trying to visualise how the various bits fit together. It is as well to look out for errors, for I fear that model engineers generally are not too good at reporting their findings to the designer or publisher. Nevertheless, it pays to buy an up-to-date set of drawings rather than use the set which Bill bought ten years ago.

Hole sizes

Drawings naturally contain details of the drill sizes that should be used to produce the general 'fixings and fastenings' holes that the model will require. If particular items are to be secured by rivets, the drawing shows hole positions or pitches and indicates the rivet size. For bolted or screwed together parts, the actual screw or bolt size is frequently not given but must be inferred from the hole sizes shown on the drawing. On the majority of drawings, these are given by reference to the series of number and letter drills which were once the standard drills for small holes. You will consequently find simple instructions such as 'Six holes, No. 34' or 'Three holes, No. 27' these being the normally accepted clearance size holes for 6BA and 4BA screws.

As a newcomer to the hobby you will most probably not have these sizes of drill available since they are now becoming the more expensive, second choice, non-preferred sizes. You will consequently have to convert these sizes into the metric equivalents in order that you may utilise the preferred range of drills. This subject is covered more fully in Chapters 10 and 11.

Materials

In many instances considerable choice is available as to the sort of materials that may be used. For example, in considering a steam engine (locomotive, traction engine or stationary engine) the cylinders may be made from brass, gunmetal or cast iron, dependent upon the duty envisaged, the size of the model or the size of the modeller's pocket – or perhaps all three. There is consequently a choice of the material to be used for the pistons to work in the cylinders, it being necessary to choose something compatible. The same is true for eccentrics and their related eccentric straps. Gunmetal straps go with mild steel eccentrics and mild steel straps with cast iron eccentrics.

Designers often acknowledge this by not specifying the materials to be used for such items, preferring to leave the builder to choose according to the size of his bank account or his personal wishes. For example, gunmetal is significantly more expensive than cast iron and modellers' suppliers frequently offer castings in both materials for such items as cylinders and their accessories, axleboxes, keeps, hornblocks and the like, for model locomotives.

Suppliers also offer materials cut specifically for the larger parts of a model, leaving the builder to decide only on the smaller items. Background reading will provide basic guidance, but there are some general rules which are quickly learned. If it needs to be strong, it should be steel, and if it's in contact with water (at normal temperatures) it should be brass, but gunmetal or phosphor bronze must be used for their better wearing qualities, for the bodies of valves and so on. Stainless steel is used for control rods and spindles for valves for both water and steam. Boilers are normally made in copper, with gunmetal or phosphor bronze bushes being fitted to the structure for the attachment of fittings. All of these metals are described in Chapter 3.

Material sizes

For many parts of a model the builder must decide for himself the size of bar material that will be used to make particular items. The drawings will show the critical dimensions of the parts and it is up to the builder to determine the method of manufacture and hence the size of material which is appropriate.

Most older designs, indeed the majority of designs, were prepared to imperial scales, and parts are most easily made from imperial materials.

At the present time, suppliers' catalogues still list imperial sizes of round and rectangular bars in the common materials. In the longer run it is likely that only metric materials will be available although it does seem that imperial copper tubes will remain in stock for the foreseeable future, thus making things relatively straightforward for the boiler builders among us.

For sheet materials and wires, older designs will be sure to use the Standard Wire Gauge (swg) nomenclature for thickness. This is the standard imperial method of specifying wire and sheet thicknesses which uses a simple number referencing system. This allocates a series of numbers to the different thicknesses or

diameters which are produced. The number series commences at zero and progresses upwards as the material becomes thinner or smaller in diameter. Several different standards were used but the British swg system was almost universally adopted in recent times in the UK, and the list is tabulated here.

Only the even-numbered gauges were widely available and these are most likely to be met as specifiers for sheet thicknesses and wire diameters on imperial drawings. The most common range of thicknesses likely to be encountered will be from 10 swg, 0.128in. (3.3mm) down to, say, 30 swg, at .0124in. (0.3mm). Below this, thicknesses are definitely in the 'shim' category and are usually specified as a decimal fraction of an inch, or simply described as 'ten thou' (0.25mm) etc.

The system may seem cumbersome, but in reality, most of us who are used to the system could readily reel off the decimal thicknesses for the very commonly used 16, 18 and 20 swg and remember 10 swg for its closeness to ⅛in. The apparent illogicality of the reversed numbering system is explained by the fact that the thinner materials have been through the rolls or drawing-down dies more times than the thicker ones.

Imperial standard wire gauges

No.	Diameter in.	Diameter mm	No.	Diameter in.	Diameter mm
0	.324	8.23	26	.018	0.547
2	.276	7.01	28	.0148	0.375
4	.232	5.89	30	.0124	0.314
6	.182	4.88	32	.0108	0.274
8	.160	4.06	34	.0092	0.233
10	.128	3.25	36	.0076	0.193
12	.104	2.64	38	.0060	0.152
14	.080	2.03	40	.0048	0.121
16	.064	1.63	42	.0040	0.101
18	.048	1.22	44	.0032	0.081
20	.036	0.914	46	.0024	0.060
22	.028	0.711	48	.0016	0.040
24	.022	0.558			

Supplies

Obtaining supplies

For most model engineers, material supplies are usually obtained from retail outlets which cater specifically for the amateur worker. Such retailers are in business to satisfy the customer who only requires material in relatively small quantities. Flats, rounds and hexagons will therefore most probably be available in 2ft (600mm) lengths, or perhaps up to 3ft (1m), standard lengths being limited by the fact that much of their business is by mail order and longer lengths are not suitable for transmission by post.

Sheet materials will also probably be available in standard sized (small) sheets, but some suppliers offer a cutting-to-size service, especially for the more expensive materials such as brass and copper. Cut-to-size pieces in these more expensive metals are often sold by weight.

Retail outlets obtain their supplies from wholesalers who are normally described as metal stockists or metal factors. If you can obtain a factor's list, you will see that every material which he sells is defined by a reference code of some sort. These codes define the available materials quite precisely, and you could therefore purchase say, riveting and turning quality brass (CZ131) as distinct from drawn brass (CZ108).

Retail outlets do not, however, normally indicate the precise type of material which they sell. This allows them some flexibility in the source of supply since they do not necessarily have to purchase specific alloys. You will be offered round brass rod in various diameters, phosphor bronze or cast gunmetal without reference to the precise material which will be supplied. This is not a problem since you can rely on the experience of the retailer to stock material which is adequate for general needs, and unless you have a specific use for something out of the ordinary, the materials you obtain from retail suppliers are entirely satisfactory. You do not need to know any detail of British Standard reference numbers, except to be aware that they do exist, and it might in due course be necessary to investigate further.

As a beginner therefore, if the drawing indicates that brass is required, just get hold of some brass and make the thing. It cannot be critical or the designer would have mentioned the particular grade he had in mind.

There are similar detailed specifications for other materials too, but the local supplier can be relied upon to stock a material that is likely to meet most common applications. So, for a start at least, just accept the colloquial names for these materials and purchase them as such.

Scrap merchants also provide another possible source of materials and should certainly be considered whenever you have no particular interest in the actual specification.

For the purchase of steel however, it is best to avoid anything for which you do not have a reasonably precise definition. There seem to be large quantities of steel produced for industrial uses which have desirable characteristics such as cheapness and good weldability which suit them to the manufacture of particular types of goods. There is no doubt that they are entirely adequate for their intended purposes, but if you 'come by' some of them you may find them hard to cut or impossible to turn to a good finish and high consumers of both good temper and time.

As a learner lathe operator you will inevitably experience certain difficulties during early attempts to cut metal on the machine and can avoid some of the likely pitfalls by purchasing your mild steel rounds in a free-cutting variety. Steel is not, in any event, particularly expensive, and the cost penalty in buying new steel is not an onerous one to bear. You will also need to have some bright-drawn mild steel available for use when case hardening is to be employed since some free-cutting steels are not recommended for hardening by carburisation.

If hardening and tempering of steel is required, for the manufacture of tools and cutters, either silver steel or gauge plate will be used. You will, therefore, need stocks of three or four distinctly different steels and should buy your requirements against the specific descriptions of bright-drawn (or black) mild steel, free-cutting mild steel, silver steel or gauge plate. This latter is also known as ground flat stock since it is supplied in rectangular cross-sections, having a ground surface finish.

An introduction to these and other engineering materials is provided in Chapter 3 which also describes heat treatment methods (annealing, normalising, hardening, tempering and carburising) which may be performed in the home workshop.

Having different materials in stock in the workshop, which might easily be misidentified, you should devise an indelible marking system so that you do not inadvertently pick out stainless steel instead of silver steel, only to find that the beautifully made tool you have just completed will not harden as you had expected. A simple system can be adopted in which one or two (generous) splashes of coloured paint mark the end of each bar according to the type of material.

One point to bear in mind is that metal is bulky, heavy stuff which attracts a not inconsiderable amount of postage and packing charges. Most of us have to order materials by post from time to time, but it is surprisingly easy to incur a charge of a few pounds in respect of transport costs. It is probably best, therefore, to assess your requirements ahead of time and plan a bulk-buying trip to a retailer who is likely to be able to satisfy your needs. If you are a member of a club, or otherwise in touch with other modellers in your locality, a joint visit in which you share the transport costs should allow access to establishments relatively far away, and if you can assess your needs for the next modelling 'season', you are likely to find this approach quite worthwhile. Even if you travel alone, you might as well spend the money on petrol and take the family out for the day, as pay the costs of having someone else transport the goods for you.

What materials to buy

It is difficult to be specific about what a beginner should buy by way of building up a small stock of materials as a start in the hobby. In the first place, tastes vary and what would be suitable for a clockmaker might remain in the workshop of a live steam man for a very long time before being of much use. In the second place, space to store the material will vary between individuals and so will the depth of one's purse, and there is always the problem posed by the location of the nearest supplier. It is all very well running out of that vital bit of material if there is a supplier in the locality, but quite a different matter to have to order by post and endure the consequent costs and delays.

Of course, with inflation being what it is, a little stock is a great comfort, especially if one is approaching retirement, but for a great many projects the bulk of the cost will be found to be in the castings and the special-to-model items rather than in the general materials and therefore it does not pay to invest large amounts of money in ordinary stock items. Should you have spare money, it is best invested in the special items that your next model (or whatever) will require, given that you already have a basic stock of commonly used materials.

So what should comprise the basic stock? In a way it is easier to be specific about what the basic stock should not be. It should not contain large quantities of the more expensive items such as brass, gunmetal and phosphor bronze. It is best to buy these in what appears to you to be sensible quantities, once a specific need has been identified. Naturally, you are likely to need some of the common sizes of round brass rod and might reasonably have stocks of the common sizes up to ⅜in. (10mm) but beyond this diameter the price begins to increase dramatically and unless you are likely to use large quantities on a particular project it is certainly not worth buying the larger sizes initially. The same applies also to hexagon brass. Large sizes are quite costly and should only be bought as required, perhaps with a small amount added on for luck.

Some hexagon brass will be useful, especially for the live steam man, but it is likely that only relatively small quantities will be required and nothing above ⁷⁄₁₆in. (11mm) or ½in. (12.5mm) across flats (AF) should be bought for the general stock.

Similar arguments can also be used in relation to phosphor bronze and gunmetal. Large amounts of these materials are not likely to be required and the acquisition of stocks can be made a low priority, certainly in the early days of establishing a workshop.

One point to bear in mind in relation to these more expensive materials is the cost of using a larger size bar than is actually required for the job. Reducing a ½in. (13mm) bar to ⅜in.(10mm) in diameter is almost the equivalent of turning away the amount of metal in a ⁹⁄₃₂in. (8mm) diameter rod, so it pays to use the size you actually need and you may need to keep a larger number of diameters in stock in order to economise on the amount of metal used. But it does of course depend upon the total amount of metal which you personally use, and for the occasional job it might be acceptable if it avoids buying a further size.

Mild steel is the cheapest of the materials which you will need in the workshop. Paradoxically, this means that modellers tend to keep a wider variety of cross-sections and sizes than they do of the expensive materials, although with cheaper stuff it is obviously not so costly to reduce a larger diameter to the size required. However, the temptation is always there to buy some more because it's good value, but do consider whether you can store it and also whether you are really likely to use it. Looking at my own stock of steel, there are certainly some sections in which I seem to have more than a lifetime's supply but since there is as yet no storage problem, no harm seems to have been done, except that money has perhaps been invested unwisely.

Steels are available in several different varie-ties and also in different sections, although the full range of sections is only likely to be readily available in what is known as bright mild steel. This is nice, clean material but is not necessarily the best form in which to purchase your steel. Items which are to be machined all over are best made from black mild steel and, if this is available, it is most definitely the best mild steel to purchase. A stock of round bright mild steel in ⅛in. (3mm) increments from ³⁄₁₆in. (5mm) to ½in. (13mm) in diameter will be likely to satisfy your initial needs and the stock can then be built up to suit your activities once the usage rates become clear.

Ground mild steel should be available from your supplier and this will provide a source of steel having an accurately sized finish. This can be used for shafts and axles for which an accurate diameter is required along the length and its use does reduce the amount of work that is needed when making such items.

Steel is also available in flats and hexagons. A small stock of flats is likely to be generally useful and might include a range of thicknesses from ¹⁄₁₆in. (1.5mm) up to ¼in. (6mm) in ¹⁄₁₆in. increments, but not in too many different widths. Up to ¼in. thick, material may readily be cut to size with a hacksaw and unless you are particularly averse to this activity you need not stock a great range of widths.

There isn't a very great demand for square and hexagonal mild steel, although this latter is useful for making the occasional nuts and bolts. This is not an activity that is likely to be required too often, but it is occasionally necessary to manufacture special fine-thread nuts and bolts for particular applications and a small stock to suit the range of spanners you use will be useful. General stock up to ½in. (12.5mm) will probably find ready use.

It might also be found helpful to have stocks of steel hexagons in the BA nut and bolt sizes, particularly the odd-numbered threads, as these are useful for making the small-head bolts which find wide use in models. If you envisage that you will require hexagon-head brass bolts or screws with these smaller-sized heads, then a small stock of the appropriate sizes might also be useful.

You are also likely to need a 'hardenable' steel for the manufacture of punches, scribers, cutters and other small tools. This material is that known as silver steel which is supplied in standard 13in. (330mm) lengths ground accurately to diameter. An initial stock of the smaller sizes, say up to ⅜in. (10mm) in diameter is certain to be useful. Another hardenable steel, known as gauge plate, is available in rectangular cross-sections but this should only be bought as required.

Chapter 3 contains an introduction to the materials that are likely to be of use in the amateur workshop, and in particular provides guidance concerning hardening and also describes the types of steel that are most suitable for the different applications that might be met. Steel is relatively cheap, as noted above, but comes in a range of different alloys, surface finishes and characteristics and it is as well to be aware of the type with which one is working. There is little point in buying 'any old steel' just because it is available and cheap, unless you actually need some rough old steel for a rough old job.

Useful oddments

You will frequently read about items 'made from the scrapbox', so it is obvious that many model engineers have one of these ubiquitous boxes. This is not to say that we produce large quantities of scrap, although this is regrettably an occasional activity. The scrapbox mostly contains material which is too good to throw away, but which is not considered to be part of the general stock. This may be because the material is not stocked in sufficient quantity to have a 'bin' of its own, or is an odd length or offcut that would easily be overlooked if kept with similar stock. It may also be that you are not exactly sure what the material is, but feel nevertheless that it may be useful. There will also be some of your mistakes which now simply represent material which might be useful, one day.

What goes into the scrapbox, and how big it is, is entirely a personal choice. It is certainly useful to segregate the materials and separate ferrous and non-ferrous boxes should prove useful in shortening any searching time when you are looking for that odd bit of material. It is probably also helpful to segregate any large-diameter offcuts since these are usually held in small quantities, and it is a nuisance if small items are mixed in with them.

If you make items which need lots of small parts, such as small-scale model locomotives, it is certainly a time-saver to segregate small offcuts of brass and nickel silver into a separate box, which need not, in any event, be very large.

If you are in a position to obtain offcuts of materials at reasonable prices, it is probably wise to concentrate initially on the more expensive ones, basically the copper alloys, brass, phosphor bronze and gunmetal, especially in the smaller sizes. Useful materials not widely sold by the modelling trade are aluminium alloys in thicker sheets, say, from about ¼in. (6mm) upwards. There is not likely to be much large-scale need for thin alloy sheets, but the thicker pieces are excellent for the manufacture of bending bars (around which material is bent – see Chapter 7) and for cutting out formers for flanging items in copper and brass.

3

Introduction to engineering materials

Iron

Iron is produced from iron ore by heating it to a high temperature, in the presence of coke and limestone, and subjecting it to an air blast. At the high temperatures achieved in the blast furnace (up to about 1800°C) chemical reactions take place that release the iron from the ore. At these temperatures, the iron is molten and it falls down to the bottom of the furnace where it is run off into open sand moulds, forming billets of iron, known as pigs. The material is therefore called pig iron.

As the molten iron falls through the furnace, it picks up impurities from the limestone and the coke, the most important of which is carbon, and the iron normally contains between three and five per cent of this element.

The rate of cooling of the iron determines the form of the carbon within the material, which may be present as graphite or as iron carbide (cementite). Slow cooling produces graphite rather than cementite and produces a coarse-grained structure which shows a characteristic grey colour if the material is fractured. Pig iron containing nearly all of its carbon as cementite is called white iron because it shows a whiter, more close-grained structure when fractured. Iron containing most of its carbon as graphite is more easily machined than white iron.

Pig iron is not used as an engineering material but is further refined in a second furnace. By incorporating special additions to the melt, a variety of irons can be produced, the two main classifications being grey cast iron and white cast iron. Since cementite is intensely hard, white cast iron is hard and durable, but extremely brittle, whereas the graphite in grey cast iron is a good lubricant and this iron is therefore softer, less brittle and easily machined. Grey cast iron is also cheap and melts at the relatively low temperature of 1200°C. It is also extremely fluid when molten thus enabling intricate shapes to be cast.

Grey cast iron has a wide range of uses, such as machine frames and machine tool beds and is used for the production of a wide range of castings, particularly for the modeller and home machinist.

Virtually all of the carbon in cast iron may be removed by reheating it in a puddling furnace and mixing millscale (oxide) repeatedly into the molten metal. This forms chemical compounds with any carbon which is present and the iron may be reduced to an almost-carbonl ss state. Chemically, it is almost pure iron, usually greater than 99%, with small amounts of other elements. This material is known as wrought iron. It is very durable, bends easily, even when cold, but resists high shock loads without permanent damage. Virtually all commercial puddling furnaces are now closed down, and it has been left to industrial museums to continue the production of wrought iron.

Steel

Production of steel

The difference between steel and iron is determined by the amount of carbon which is present. Wrought iron has virtually no carbon whereas cast iron contains between 2 and 5 per cent. Steels lie between these two extremes, normally containing between 0.1 and 2 per cent carbon. Although other elements may be present, the carbon is by far the most important constituent.

Consequently, plain (non-alloy) steels are normally classified according to their carbon content, the commonest material being mild steel, which has a carbon content of between 0.1 and 0.25 per cent. Medium-carbon steels contain from 0.2 to 0.5 per cent carbon and high-carbon steels have more than 0.5 per cent.

Black and bright-drawn mild steel

The most-frequently used steel is that known as mild steel. This has between 0.1 and 0.25 per cent carbon. It is widely available in sheet and bar form, making rounds, flats (rectangles and squares) and hexagons all readily available, in addition to sheet, in a wide range of sizes and thicknesses.

At the steelworks, the basic material is produced in large billets which must be rolled, pressed, drawn or otherwise worked down to the sizes and cross sections which are required. This working down may take place with the material in a hot condition, in which case the finished steel has a characteristic black surface scale and relatively poor surface finish. Having been hot worked, the material is, however, relatively free from internal stresses.

For the most part, model engineers' suppliers stock what is known as Bright-Drawn Mild Steel (BDMS), sometimes simply referred to as Bright Mild Steel (BMS). This has a bright, scale-free surface since the sections are produced by cold-drawing billets of steel progressively through a range of dies to produce the required sections. The cold drawing (cold working) stresses the steel and although what is called the tensile strength actually increases, the ductility (the flow capability, or the ability to bend) is considerably reduced. This means that BMS should be heat treated before any cold bending operations are performed, otherwise the material may be overstressed and a fracture will result.

The locked-in stresses in BMS are also a problem if large parts of a rectangular cross-section bar are removed to form slots or holes, or if large sections are removed to produce a tapered and fluted connecting rod, for example. Although stressed by the cold-drawing process, the rectangular bar is in equilibrium and will remain in the same shape if no metal is removed. Taking material out of a bar which is in this stressed condition, removes a vital part, the initial presence of which balances the stresses in adjacent parts. The stresses in the remaining material are therefore unbalanced, and distortion is the result.

BMS is not the ideal material for use when making parts of complex shape. However, the internal stresses can be relieved, and stress relieving of rectangular cross-section BMS (particularly) should become the rule rather than the exception. Relieving the stresses in round or hexagonal BMS is not normally undertaken, since turning tends to remove metal in a symmetrical fashion. Stress relieving, or normalising, is described below.

BMS was originally introduced for use in situations where a clean surface finish was required without great accuracy, so if a particularly accurate overall size is required, machining of the outside will be necessary and the next larger standard size of steel will naturally have to be used. There is, therefore, no disadvantage in the use of black mild steel and it does have considerable advantages where large amounts of metal need to be removed from rectangular cross-sections or where cold bending will be performed.

If accurately sized circular rods of mild steel are required, Precision Ground Mild Steel (PGMS) is readily available from modellers' suppliers. This is available in increments of ⅛in. from ⅜in. to 1⅛in. and in some metric sizes, from 6mm to 40mm in diameter. This material is ideal for model locomotive axles, crankpins and the like. Many modellers seem not to appreciate the ready availability of PGMS and tend instead to use an alternative steel, also having a ground surface finish, known as silver steel (see below) but PGMS is preferable if load carrying is required.

Free-cutting steel

Mild steel is fairly easy to machine to a good surface finish, but the removed material usually comes off in long, spiral curls which fly all over the workshop. In industry, where high-speed turning and drilling are naturally adopted as a means to reduce costs, the removal of the long curls of swarf from the machine tools poses significant problems. Consequently, steels have been developed from which the removed material comes off as small chips rather than long curls. The breaking up of the curls is caused by adding 0.1 to 0.2 per cent each of phosphorus and sulphur to the steel. This induces a certain degree of brittleness which causes the removed material to come off in the form of chips. Steel which has been 'doped' in this way is known as Free-Cutting Mild Steel, sometimes abbreviated to FC Mild Steel. It is easier to produce a smooth machined surface on free-cutting steel, hence its usefulness in the home workshop.

Carbon steel

Steels containing more than 0.5 per cent carbon are known as high-carbon steels. They are extremely useful since they are supplied in a soft condition but may be hardened to different degrees of hardness without elaborate equipment. Two forms of high-carbon steel are of interest to the amateur because of their wide availability – silver steel and gauge plate.

Silver steel is a high-carbon steel suitable for hardening, intended for the manufacture of cutting tools, punches, scribers etc. It is most commonly supplied in 13-inch (330mm) lengths of ground rod manufactured by the firm of P. Stubs. These have a ground finish accurately sized to within .00025in. (.00635mm). The rods carry the imprint 'STUBS' at one end which distinguishes them from other bright steels.

Stubs silver steel has 1.1 or 1.2 per cent carbon and small additions of manganese, chromium and silicon. It has a slightly higher tensile strength than mild steel but is much less ductile and it should not be used for load-bearing pins and suchlike. It is frequently recommended for such duties as crankpins due to its accurate grinding to size during manufacture, but it is more difficult to machine than mild steel and its wearing properties are no better. It therefore offers no mechanical advantages over mild steel and its higher cost is disadvantageous, to say the least.

Silver steel should also not be considered for use in its hardened state for such load-bearing duties. If a harder, wear-resisting surface is required, this may be obtained by case-hardening mild steel. This produces a hard surface with a softer, more-resilient core, whereas hardened silver steel is much more brittle (right through) and is therefore more liable to fail in service.

Silver steel is very difficult to machine to a smooth surface finish and a free-cutting carbon steel is becoming available from some suppliers which has the same properties of 'hardenability'. This is not yet available in such a wide range of sizes as silver steel but offers a big advantage in terms of ease of machining. It is also cheaper than the ground-finish Stubs silver steel.

Hardening and tempering of silver steel for the manufacture of tools, for which the material is intended, is described under the heading 'Heat treatment of steel' below. Case hardening of mild steel is also described later.

Gauge plate, or ground flat stock, to use its other name, is also a high-carbon steel, but is supplied in rectangular cross-section bars, 18in. (500mm) long, accurately ground to thickness. Composition varies, but generally about 1 per cent of carbon is present, together with small amounts of manganese, chromium, tungsten and vanadium.

As its name implies, gauge plate was introduced for the manufacture of gauges. Like silver steel, the material is supplied in the soft state but may be hardened after machining to the required shape. It is also suitable for making cutting tools and is especially useful for making form tools, which can be a great help when a 'production run' of small turned components is required. Gauge plate is very expensive (7 to 10 times the price of mild steel) so should only be bought as required.

Hardening and tempering of gauge plate is performed in a broadly similar fashion to that used for silver steel, but the required technique is subtly different and varies from one manufacturer to another (see 'Heat treatment of steel' below).

Silver steel and gauge plate are both available in imperial and metric sizes.

Stainless steel

It is possible to create an alloy steel which is virtually rustless. This is accomplished by utilising chromium as an alloying component with iron, together with small percentages of other elements. The chromium produces a bright, silver-coloured alloy and when the percentage of chromium exceeds 10 or 11 per cent, the range of materials known as the stainless steels is produced.

Chromium is commonly used as the plated finish on steel or brass, either for decoration, as a means of avoiding corrosion, or to produce a hard surface. Chromium combines readily with oxygen and forms an oxide skin which effectively precludes further chemical action. Alloyed with iron, it confers its 'rustless' quality on the resulting alloy, but it is a hard material and this makes most stainless steels difficult to machine.

Stainless steels do however vary in their compositions. The principal additional component to these alloys is either nickel or carbon and if one is present, the other occurs only as a trace impurity. Quite large proportions of nickel may be used (up to 20 per cent) with between 8 and 20 per cent of chromium. However, if carbon is used in the alloy it does not normally exceed 2 per cent. The presence of carbon is vital if the stainless steel is to be subjected to heat treatment but this is not normally required by the amateur worker.

Nickel forms one of the most versatile and important elements used in the production of alloy steels (of which the stainless types are only one variety) since, depending on the percentage of nickel which is present, slow cooling, or annealing, can leave the steel in various conditions and special non-magnetic irons and steels can be manufactured.

Many stainless steels are therefore either non-magnetic, or only mildly so, but some may require annealing to achieve the non-magnetic state since cold working tends to induce a degree of magnetism.

Due to the presence of large percentages of chromium, stainless steels can be difficult to machine. To overcome this problem, free-cutting or free-machining alloys are available. These are created by the addition of small amounts of sulphur, as for free-cutting mild steel, but for stainless steels, selenium is sometimes used instead of sulphur.

Only very small percentages of sulphur are required, a typical analysis showing only 0.3 per cent of this element. One particular free-cutting alloy is very close to the universally known 18/8 stainless steel (18 per cent of chromium with 8 per cent nickel) having 17 to 19 per cent chromium, 8 to 11 per cent nickel and up to 1 and 2 per cent of silicon and manganese. It is described as a non-magnetic, free-machining alloy and contains a maximum of 0.12 per cent of carbon.

Model engineers' suppliers normally have available tubes in stainless steel in a small range of sizes up to about 0.5in. (12mm) in diameter and circular rods up to 1in. (25mm). A small range of rectangular sections is also available from suppliers catering for the live steam enthusiast as it is used for superheater return bends and for the manufacture of fire grates. Ground rods are available from some sources.

Heat treatment of steel

Types of heat treatment

Heat treatment is an operation, or combination of operations, involving controlled heating and cooling of a metal or alloy while in the solid state. The purpose of the heat treatment is to bring about some change in the physical properties of the material with the object of changing its hardness, increasing its strength or making it more ductile or malleable. Most metals and alloys react to some form of heat treatment.

Heat treatment of steel is important since the carbon which is present in all steels is capable of being transformed into hard or soft varieties by heating and cooling an item in a precise, but simple, manner. The main forms of heat treatment performed on steel are as follows:

Annealing
This is a process of softening a previously hard metal, usually so that it may be reworked in the soft state.

Hardening
Steel may be hardened to increase its resistance to wear and abrasion, or it may even be rendered sufficiently hard to cut other metals.

Tempering
Steel which has been hardened may be extremely brittle, making it useless for most applications. Some of the ductility may be regained by sacrificing some of the hardness, the process being known as tempering.

Normalising
Normalising is a process which restores the structure of a metal to its normal condition. If a metal has been worked (hammered, drawn or rolled) while in a cold state the grain structure becomes distorted.

This can also occur if the metal has been subjected to prolonged heating, when grain growth, or swelling, takes place. Grain distortion is undesirable and normalising is used to restore grains to normal size. It is therefore sometimes referred to as grain refining.

Hardening of carbon steel

Steel is effectively pure iron to which small quantities of carbon have been added. This results in the formation of iron carbide, or cementite, making the steel progressively harder and tougher as the carbon content increases up to about 1.5 per cent.

Steel which has 0.87 per cent of carbon is known as eutectoid steel. Its structure comprises thin particles of cementite alternating with similar particles of ferrite, in the ratio of 87 per cent ferrite to 13 per cent cementite. This particular structure is known by the name Pearlite.

If eutectoid steel is heated to 700°C, the pearlite undergoes a change as the cementite and ferrite merge together to form a solid solution of carbon in iron, called austenite. This is non-magnetic and the transformation may be confirmed by bringing a magnet close to the heated steel.

A similar change takes place for all steels, except that, for other percentages of carbon the change takes place in two stages. At the lower critical point (700°C) the change in structure starts to take place but only for eutectoid steel does the change completely occur at this point. If there is more or less carbon than is found in eutectoid steel (0.87 per cent) the change to austenite is not completed until what is known as the upper critical point is passed. This varies according to the percentage of carbon which is present. For silver steel (1.1 or 1.2 per cent carbon) the upper critical point is 800°C.

The change to austenite, during heating, does take time to occur and the steel must be held above the upper critical point for a reasonable time. 'Reasonable' is generally taken to be roughly one hour per inch of thickness. This means 15 minutes for work which is 0.25in. (6mm) thick, or 7 to 8 minutes for work which is 0.125in. (3mm) thick. These times should enable the maximum possible hardness to be achieved.

The most important feature of austenite is that it can only revert to pearlite if it cools slowly. Rapid cooling prevents this change and instead 'freezes' the structure in a modified austenitic condition which takes the form of a glass-hard constituent called martensite.

The more rapid the cooling of the austenite, the greater the quantity of the glass-hard martensite which is created and the harder is the resultant steel. However, extreme hardness brings with it extreme brittleness.

Rapid cooling is arranged by plunging the heated material into water, or better still, brine (¾lb of salt per gallon of water, 500g per 6 litres) which is at room temperature. The item (tool) must be plunged into the liquid while still above the upper critical point, pointed or 'business' end first, and should be given a gentle up and down motion to encourage as rapid a cooling as possible.

Brine is a better quenching bath than water due to the fact that, in water, bubbles tend to form on the surface of the tool as the water boils locally when the tool tip is plunged in. The bubbles insulate the metal from the cold water and therefore do not provide such a rapid quench as is possible with brine which does not exhibit this effect. Nevertheless, little trouble is likely to be experienced with water quenching if the tool is given a gentle motion, as described above.

A similar procedure is also used for gauge plate except that this is described as 'oil hardening' and should therefore be quenched in oil. A clean engine oil, viscosity about 20SAE, can be used for quenching, but it is inclined to flame-up and special quenching oils are available which avoid this problem.

Hardening temperatures for gauge plate do vary from manufacturer to manufacturer as the critical points are modified by the small additions of manganese, vanadium etc. which are usually present. The manufacturer's recommendations, which should be available with the purchased material, should be followed, but as a general rule, a temperature of between 810°C and 840°C will be found satisfactory for hardening.

Estimation of temperatures for hardening

For hardening, temperatures of the order of 700–800°C must be estimated; broadly in the 'red hot' category. However, there are degrees of redness and although the degrees are highly subjective, they do give a guide to the temperature which has been achieved. The discernible red colours commence at a temperature of just over 500°C with the 'barely red' description and continue through to 'bright orange-red' which occurs at 1000°C. The full list of temperatures, together with their usual descriptions, as seen in subdued daylight, is as follows:

Colour	Temperature (°C)
Barely red	520
Dull red	700
Blood red	750
Cherry red	800
Bright cherry	825
Red	850
Bright red	900
Very bright red	950
Bright orange-red	1000

The above descriptions are highly subjective but you should be able to experiment by heating a steel sample and watching the colour change progressively as the temperature increases. For the hardening of silver steel and gauge plate, temperatures in the range 800–850°C are required, putting the colour range in the region of cherry red to bright red.

Although it is necessary to ensure that the temperature of the whole sample has reached a value above the upper critical point, and has remained there long enough to convert all of the material to austenite, there are dangers in holding the work at high temperatures for long periods. The first problem is decarburisation which is the 'burning out' of the carbon from the heated surface of the metal. This reduction of the carbon reduces the hardness which may be achieved and therefore destroys the very quality which is desired.

The second problem is caused by oxidation of the metal surface when held at high temperatures. This produces a scale on the surface which degrades whatever surface finish the item had prior to commencing the hardening process and may make it unusable in some instances.

If the tool is to be ground after hardening, then neither effect is likely to be a problem since the scaled or decarburised material will be removed. If this is not the case, the method used to overcome the problem is to coat the point (business end) of the tool in a protective material such as a paste made from chalk and water. Sometimes, it is recommended that iron wire should be wrapped around the sample in addition to the use of chalk paste, but the wire must be removed before quenching otherwise the surface may not be cooled sufficiently rapidly to produce the required degree of hardness.

Tempering

Once the hardening has been performed, the material is extremely brittle and the tool may shatter if dropped. The material must therefore be 'let down' from this very hard but brittle state so that it is just as hard as is needed for the particular service. The letting down, or tempering, also relieves the internal stresses caused by the sudden cooling during quenching.

Tempering consists of heating the hardened steel to a certain temperature and then cooling it. The hardened steel consists largely of martensite, but if this is reheated, further changes in the structure occur to create softer, tougher materials.

The higher the temperature to which the martensite is raised, the softer and more ductile is the steel structure which is formed. Reheating hardened steel to just below the 700°C lower critical point produces a modified cementite which has minimum hardness and maximum ductility. Reheating to lower temperatures leaves progressively harder compounds as the reheating temperature reduces. Clearly, no reheating leaves the martensite unaltered and therefore at maximum hardness. A range of structures of differing hardness can be created by reheating hardened carbon steel to a range of different temperatures.

Estimation of temperatures for tempering

For tempering, the actual temperature to which the steel is raised is very critical, ranging from 200°C to a little over 300°C depending on the service to which the tool is to be put. When it is realised that this limited temperature range covers the full range of hardness which is likely to be required, from cutting tools to 'softish' springs, the degree of control of temperature can be appreciated. Fortunately, these temperatures can also be estimated by observing the colour of the tool as it rises in temperature.

If steel is heated in an oxidising atmosphere such as air, a film of oxide forms on the surface. The oxide changes colour as the temperature increases and the colours are used as a means of judging the temperature. This is, again, a subjective test, but the practice is so well established that little difficulty is likely to be encountered.

Colour	Approximate Temperature (°C)	Uses
Pale yellow	210	Turning tools for brass
Pale straw	230	Turning tools for steel, taps and dies
Dark straw	250	Drills, milling cutters
Yellow-brown	255	Cutters for hard wood
Brown-red	265	Wood boring tools
Brown-purple	270	Twist drills, coopers' tools
Light purple	275	Axes
Purple	280	Bone and ivory saws
Dark purple	290	Punches and cold chisels
Full blue	295	Screwdrivers
Dark blue	300	Springs

Lists of tempering colours and their related temperatures do vary from one authority to another, as might be expected for such a subjective test. Some lists are very abbreviated, showing only six or seven colours, but a relatively comprehensive list is given here, together with the approximate equivalent temperatures and the normal service for which the tempered steel is used. In normal references to tempering, little attention is paid to the actual temperatures required, the usual advice being to 'temper to dark straw' or whatever.

The best way to reconcile the list of colours in your mind, and to practise the procedure, is to carry out an experiment for yourself. Take a piece of mild steel, about 0.25in. (6mm) in diameter, and polish a couple of inches at one end with emery cloth until it has a nice bright finish. The colours of the oxides of steel do not develop correctly within a flame, so the end of the rod (point of the tool) cannot be heated directly. Instead, heat a point about 1.5in. (40 mm) from one end with a single small flame from a blowtorch. As the steel heats up, the range of oxide colours will be seen progressively, from pale yellow, through straw, brown, purple and blue as the temperature increases.

If the flame is held stationary, 1.5in. (40mm) from the end, a band of the colours will be seen to progress along the rod, away from the flame, the coolest, pale yellow, leading this band towards the end. Within the flame, these colours cannot be seen, but they will be evident, travelling both ways from the 'hot spot' as the steel heats up.

Take the flame away when the pale yellow reaches the end of the rod but observe that the travelling band of colours still keeps on moving, the end of the rod becoming progressively straw, brown, purple etc. as heat from the hottest part is conducted away. The tip of the rod thus continues to heat up even after the flame is removed.

If you are attempting to temper to dark straw, say, you need to take account of the fact that the end of the tool will continue to heat up after the flame is removed. You must stop this happening, otherwise the tool will be rendered too soft as the 'hotter' colours reach the end. The answer is to remove the heat in the tool as quickly as possible once the tempering colour has reached the end. This is achieved by removing the tool from the flame and immediately quenching it in water or brine at room temperature, as for hardening.

The above tempering procedure inevitably raises the body of the tool to a higher temperature than the point, therefore producing a soft body and a hard tip to the tool. For punches, scribers and so on, this is quite acceptable, but if a cutting tool having cutting edges all along its length is being made, such as a broach or a tap, its whole length must be tempered. In these instances, it is most convenient to lay the tool in a small tray of dry sand and heat it indirectly by means of a blowtorch until it achieves the required colour.

The alternative is to immerse the item in a liquid held at a suitable temperature. Molten solders can be used, or alternatively, an oil bath, for which a normal oven thermometer will provide a sufficiently accurate measurement of temperature. The tool should be allowed to soak for a sufficient time for it to heat right through. Once again, engine oil of about SAE20 viscosity may be used, but quenching oil is to be preferred. After tempering in oil or sand, the item should be quenched in water or oil in the normal way.

Annealing hardened and tempered carbon steel

Carbon steel tools which have been overheated in use can be rehardened and tempered again provided that the material is first restored to its unhardened (pearlite and cementite) state. This process is known as annealing. If annealing is not first carried out, there is great danger that cracking or distortion will occur when an attempt is made to reharden.

For complete annealing of carbon steel, the tool must be heated slowly to just above the upper critical point (800°C) and held there for one hour per inch of tool thickness, as for hardening. The subsequent cooling must also take place slowly and if the annealing is performed in a furnace, the tool is normally left inside and the whole allowed to cool naturally. If the annealing is done in an open hearth, as is likely in the home workshop, the object should be covered with a good layer of hot ashes or surrounded with as much hot firebrick as can reasonably be arranged. Once annealing is complete, hardening and tempering may be undertaken normally.

Case hardening

Mild steel cannot be hardened in the same manner as carbon steel since it contains too little carbon and insufficient martensite is formed to give the material any significant hardness. However, the process known as case hardening can be used to create a hard surface on mild steel, thus providing an extremely hard (high-carbon) surface, but a soft, tough inner core. Case hardening is not recommended for free-cutting mild steel and components which are to be given a hard surface by this means should be made from black or bright-drawn mild steel.

To create the hardened skin, a way has to be found to convert the surface mild steel into a high-carbon steel. This can be achieved by placing red hot mild steel in contact with a high-carbon material and allowing additional carbon to be absorbed into the surface. Deep penetration by the carbon is not normally achieved, but a hard skin, or case, up to $\frac{1}{16}$ in. (1.5mm) thick is perfectly possible.

The high-carbon steel surface may then be hardened by heating above the upper critical point and quenching, as for silver steel or gauge plate. This produces a 'dead hard' surface but leaves a soft core and in many applications the ability of the material to flex under load is extremely advantageous, especially considering the very hard surface which can be produced.

For case hardening of the small quantities of relatively small components likely to be of interest to the amateur, special carbon-bearing substances are available commercially. The object to be hardened is heated to bright orange-red (about 1000°C) and dipped into a tray of the compound. This is supplied as a powder which melts on contact with the heated sample and sticks to its surface.

A good coating of the preparation (technically known as a carburiser, but more commonly called case hardening compound) is built up on the sample by repeated heating and dipping. When the article is adequately coated, it is heated once more to 1000°C and then quenched in water. The remaining carbur-

iser may then be removed and the article polished or cleaned up to suit the application.

Should it be necessary, the hardened case may be rendered soft by following the procedure described above for softening hardened and tempered carbon steel, except that there is no need to 'soak' the sample for such a long period since only the surface is hardened.

Normalising mild steel

Normalising is intended to ensure that the internal structure of steel takes on a uniform, unstressed condition in which the 'grains' of steel assume a uniform size. It is therefore a distinctly different process from annealing which is designed to reduce high-carbon steels to a soft, or machinable, condition.

Low-carbon steels consist of pearlite and ferrite, neither of which is intrinsically hard, so what is generally described as mild steel is always sufficiently soft to be in a machinable condition. However, the internal structure of the steel may not be in the ideal state. If the steel has been cold worked (drawn, rolled, hammered, guillotined from large sheets etc.) the structure of the metal has been distorted and there will be unequal stresses in the material. These stresses sometimes produce a distortion in the shape of the steel, but not always.

For example, a piece of bright mild steel sheet, guillotined from a large sheet usually shows a large-radius 'bow' from end to end, and perhaps even a twist along the length. Sections of bright mild steel, on the other hand, are normally quite straight as bought, but will distort significantly if large sections of the material are removed by machining.

The cause of the distortion is unbalance of the stresses that have been set up by working the material in the cold state. Normalising is the method adopted to remove the residual stresses and for low-carbon steel (mild steel) places the material in its best condition for machining. Removal of the stresses also reduces the likelihood of distortion occurring during case hardening.

Normalising simply consists of heating the steel to just above the upper critical point and then allowing it to cool in still air at room temperature. For mild steel with 0.3 per cent carbon, this temperature is about 800°C (bright red). The sample should be heated up slowly to ensure that an even temperature is achieved, and then allowed to cool in still air. It is vital that cooling is uniform, and draughts should be excluded.

There is one other situation in which grain distortion can be caused – this is by prolonged heating of a steel sample, well above the upper critical point. This can occur, for example, where much hot forging of the material has taken place and several reheatings to quite high temperatures have been required. This causes grain growth or swelling to take place.

Normalising – sometimes referred to as 'grain refining' – tends to restore the grains to their normal size and refinement and therefore removes any internal stresses resulting from prolonged heating.

Copper

The major source of copper is the ore known as Pyrites, which normally contains more than 30 per cent copper. The copper is released from the ore by use of a blast furnace, but the ore requires various stages of refinement prior to smelting in order to produce a relatively pure copper. Even so, further refining is required to produce commercial grades of copper.

One way of refining copper is to use electrolysis to isolate pure copper from ingots of impure smelted copper. This produces a high-purity copper, known as electrolytic copper, which is used for preparation of copper-based alloys.

Pure copper possesses valuable properties which make it an extremely useful material. It has a high conductivity to both heat and electricity and has the added advantage that it is both ductile and malleable. This means that it can be drawn down to produce very fine wire and can be pressed, forged or spun into complicated shapes without cracking. Copper is also corrosion resistant and may be joined by soldering, brazing or welding.

Copper is supplied in sheet, bar and tube forms, bars being available as rounds and flats (rectangular cross-sections). Its hardness varies according to the way that it has been worked or heat treated. Basically, it may be in the hard or soft condition. Bars or tubes which have been drawn or extruded, without subsequent heat treatment, will be in the hard condition, but tubes (particularly) are also available in the soft state, and are much used, for example, in the refrigeration industry. Soft tubes are normally supplied in a coil, but hard copper tubes, e.g. for plumbing, are in the straight condition.

Tubes are available having a specified bore, or alternatively having a specified outside diameter, so it pays to be careful when ordering. Model engineers' suppliers stock by outside diameter.

Copper strips are available in some sizes in the soft condition, particularly those with small cross-sectional areas, but sheet is not normally stocked in the soft state since large sheets are susceptible to damage and therefore difficult to handle. Sheets are therefore normally hard or are in an intermediate condition between hard and soft. The processes under which these various tempers are produced are described under the heading 'Working copper and brass'.

The principal model engineering use of copper is for boilermaking, for which its ductility, corrosion resistance and solderability make it ideal. The flanging and forming of copper (and brass) is described in Chapter 7. The softening of hard copper is described under the heading 'Working copper and brass' below.

Copper is also extremely important because of its wide use in the preparation of alloys, since it is the basis of brass, bronze, gunmetal and monel metal, among others. An alloy is a combination of two or more metals which are smelted together to form a combination material.

Alloys generally may be harder and stronger than the pure metals which are their constituents, and may have melting points lower than any of their constituents. There is, therefore, much scope for development of new alloys having special characteristics. This is true also of steel alloys, but these are of less interest to the amateur than the copper-based materials.

Alloys of copper

Brass

The most common alloy of copper is brass. This word is used to describe many actual alloys but it is too vague a term to have any scientific meaning, although it is in general use. Most brasses are alloys containing only copper and zinc and are known as binary alloys since they contain only two components. However, there are some brasses which contain other metals, conferring special properties on these alloys.

When less than 36 per cent of zinc is present, brass is very ductile and can be cold worked into complex shapes e.g. cartridge cases. Brasses with less than this percentage of zinc are termed Alpha Brasses. Nearly all sheet brass is of this type.

As the zinc content of brass increases, it becomes more brittle (zinc is itself brittle) and brasses with between 46 and 50 per cent of zinc have a quite different structure from alpha brass, being harder but very brittle. These

alloys are known as Beta Brasses and their properties make them suitable only for use as brazing materials.

If the percentage of zinc is between 36 and 46 per cent, the material is known as an Alpha-Beta Brass since it contains both types of alloy. These brasses are harder and stronger than the alpha brasses and most bar stock is of the alpha-beta type. Bar stock brasses frequently contain other alloying elements to improve the strength or machinability. Aluminium increases the strength, and high-tensile brass contains 3 per cent aluminium in addition to iron and manganese. A small percentage of lead may be added to improve the machinability of the brass, but large quantities tend to weaken the alloy significantly so only small amounts are usually added.

Like copper, brass is available in sheet, bar and tube form, but square-section tubes are available in addition to circular section, and thick-walled tubes can also be obtained. These are described as 'hollow rod'. In addition, brass is available in equal and unequal angles, tee, U and half-round sections. Brass suffers a progressive breakdown at high temperatures and is not suitable for making boiler fittings, other than for 'toy' boilers operating at low pressures.

Manganese bronze

A useful variant on the brass theme is a material known as manganese bronze. This is basically an alpha-beta brass to which up to 2 per cent of manganese is added together with up to 1 per cent each of aluminium, iron and tin. Usually, 58 to 60 per cent of copper and 39 to 41 per cent of zinc form the basis of the alloy. This material is as strong as mild steel, highly corrosion resistant and is excellent for bearings and steam fittings. It is also easy to machine.

Bronze and gunmetal

Bronze is the general name given to copper–tin alloys but commercial bronzes are usually composed of copper, tin and zinc. Bronzes contain 70 to 90 per cent of copper, 1 to 18 per cent of tin and 1 to 20 per cent of zinc plus other elements in varying proportions depending on the properties required of the alloy.

The most common bronze likely to be used in the home workshop is that generally known as gunmetal. This name is given to a group of bronzes originally used for the making of cannons. It resists corrosion, is relatively strong, casts well and is widely used for pump bodies, steam fittings and for castings that are subject to pressure or shock loads.

The alloy is normally used as cast (with appropriate machining), specific castings being provided for steam fittings and valves, cast bushes for model boilers, regulator bodies and water pumps. Gunmetal is also frequently used for steam engine cylinders and pistons, but in this case two distinct alloys will be supplied, in order to ensure adequate life, since adjacent sliding or rotating surfaces formed of identical materials suffer high wear.

Gunmetal is also available in continuously cast sticks, sometimes called continuously cast bronze. This is available as solid or cored (hollow) round bars which are ideal for the manufacture of bushes or bearings.

Phosphor bronze

The description phosphor bronze is generally applied to a bronze containing 10 to 14 per cent of tin with 0.1 to 0.3 per cent of phosphorus. This alloy is stronger and harder than gunmetal and it makes a good bearing material, particularly for components which must support heavy loads. Commercially, it is frequently used in the form of castings because it is very fluid when molten and may be cast into intricate shapes. In the home workshop it is most likely to be used in the form of drawn bars, for use as bearings, or for the manufacture of steam fittings for model boilers. Rounds, squares, hexagons and flat bars are widely available in this useful material, but it has the disadvantage that it is not easy to machine.

Hard-drawn phosphor bronze wire can also be obtained. This is quite springy and yet can be bent relatively easily for the manufacture of non-corrosive springs without subsequent heat treatment. Rolled strip is also available, which also finds use as springs. It has relatively good electrical properties and is used for spring-contact electrical connections, as also is hard-drawn wire.

Monel metal

Monel metal is primarily a nickel-copper alloy, usually in the proportions of 2 parts nickel to 1 part copper. Small percentages of manganese, silicon, iron, sulphur and carbon are normally present, together making up less than 3 per cent. This alloy came about because in Canada, deposits of ores were discovered containing 2 parts nickel to 1 part of copper and a great deal of expense, and very elaborate smelting processes were required to separate the two elements for use. A smelter, Ambrose Monell, experimented by smelting the combined ore directly, thus producing a natural alloy of the two elements. This was the inception of monel metal, a first quality, corrosion-resisting alloy. Monel is one of only very few alloys that are smelted directly from mixed ores.

Monel is used extensively where relatively high mechanical strength is required at high temperatures (it will withstand temperatures up to 500°C) or where its high corrosion resistance is required. It therefore finds much use in hospital, laundry and food handling machinery.

For the model engineer, it is used for valves and valve seats and for stays in model boilers. It is an extremely tough alloy but can be machined readily if sharp tools are used.

Nickel silver

Nickel silver is an alloy of copper, zinc and nickel. Normally, 20 per cent of zinc is present together with 55 per cent of copper and 20 per cent or so of nickel. Small quantities of cobalt, lead and iron are also usually present.

Nickel and, to a lesser extent, zinc are shiny, white metals. When these are mixed with copper, the resultant alloy has a bright, silvery appearance. The alloy is generally known as nickel silver but is also referred to as German Silver, the names arising from the bright colour of the alloy rather than from the actual presence of silver.

Nickel silver has poor electrical conductivity and is used for making electrical resistance wire. It solders well, takes paint better than brass and has relatively good corrosion resistance. It is popular as a material for use in the construction of small-scale model locomotives and is available in sheet form, from 'shim' thicknesses such as .005in. (0.13mm) to ¹⁄₁₆in. (1.6mm) or so, in addition to rounds and flats. It machines easily also (somewhat like brass) and is useful as a substitute for steel for small model fittings which will not be painted but require some better corrosion resistance than steel fittings would provide.

Nickel silver also casts well and is used extensively as a steel look-alike in the small-scale model railway field. In the larger scales it is frequently used for live steam locomotive crossheads due to the ready machinability of the castings and steel-like appearance.

Working copper and brass

Work hardening

Materials which are malleable and ductile have an internal structure which allows individual grains within the material to slide over one another without a great deal of energy being absorbed. Such materials are therefore very easy to bend, or squeeze into new shapes. The easy slippage between the grains does not continue unrestrained however, as the material is worked. Only limited slippage of the grains can occur, since once a grain has slipped to take up the available adjacent 'space' it becomes locked to its neighbours and is no longer free to slide.

As the material is worked in the cold state, the grains in the region being worked progressively take up their 'free' movement and eventually will not slip relative to one another. The material therefore appears to become much harder as it is worked and the process is known as 'work hardening'.

Copper is one of the common materials which is malleable and ductile but which suffers work hardening. Since copper forms the major constituent of most brasses, they also exhibit this characteristic, but it is only the alpha brasses (less than 36 per cent of zinc) which are ductile and likely to be formed into complex shapes.

Any cold working of the metal distorts the grains within the structure so that simple cold rolling of copper and brass, to produce commercial sheets, also causes work hardening.

Annealing

Once the material has become hard it is very difficult to work and may fracture if further deformation is attempted, so it must be softened before further work can be done. This process, known as annealing, consists of raising the temperature to a sufficient level to allow rearrangement of the individual atoms into their preferred shape and juxtaposition, thus allowing re-establishment of the 'free spaces' and the rounded and even grain structure. Grain refining also takes place, as it does for carbon steels, and since smaller grains give greater strength and better shock resistance, annealing brings the metal to the best possible condition.

The temperature required for the annealing of copper and brass is in the range 400–600°C. Industrially, where annealing is carried out in temperature-controlled furnaces, 400°C is normally used. For the home workshop, something more 'visible' is required and heating to a dull red will be found satisfactory. This is equivalent to 700°C and therefore above the required temperature, and since the structural change only takes a short time to occur, excessive heating is not required. Viewing the heated metal in subdued daylight will normally allow redness to be seen at 600 to 650°C, which is ideal.

There is no absolute need to quench copper or brass to achieve a satisfactory anneal, but it does help to remove any scale which forms. It does have a tendency to distort largish sheets since the rate of cooling is not usually uniform, but if this is not a problem, clean water at room temperature may be used.

Overheating of brass and copper for long periods causes grain growth to occur, as it does for steel, thus rendering the metal weaker and less able to withstand shocks. This is not a problem when annealing but should be borne in mind when silver soldering is carried out as temperatures up to 800°C may be required.

Degrees of temper

Hot working of copper and brass does not cause work hardening since the atoms have sufficient energy to rearrange themselves into the preferred structure. It is, therefore, possible to produce copper and brass in the soft condition. This is available commercially in some forms (tubes, for example) but is susceptible to damage in sheet form and therefore only available in a limited range of sizes.

However, since these materials are hardened by cold working, it is possible to complete the manufacturing process by cold rolling to final thickness. Depending upon the amount of cold working, the metal may be in any state between soft and fully hard. The degrees of hardness, or tempers, which are available are usually soft, quarter, half and fully hard and metal stock holders will be able to supply copper and brass to these degrees of temper. Half hard represents a compromise for most purposes and is the most commonly available material. It is sufficiently hard to permit ease of handling, marking out and cutting and yet may be rendered soft by a simple annealing process.

Aluminium and its alloys

Aluminium is an important engineering material due to its light weight, corrosion resistance and good electrical and thermal conductivity. Its most serious disadvantage is its relative weakness (it has only about one-third the strength of steel). This means that careful design is required to place appropriate sections in the correct orientation with respect to the anticipated loads. Fortunately, aluminium is very workable and the production of the required shapes is quite straightforward, whatever the method of manufacture involved.

Since aluminium in its pure state is soft, it is seldom used in this condition. It does, however, form the basis of many extremely useful engineering alloys, being alloyed with copper, manganese, magnesium, or silicon (or even all four) in various proportions, to produce specific characteristics. Aluminium has good corrosion resistance since, like chromium, it rapidly forms a tough oxide skin on contact with air, which prevents further corrosion. When alloyed with other metals, oxide formation is naturally not so readily induced, but nevertheless, most aluminium alloys do exhibit good corrosion resistance.

Like copper, pure or near-pure aluminium is a work-hardening material and it is therefore available in different degrees of temper that characterise the sheet forms of copper and brass. Some alloys of aluminium, particularly alloys of aluminium and copper, are susceptible to a process of age hardening i.e. they may be rendered soft by heat treatment but will then reharden over a period of time. Rehardening may take place at elevated temperatures or may occur at room temperature.

Perhaps the best-known name among aluminium alloys is Duralumin (or Dural); an alloy with 4 to 5 per cent of copper and small amounts (less than 1 per cent each) of silicon, iron, manganese, magnesium and zinc. Duralumin was the original 'strong' alloy, having roughly twice the strength of commercial aluminium, or about two-thirds the strength of steel.

Duralumin is available as extruded rounds, hexagons and flats from most factors or alternatively these are available in alloys containing only 0.1 per cent of copper, but greater quantities of magnesium, quite distinct materials being available having 0.5 to 1, 1 to 2.5 or 4 to 5 per cent of magnesium.

Extruded angles, tees and Us are normally only available in magnesium-rich alloys, rather than dural, and the same is true also of drawn and extruded tubes.

Sheet, plate and strip are supplied in commercial aluminium (99 per cent pure), magnesium-rich alloys or as the copper-rich dural. In model engineering, aluminium alloys tend to be used in just those situations in which their light weight is advantageous, especially since relatively strong alloys are available. Model aero engines use these alloys extensively as they are ideal for pistons, crankcases, cylinder barrels and cylinder heads, sumps, timing covers and so on. Since the use of these alloys is now well established, they are used for major components for small internal combustion engines generally. Where adequate strength can be achieved, these alloys are also used for castings for parts of some tools for home machining, mostly on account of the relatively low cost and ease of manufacture.

Heat treatment

Dural exhibits the age hardening referred to above and since this occurs at room temperature, the material is naturally normally in the hard state. In this condition it cannot be bent through more than small angles without fracturing and sheet dural must be annealed before any significant bending operations can be carried out.

One of the problems of annealing dural is estimation of the temperature of the sample. Since the alloy melts at about 650°C and a propane torch will comfortably exceed this, it is all too easy to melt the sample and reduce it to a pool of scrap. The usual way to assess the temperature is to heat up the sample (keeping the torch well on the move) and to test its temperature periodically by rubbing the end of a 'dead' match along the surface. When the dural is hot enough for the match to ignite, the annealing temperature has been reached (500°C) and the material must immediately be quenched in water at room temperature. This renders the dural fully soft and it will remain in this condition for a while, progressively age hardening at room temperature over a period of a few hours. Any bending operations should therefore be carried out as soon as possible after completion of the anneal.

Casting

For the production of castings, aluminium alloys having relatively high levels of silicon are used since this gives greater fluidity to the alloy in the molten state. If particularly thin sections are to be cast (where greater fluidity is required) more silicon is generally utilised.

The addition of silicon to the 'casting' alloys of aluminium tends to increase the difficulties of machining since the silicon is itself abrasive and tends to form particularly hard compounds within the material. Cast aluminium is, therefore, frequently much more difficult to machine to a good, bright finish.

The low melting point of aluminium also makes it suitable for casting in the home workshop. This makes it possible to make castings from one's own patterns, for items that are not too highly stressed, and can provide a more than useful introduction to the patternmaker's and foundryman's trades.

4

Setting up a workshop

Introduction

Much of what is needed in a workshop may seem at first sight to be enormously expensive. There are several points to bear in mind however. First of all, it is highly likely that you already have some interest in manual skills and will have a 'basic' set of hand tools, perhaps even a bench and vice, and somewhere to work. Secondly, it is not necessary to acquire the contents of a complete workshop immediately, and thirdly, many of the smaller items (and even some quite large ones) can readily be made once the basic machinery is available. It is also worth bearing in mind that the best models are not necessarily made on the newest or most sophisticated machinery, and new items are not, therefore, vital to success.

Given that a bench and vice will naturally be required, together with an area set aside for assembly work, the items needed can be broadly grouped under the following headings:
- a drilling machine
- a lathe and its attachments
- measuring and marking-out equipment
- hand-cutting tools (saws and files)
- threading equipment (taps and dies)

- some smaller hand tools (scribers, punches, clamps etc.)
- some means of sharpening cutting tools.

If much soldering or brazing (silver soldering) is likely to be required, at least a rudimentary brazing hearth and the necessary torch (blowlamp) and gas supply will also be required.

The provision of space for the workshop should be the first consideration. Once this is available, the purchase of a lathe is likely to be the next step, if only for psychological reasons, since it is the possession of this versatile tool which turns one from a 'hobbyist' into a 'model engineer'. The general characteristics of small lathes are described in Chapter 12 and the purchase of a secondhand machine is considered in Chapter 17.

Space for the workshop

General considerations

Space (floor area) is a fairly expensive commodity, at least in the South-east, where a 3-bedroom house offering some 1500 square feet

(150 square metres) of floor area is currently valued at around £70 per square foot (£700 per square metre). The provision of the space for a workshop is therefore likely to be the major item of expense, since some dedicated area is essential to house the machinery, benches, and so on which are required.

Several possibilities exist, however, from the spare room to the garden shed, not forgetting the use of part of a garage, or a purpose-built extension. In reality, a combination of locations can be more convenient and may well allow quite a small floor area to be used for the basic workshop, without the necessity for it to accommodate all of the items required.

The essential requirements for the workshop are that it should be dry, warm in winter and in a location or situation which makes it available for the 'odd half hour of modelling' for, try as we might, many of us cannot ordinarily spend long periods away from family and social commitments. Above all, the workshop must be a suitable location in which to store relatively expensive machinery and valuable models, so it must be secure.

To achieve these desirable characteristics, the workshop must be substantially built, well insulated and incorporate at least a modest degree of heating and ventilation, as a means of removing excess moisture from the atmosphere. In these respects, a spare room in the house may seem desirable, but there are some disadvantages to this. First of all, there is the noise problem. Some processes, even simple sawing, are by their nature noisy, and a workshop within the house may impose some restrictions in this respect. There is also the problem that the close proximity to the living quarters increases the likelihood that swarf, oil and general dirt will spread around the house.

A spare room on the first floor is also not ideal for the installation of heavy machinery (it is fine once up there, but difficult to install in the first instance). The first floor is also unlikely to be suitable if the model you will build is likely to be heavy. Even a modest 0-6-0 tank locomotive in 5-inch gauge may well weigh 80 to 100 pounds (36 to 45 kilos) when complete and may be difficult to get downstairs at the end of the day – or 10 years, depending upon how fast you work!

If a spare room is available on the ground floor, perhaps a utility room or playroom which can be dispensed with, this is altogether a different proposition. It meets the essential requirement of being instantly available, will most likely have some form of heating already installed, and is certain to be provided with ventilation. Provided that noise is not likely to be a problem, it is socially more acceptable to be within the house, since you are available should the family need your attention. You can also readily be supplied with refreshment and can easily 'down tools' for a social break over a cup of tea or coffee.

Bearing in mind the need to store steel and iron components and materials, a positive approach to rust prevention must be adopted. The essential requirement is to keep the atmosphere dry. This can be by use of ordinary heating and ventilation or by use of one of the small dehumidifiers which are now becoming available. Unfortunately, these are expensive and also take up significant space, thus consuming commodities which are generally in short supply. The alternative is likely to be an electrically powered fan heater or a tubular heater, allied to ventilation.

If the workshop is within the house, the domestic heating system might satisfy this essential requirement, which is very convenient, but even nice, dry heating provided by radiators cannot be expected to prevent rusting entirely since there are inevitably spaces which do not have sufficient air flow, and moisture-laden air can accumulate here and there. Under no circumstances should moisture-producing heaters such as paraffin burners or gas heaters be used, unless provided with a properly installed external flue.

Space requirements

It is difficult to be specific about the actual floor area required since it depends a great deal on the size and type of your machinery, the type of model engineering which is done and whether some additional space is available outside the main working area. As an example, my workshop is roughly 8ft (2.4m) square which is adequate for model locomotive building except for the fact that three of the walls include a door, all three of which are used from time to time, so models, some materials and brazing equipment have to be stored elsewhere.

When originally established, the workshop housed a small bench, with 4-inch (100mm) vice and drilling machine, the lathe on its own bench, a small bench used more as a table, and two wooden bookcase-type units that were used for general storage. These were roughly the same height as the small bench and about 9in. × 30in. (230mm × 760mm) in plan. Each bookcase unit was just about large enough to accommodate a 5-inch gauge 0-6-0 tank locomotive. The basics of the workshop were thus housed reasonably well in 64 square feet, or about 5.75 square metres.

Recently, the small bench/table has been replaced by a smaller one, and one of the bookcase units has been discarded, permitting the introduction of another small bench which provides a mounting for a small mill/drill, but this has so cramped the workshop that a visitor is now difficult to accommodate, and the horizontal table on the mill/drill must be traversed to one end or the other of its travel, according to whether one needs to sit at the bench/table, or gain access to the adjacent garage.

Even so, with only a single door, this space would also house the second bookcase unit, which would be valuable for assembly work and for further storage, or would accommodate larger models. Even an 0-6-0 tender locomotive in 5-inch gauge would be an embarrassment at present.

Garden shed workshop

Introduction

For many, the possibility of an indoor workshop is fairly remote, if only on the grounds of cost. For this reason, the garden shed is often the only choice and is the traditional location, at least in the UK. Such a location does bring with it certain advantages – noise and dirt are unlikely to be a problem to the rest of the family and the enforced walk between workshop and house provides an opportunity for dust and swarf to fall off clothing and footwear. This is very advantageous as it is amazing how far from the workshop 'sticky' swarf such as copper and phosphor bronze will travel before deciding to embed itself in the carpet. One acquaintance relates how wheel turning, or other cast iron machining, always produces a rusty patch outside his garden workshop, just where the dust tends to fall away.

The cost of an 8ft × 6ft (2.4m × 1.8m) wooden garden shed is around £250 at the present time, depending upon the finish and quality (type) of timber which is used. It is a simple matter to design and build a workshop to one's own specification and is a cost-effective way to provide the workshop space. The overall size can be tailored to suit the site, and the internal layout and timber sizes can be arranged to suit the type of insulation which is to be installed (and this is highly recommended). Positioning of the internal timber 'studs' can also be arranged to suit the wall covering to be used (again, highly recommended) and the result will certainly not be inferior to commercial products.

Commercially, frames made from 50mm × 75mm timber are used, or sometimes even lighter material, and this is certainly strong enough, although if it is envisaged that the walls will be used for supporting large amounts of shelving, or heavy cupboards, 50mm × 100mm frames might be preferable. The basic structure can be arranged as a series of modules comprising a floor unit, two sides and two ends which are bolted together using coach bolts. Simple, bolt-together roof frames, one for each end and intermediate ones at one metre intervals can be used to support the roof, which can conveniently be covered with pre-felted chipboard, the ridge afterwards being overlaid with roofing felt.

An examination of a few commercial products will show the general type of construction which is adopted, and will indicate where the overlaps must be arranged in the external cladding in order to provide a weatherproof structure. The only likely problems are in the provision of windows and the door, but commercial items can be utilised. The need for ventilation and security should not be forgotten, but it is not advisable to purchase a door which is too heavy to be supported by the framing.

The workshop must stand on a firm base. Ideally, this should be a solid concrete foundation, but it is acceptable if paving slabs are laid onto well-consolidated ground to provide the base. A floor construction comprising 75mm × 50mm or 100mm × 50mm timbers, standing on edge, with floorboards or chipboard flooring nailed to the timbers will serve adequately, but if chipboard flooring is used, it is essential that it is a moisture-resistant type.

External cladding to the walls and ends should overlap the flooring, but not reach the foundation level, so that good ventilation is provided below the floor. All timber used should be liberally treated with preservative during construction and an external finish adopted which will be easy to maintain.

Insulation

Provided that the workshop is well insulated (even the garden shed type), the provision of background heating need not be too costly, and a thermostatically controlled fan heater, for example, set to maintain the temperature at 10°C will provide the most useful form of overall heating as well as being capable of bringing the temperature up to a comfortable level when the 'shop' is occupied. The heat losses through the structure can easily be calculated, once its size, and the type and thickness of insulation are known, and the local library will provide a ready source of reference material to enable this to be done. As a guide, the theoretical heat loss from my workshop is approximately 700 watts assuming inside and outside temperatures of 10°C and 0°C respectively.

Nowadays, a wide range of materials is readily available for insulation, and the final choice often rests on those materials which are easy to use. For walls, expanded polystyrene is one of the most convenient, and is also one of the cheapest. It is a high-void material comprising very small pockets of air entrapped in polystyrene. It is made in 25mm and 50mm thick sheets, in a grade which is treated with a fire retardant and therefore safe to use in buildings. It is light in weight, cuts easily and is a very good insulator.

In addition to providing insulation, the installation also has to consider the effect of moisture in the air. If warm, moist air is in contact with a cold surface, it cools and releases some of its moisture and condensation occurs at the cold surface. In a modern house, condensation is most frequently seen on single-glazed windows, but seldom on walls, since their cavity construction is designed to mitigate the condensation problem. If the interior of a workshop is to be heated, the possibility of condensation occurring must be considered.

Figure 4.1 shows the problem diagrammatically. Warm, moisture-laden air on the inside of the structure is insulated from the cold exterior by insulating material and the external wall material. However, there is a temperature gradient across the insulation, and if moisture can pass through this, condensation occurs on the internal surface of the wall. This cannot be permitted, since the insulation will become saturated and degradation of the structure will occur. The answer is to install a vapour (or moisture) barrier, the most common material for which is polythene sheet. This must be positioned between the insulation and the warm interior in order to prevent water vapour coming into contact with the cold external wall.

One important point to remember; the vapour barrier can only be effective if it is completely impervious. This means that care has to be taken to ensure that no tears or cuts are made in the polythene and if joints need to be made, they should have adequate overlaps and be doubled over and sealed with an impervious tape.

The use of a vapour or moisture barrier between the moist room air and the cold wall of

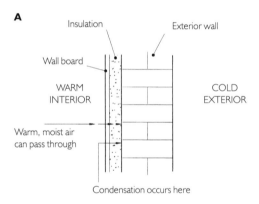

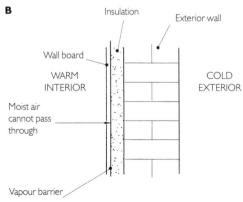

Figure 4.1   The need for a moisture barrier in an insulated building.

the structure means that a wall covering of plasterboard or insulation board is required. Plasterboard is preferable since it presents a harder and more durable surface and is more readily decorated than insulation board which has a soft and very absorbent finish. If the vertical timbers are positioned to correspond with the joints between the sheets of board, it is a simple matter to cut the sheets to the correct height and nail them to the timber. It is customary to nail plasterboard into position with a ½in. to ¾in. (12mm to 20mm) gap between the floor and the lower edge of the sheet. This is to prevent the absorption of liquid which might otherwise degrade the plaster core, should the floor be washed or any liquid be spilled, and the same consideration applies also to the use of insulation board. The gap must be allowed for when cutting the sheets and is readily arranged when fixing a board by use of suitable packing to stand the board on when it is being nailed into position.

Several methods can be adopted for concealing the gaps between the boards. A paper strip, with central or longitudinal corrugations, or ribbing, can be stuck over the joints with heavy-duty paperhangers' paste – this is perhaps the old-fashioned way. Alternatively, a non-corrugated strip can be 'plastered' over the joint with a shallow wedge of board-finish plaster applied to each side thus making a wide ridge which is virtually invisible after painting. A variation of this method is to use a ceiling texture material such as *Artex* to form a feather-edge strip about 12in. (300mm) wide spanning the joint between two plasterboard sheets. The jointing should be reinforced using a 4in. (100mm) wide cloth scrim which is first applied to a thin coat of *Artex* which spans the joint, the whole being afterwards covered with the feathered strip. This is the method used for joints in ceiling board and this, or the plain paper strip, is also used in houses which have what are called 'dry liner' wall finishes.

Once the plasterboard has been installed,

and the joints made good, a standard skirting can be nailed into position to protect the lower edge of the board and close the gap between board and floor.

The positions of the timber studs to which the plasterboard is fixed should be evident, or have been marked, and electrical outlets can therefore readily be installed for the machine tools, and for a wall-mounted fan heater. This can conveniently be of the type which is fitted with a thermostat to allow selection of the set temperature, and while this may seem extravagant, it certainly provides a cosy working environment. Good insulation allows reasonable temperatures to be maintained at not too great a cost, both for storage and use, and the insulated-and-heated environment affords protection to what are becoming valuable machine tools and models.

Garage workshop

Isolation of the workshop

Unless a garage can be dedicated exclusively as a workshop, it does not come very high up the scale of desirable locations. Firstly, it may very well be damp, due to the lack of a damp-proof membrane and damp course in its floor structure, but even if this has been attended to, which is unlikely, the regular presence of a hot and damp vehicle is sure to create a humid atmosphere which rapidly promotes rusting. An unmodified garage cannot be recommended as an ideal location in which to store ferrous parts, and this includes many types of model, most machinery and much stock material.

Unless the machinery you have, or intend to purchase, is exceptionally large, a garage also represents a very large space and a better environment may well be more readily provided by partitioning off one end to create an isolated space. The partition between garage and work-

shop can be constructed by using a timber frame of 50mm × 75mm or 50mm × 100mm prepared timber incorporating a standard commercial frame and door, if one is required. Garage side cladding may be by nailing on a thin plywood cladding, using an exterior (waterproof) grade if dampness is likely to be a problem. It is a good idea, in any event, to use a damp-proof material below the timber frame and a standard roll of damp proof course (DPC) felt can be used for this.

Some form of heating for the workshop is thoroughly recommended and this becomes more effective (and cheaper to run) if the walls, floor and ceiling are insulated. Expanded polystyrene, as recommended for the garden shed workshop, should be used for the isolating partition, and also for the walls, if possible, and the installation of a ceiling, with a glass fibre or mineral wool blanket above, is also advantageous in reducing the heating costs. For the walls and isolating partition, a moisture barrier must be installed, and a plasterboard lining, as described above will complete the installation, as for the shed workshop. Provided that there is a large air space above the ceiling, with good air circulation, condensation will not be a problem in the roof space.

Floor insulation

If a workshop can be provided by segregating part of a garage, some floor treatment is desirable since a garage floor does not normally have a damp-proof course. The choice lies between a professionally laid damp-proof membrane, with concrete screed floor, or a floor laid on wooden joists resting on the existing concrete, with a damp-proof course between. A construction of the type shown in Figure 4.2 is suitable, in which 50mm × 75mm joists are used, with the gaps between the joists filled with mineral wool or glass fibre insulation. This type of construction loses relatively little heat in comparison with a concrete floor (in a ratio of about 1:4). It therefore assists retention of heat and reduces heating costs.

Two moisture barriers are required – one to prevent the ground-borne moisture from permeating the insulation (and the supporting timbers) and another to prevent water vapour in the room air from penetrating the structure and causing condensation on cooling at the cold concrete surface. However, the timber and insulation are not sealed into a cocoon since the upper and lower polythene sheets are not sealed at the edges. The lower sheet is laid on the floor and carefully folded at the corners to lie up the wall a few inches above the damp-proof course. Prior to laying the polythene, strips of conventional DPC must be placed on the oversite concrete, below where the joists will lie, to prevent the polythene being punctured by the rough finish.

Once the joists and insulation are in position, the upper moisture barrier is positioned in the same way, to lie up the walls, but is not sealed to the lower layer. The below-floor space can therefore 'breathe'. Both moisture barriers must extend above the DPC to prevent moisture entering the enclosed space.

The type of floor described is frequently recommended where good insulation is needed, not only with false joists, as described, but also with floors having expanded polystyrene as the insulation, either screeded over with concrete or covered with flooring-grade chipboard. Both surface treatments act to spread the load applied to the floor so that it may be supported by the polystyrene, but the construction is suitable only for relatively light loads. Although polystyrene is normally impervious to moisture, the fact that it must be laid in relatively small sheets means that it does not provide an unbroken moisture barrier, and two polythene sheets are recommended for use in floors.

Workshop electricity

The workshop will need several electrical outlets (sockets) so that the machine tools, lamps, soldering irons, and so on, can be connected to a source of power. The electrical supply available in the home is what is known as a single-phase supply, having just live and neutral connections, with an earth. To match this supply, the machine tools which are bought should be fitted with single-phase motors. This makes them suitable to be plugged into a normal 13-amp outlet. Lathes, milling machines and mill/drills of light construction, or those intended for amateur use, are likely to be fitted with single-phase motors.

If larger machine tools are purchased, or those intended for industrial use, they may be fitted with three-phase motors which cannot be connected to a domestic 13-amp outlet. In these cases, a special three-phase supply will need to be wired into the workshop, or an adaptor purchased which produces a three-phase supply (or its equivalent) from a domestic supply. An adaptor is the cheaper alternative to the provision of a three-phase supply, since the connection of a three-phase supply requires a new electrical

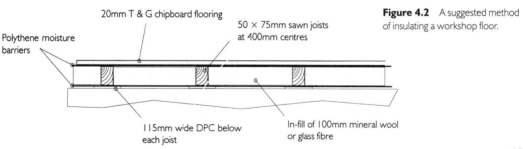

Polythene moisture barriers

20mm T & G chipboard flooring

50 × 75mm sawn joists at 400mm centres

115mm wide DPC below each joist

In-fill of 100mm mineral wool or glass fibre

Figure 4.2 A suggested method of insulating a workshop floor.

feed into the property from the electricity company's cables in the locality.

If the workshop is established in a purpose-built location, or is an adaptation of part of a garage, for example, it may require a completely new electrical installation. Any electrical work, even minor modifications, should only be carried out by a competent electrician. It is important that all of the tools are correctly installed, from an electrical point of view, and are correctly earthed. Only a competent person can ensure that this has been done, and only a person with the appropriate knowledge and test instruments can check an installation to confirm that it conforms to the regulations and is safe to operate. It should be particularly noted that the earthing requirements for a garden shed workshop may require special consideration due to the distance from the house, and this again dictates that the installation is inspected professionally, prior to use.

Fitting out

Benches

Like a good kitchen, a good workshop needs plenty of work surface, plenty of storage and somewhere to sit down to drink a cup of coffee. The work surfaces are generally the benches, the storage spaces are boxes, cupboards and shelves, and the place to sit for a cup of coffee might also equate with another bench. Benches are thus required in profusion and the following specific areas are probably desirable, if you can find the space:

(1) a dirty bench, very sturdy, and fitted with a large vice. This is where sawing and filing processes are performed.
(2) a fairly clean bench which can be used for fine sawing and filing, polishing and general finishing off.
(3) a clean assembly area, remote from sawing,

filing and machining.
(4) a storage bench for items under construction. This might need to be large if a large-scale tender locomotive is being built.

Needless to say, few workshops will have the space for all of these separate areas and some compromises must inevitably be made by combining functions. Unless very large models are being built, assembly work generally amounts to the equivalent of 'instrument making' in industry. Many people find that this is most conveniently undertaken in a sitting position, and this also applies to the finer sawing, filing and polishing activities, so a desk-type bench can conveniently satisfy the need for both the clean and fairly clean activities provided that they can be separated in time. A small clamp-on vice and clamp-on vee-notch table for the piercing saw (see Chapter 6) allow an ordinary table or a secondhand desk (provided it is rigid enough) to be used for smaller work of this type.

For heavier sawing and filing, it is generally recommended that the top of the vice should be at elbow height, so that the forearm is horizontal during these activities, therefore it is a good idea to obtain a vice (or at least measure the height) before designing and building the heavyweight bench. For a work surface at table height, where only light work will be performed, a comfortable height for writing is probably the major consideration, but if you have particular eyesight problems, you may need to consider these. The height required thus depends upon you and the chair you will use.

Since drilling produces volumes of swarf, the bench to which the vice is attached might also accommodate the drilling machine if it is of the bench-mounting type, thus combining two of the essentially dirty processes.

Unless the lathe is provided with its own stand, it too will need a bench. This must be rigid and substantially built in order that the machine can be set up correctly and maintain its accuracy. Naturally, as a swarf producer, the lathe needs to be located at the dirty end of the shop.

The bench vice

The main vice in the workshop should be the largest you can afford or accommodate, and it must be mounted to a substantial bench. A good quality, 4-inch (100mm) vice will generally be sufficient for the general run of modelling work, something smaller being provided for use at the table bench, in the form of a clamp-on vice, or by making or purchasing a small-piece vice for clamping into the bench vice.

The important feature of all vices is that they should have smooth jaws which remain parallel throughout their gripping range. Most bench vices are manufactured with hardened, serrated jaws, but unless you are interested in gripping only full-size, agricultural items, such jaws are useless. They naturally grip well, but only by making impressions of their jaw serrations in the work, and this is definitely not required in the modeller's workshop. Serrated jaws should therefore be removed and replaced by home-made versions made from ordinary mild steel.

An alternative to replacing the jaws is to purchase or make up slip-on soft jaws. These comprise steel pressings to which are riveted strips of soft, composition material, making the jaws suitable for holding soft materials without marking them. Something of this sort, or loosely inserted pieces of soft packing, is frequently required when working with soft materials, since even mild steel jaws will mark or squash soft materials such as copper, brass, nickel silver and aluminium alloys. Soft jaws of the type described are not good for holding very small items, since they compress due to the localised loads.

A clamp-on vice for the table bench does not need to be larger than a 2-inch (50mm) jaw. Some vices of this type are provided with bolt-on jaws, frequently soft plastic mouldings, but unless you envisage a great deal of work in soft materials, or with work that is very easily damaged, such as printed circuit boards, it is better to use the type without removable jaws,

Figure 4.3 A small vice, with integral clamp, fitted to my second bench.

such as that shown in Figure 4.3, since very soft jaws are easily damaged by gripping small items.

Seeing and looking

Good lighting is necessary to most of the operations carried out in the workshop. It is vital for all machine work, since there is a repeated need to bring the cutter up to the work to detect the point at which it first starts to cut. When cutting or filing to a line, good lighting is essential. It is vital too, during marking out or any operation which requires that a rule or vernier be read to the smallest divisions on the scale.

Since the locations of these different operations are likely to be different, it follows that several small sources of light are required. These must be movable so that the light can be brought to bear from different angles and are ideally provided by the common type of counterbalanced reading lamp. The only disadvantage of these lamps is the large size of reflector fitted to those types which use conventional, tungsten-filament bulbs since they tend not to be very convenient when used on the lathe. The type of lamp which uses a low-voltage tungsten-halogen light source is preferable, since this offers a significant reduction in the size of the bulb, and hence, the size of the reflector. There is a cost penalty, however, and this may dictate the choice.

Once the lighting is installed, attention can be turned to the problem of seeing. It may be quite impossible to separate the adjacent engravings on a rule with the naked eye, and so some assistance is necessary in this direction. The first tradesmen to meet the need for such assistance were the watch, clock and instrument makers involved in the development of ever-smaller precision mechanisms. These craftsmen were doubtless associated with early optical experiments and were thus quick to understand the usefulness of a magnifying glass.

However, the 'Sherlock Holmes' type of magnifying glass was impossibly large to cast, grind and polish and early lenses were consequently very small. The habit grew of using these small magnifiers close to the eye, and they were mounted into holders small enough to be held in the eye socket and were known as loupes.

The type of lens used is the same as those used by schoolboys for focusing the rays of the sun onto a piece of paper and burning a hole in it. The lens is a bi-convex, and it brings the sun's image to a sharply defined spot on the paper when it is placed the correct distance away. This distance is known as the focal length. The magnification which is provided when a bi-convex lens is placed between the eye and an object is related to this measurement, and the lenses are described by their focal length.

The shortest focal length gives the greatest magnification and the range of lengths sold is from 1½ inches (38mm) to 4½ inches (115mm). A 4½ inch glass does little more than concentrate the view, but what it does do is to allow the eye to focus on something roughly 4 inches away (100mm), or 3 inches (75mm) in front of the lens. A 1½ inch glass allows the eye

Figure 4.4 Watchmakers' eye glasses.

to approach to within about 3 inches (the object is 1½ inches in front of the lens) but has the disadvantage that the field of view is narrow and only a limited range of objects is in focus.

The glasses are sometimes described by the nominal magnification which is provided, a 4½ inch giving 2 times, a 3 inch giving 3.5 times and a 1½ inch giving 6 or 7 times. A 3 times is a useful size for general use when marking out and centre punching and the only other consideration is whether you can hold it conveniently in the eye socket. A 4½ inch and a 6X (described as a 42mm) are shown in Figure 4.4 together with one type which clips on to spectacle frames. The two glasses are significantly different in size and this might be the determining factor when making a purchase.

There are several types which fit onto spectacle frames. The one shown in Figure 4.4 can be swung out of the way when not required, but others may not incorporate this feature – which type to buy depends on how you envisage its use.

If relatively long-term use of a glass is needed, for example when painting small, intricate work, a binocular magnifier which fits to a headband might be preferable. These are again available giving different magnifications and are essentially designed to work at different ranges, from something like 1½ times magnification at 20 inches (500mm) to 3½ times at 4 inches (100mm). A compromise of something like 2 or 2½ times working at 8 to 10 inches (200–250mm) is correct for an initial trial.

A quick and simple way to check the focal length when purchasing any magnifier is to project a picture of a nearby window onto a sheet of paper. If you retreat a little from the window and hold the lens up, with the paper behind, facing the window, and vary the distance from the lens to the paper, you will see an image of the window appear on the paper. When this is sharply focused, the lens is approximately at its focal length from the paper. If a sheet of paper is not available, it is quite possible to project onto your hand, and if you are out of doors, then you can focus the sun, as the schoolboys do, but in this case, use someone else's hand! Indoors, without a bright window to use, the image of a light can be focused in the same way, but a reasonably distant one needs to be used or the result is not very accurate.

Storage

General considerations

In comparing the requirements of the workshop with those of the kitchen, the general suitability of secondhand kitchen units should be

Figure 4.5 The traditional type of toolmaker's cabinet.

borne in mind. While base units are hardly sufficiently robust to make an ideal bench for sawing and filing, they are provided with substantial chipboard work surfaces which are easily cleaned, and have drawer units, shelves and so on that provide under-cover storage which is readily organised to suit the modeller. Wall cupboards and midway shelf units might also find ready uses.

One advantage of being able to utilise kitchen units is that much storage will be under cover and therefore protected from dust, swarf and general dirt which is apt to accumulate. This is immensely preferable with one exception – ventilation is a problem inside a closed cupboard, and this can lead to corrosion on iron and steel due to stagnation of moisture-laden air in proximity to cold metal. This is a serious problem, even inside a centrally heated house, and it can also affect cutting tools such as taps, dies and reamers if they are left for long periods in closed boxes, without being disturbed. So, you need to open the cupboards regularly, and perhaps even move the stock about. As far as tools are concerned, the solution is obvious – just get out to the workshop and use them!

Tool storage

Tools divide broadly into the large-and-dirty and small-and-handy categories and their storage can be considered in these two ways. The larger tools are those associated with sawing and filing. In a small workshop, nothing is ever likely to be far from where it is needed, but it is naturally convenient if saws and files are located adjacent to the bench vice. The dirty bench should incorporate drawers to hold these, together with hammers and mallets. Other large hand tools such as pliers, pincers, tin snips and screwdrivers can also be located here. If drawer space is at a premium, a rack on the end of the bench, or on the wall behind, might be preferred.

The traditional storage for a craftsman's small tools is the toolmakers' cabinet, one type of which is illustrated in Figure 4.5. These cabinets provide one or two long drawers and six or eight small ones, according to the design, and some, like the one illustrated, have a top section covered by a lift-up lid. They are designed, and therefore ideal, for the storage of small tools which might be used at the assembly or fitting stages, and can store the smaller taps, dies, reamers, tap wrenches, spanners, pliers and cutters which are used at those stages. If a shallow, long drawer is provided, this is ideal for the storage of rules, protractors, a depth rule or perhaps combination square and vernier caliper.

When determining methods of storage, the other scarce commodity (apart from space and money) should also be borne in mind – time. There is never enough of this, so that which is available must be maximised. One way to achieve this is to reduce searching time by storing small items so that they are immediately available and identifiable. The storage of drills is a case in point. It is all very well having a large box full of drills covering the range from ¼in. to ½in. (6mm to 12mm), but quite a different matter to find the one to drill a hole .256in. or 6.5mm in diameter. The selection process can be long and tedious, measuring drills which look right, repeatedly, with a micrometer until the particular one is found. Drills are much better stored in drill stands or boxes of the types illustrated in Figure 4.6, although the open stands are dust collectors if they are stored in the open.

A similar argument can be applied to small taps. A drawer full of BA taps, which are in any event small, might contain over 30 taps if you have three of each of the common sizes (the normal set). The 0 and 2 sizes are readily distinguishable, perhaps also 10 and 12, but if the

Figure 4.6 Three types of drill stand.

need is for a particular type of size 8, several 6s, 7s and other 8s are sure to be examined before the one required is found.

One way to overcome this is to make up drilled blocks designed to accept the three taps in a set, perhaps also with the tapping and clearing drills for the particular size. These blocks are pleasant to handle if made up in hard wood, but oak should not be used since it causes corrosion. Depending upon the shelf or cupboard storage which is available, larger blocks holding several sizes might be more convenient. This type of storage is not required for dies since, in any event, there are fewer of them for a given thread form (only one of each size) and they are generally more readily identifiable.

Stock storage

Everyone has his or her own ideas about how much stock to lay in and how best to store it. However, one soon begins to build up a large amount, and its storage can become a problem. One way to solve this is to store rod and bar materials in lengths such that they can be accommodated within the depth of a standard bench and to build into one of them a long pigeon-hole rack. My rack has holes roughly 2ft (600mm) long and 2in. × 2in. (50mm × 50mm) or 2in. × 4in. (50mm × 100mm) cross-section. Accordingly, the rack holds the standard 2ft (600mm) lengths which are sold by most modellers' suppliers, and it is only the occasional longer lengths of bar which cannot be accommodated. These stand in some convenient corner until actually needed.

A simple solution to the problem is to utilise square-section rainwater downpipe, as some retailers do for their exhibition stands. If you are a woodworker, and have some suitable plywood available, a rack can easily be constructed.

The pigeon hole or tube arrangement works very well provided that the lengths of material are reasonably long, but short ends and odd stubs of material tend to get lost among the long pieces and inevitably get pushed to the back. A box in which to store these oddments is a useful adjunct to the main store and can possibly equate with the 'scrapbox'.

Storage of sheet materials is not so likely to require much space, since a large number of different types is not required and most sheet is purchased for specific jobs. A single shelf in a cupboard, or a stand-up slot in a bench might conveniently hold sheets up to 2ft (600mm) square and quite a lot can be packed into a space 9in. (230mm) high, or wide, depending on whether the sheets lie down or stand vertically. Once again, smaller offcuts tend to get overlooked among the larger pieces, and an oddment box or subsidiary store is very useful.

Any castings which have been purchased for a particular model can be stored in suitable boxes, but it is preferable if these can be closed up so that the contents remain clean and dust free. Cardboard boxes are perfectly satisfactory for this, provided there is not a need to carry them about frequently and nothing elaborate (or expensive) needs to be arranged. The boxes do need to be fairly strong since the wheel and cylinder castings, even for a modest 3½-inch gauge locomotive amount to quite a weight.

Storage of cutting tools

Many of the items in the workshop are designed for cutting, a process which can only be performed satisfactorily by a sharp tool. Tools which become blunt must therefore be resharpened, or be replaced if sharpening is not possible. Cutting tools naturally wear away and become blunt through use, but they can also be blunted by careless storage. If all of your BA taps are kept in one box and carelessly taken out or dropped back in, or the box is handled so that the taps rattle against one another, they become blunted even though they are not in use. This can seriously shorten a tap's life, putting more pressure on that rare commodity, money. If loose storage is more convenient, then the taps can be protected by slipping on short lengths of plastic tubing.

Reamers, end mills and slot drills, which have cutting edges all along their working lengths, are especially prone to damage if carelessly stored, and are best kept in compartmented wooden trays or drawers so that they cannot damage one another. Many of these items are nowadays supplied in plastic boxes, and if bought new, will have ready-made protection.

Some cutters are supplied with protection in the form of wax-dip sleeves. These can be removed by cutting carefully down the length of a flute with a sharp knife and removing the sleeve by a combination of peeling and unscrewing. The sleeve can usually be removed intact by this method and can be replaced on the cutter by reversing the action. Since the cutters are coated with oil before immersion in the wax, the retention of the sleeve is also useful as a rust-prevention measure. If the sleeves become damaged, plastic tube can be used as an alternative.

Storage of small items

For the storage of small items such as nuts, bolts, washers, rivets and so on, an extensive collection of small boxes is required. From the point of view of time saving, it is vital to be able to access a given size of nut or bolt, or whatever, rather than having to scrape about in a box which contains all sizes. It pays, therefore, to segregate different sizes into their own boxes and it might even be beneficial to store nuts, bolts and washers separately, for a given size, or to segregate hexagon-headed bolts and screws from other types. An extensive collection of small boxes is therefore required.

Traditionally, small-item storage has been in tobacco tins, but with the reduction in the number of smokers and the use of plastic packaging, this source of small boxes has almost dried up. However, if you know a smoker, (or an ex-smoker), particularly if he favours (or favoured) a pipe, cultivate him. There may be a stock of tins which is no longer required and they can readily be labelled and will stack nicely.

Failing the availability of tobacco tins, the modern equivalent has to be the plastic box. You may be lucky and be able to locate a source of supply, since much packaging is nowadays simply discarded, but failing that, an extremely useful range of interlocking boxes is available from suppliers of electronic components. The system is based on a standard module comprising a plastic sleeve, about 2in. × 2in. (50mm × 50mm) in cross-section, into which is fitted a clear plastic drawer. The sleeves are moulded with external dovetails so that they can be interlocked to form a nest of units and they may be wall-hung by using keyhole slots moulded into the outer sleeves.

The range of drawers uses one-module and two-module wide drawers, but other sizes are also available from some sources, for example, 2 wide × 2 high and 4 wide × 2 high. While not perhaps the cheapest available, there is great merit in being able to rearrange the unit, or add to it as time goes on, and the ability to hang the sleeves on the wall is also a useful alternative to providing shelves.

Hand tools

General tools

Many of the hand tools used by the modeller are not different in any way from those used ordinarily and anyone interested in things practical is sure to have available pliers, screwdrivers, hammers and the like, perhaps also wire cutters, soldering irons, some drill bits and a hand drill to use them in. With the general availability of power tools these days, a DIY electrically powered drill is almost certain to be available also.

What may not be available are the tools needed for small-scale modelling; small screwdrivers, usually called watchmakers' screwdrivers, smaller spanners, especially for the BA sizes of threads, including the odd numbers, which are quite widely used in modelling, and some of the more specialised items such as pin chucks and pin vices, centre punches and riveting tools. Taps and dies for the common threads used in model engineering may also need to be obtained, together with the relevant tapping and clearing drills. Precision measuring equipment will also most likely be needed.

The more specialised items which are

required are described in succeeding chapters. Use of the important hand tools for marking out, sawing, filing and cleaning up or polishing is described in Chapter 6 and the more usual devices for performing the important measuring functions are described in Chapter 5. Hole drilling is described in Chapter 10 and the production of screw threads using hand methods, in Chapter 11. Riveting, and the tools needed, are described in Chapter 8.

Soft-faced hammers

One operation which occurs very frequently is the mounting of work into holding devices associated with machine tools. These are principally machine vices and workholding chucks for the lathe. When placing work in these devices, it is usually essential that squareness is achieved, and that the work 'seats' correctly against the relevant reference faces, when this is appropriate. Correct seating is achieved by partially tightening the workholding clamp(s) and then tapping the work down onto the seating face.

One way to do this without marking the work is to use a normal (hard-faced) hammer but to interpose a soft block between it and the work. It is more convenient to use a soft-faced hammer, however, and two or three of these useful tools should be included in the tool kit. Several types (and different weights) are available, and the governing factor in making the choice is the size of the machinery and the likely size of the work. Figure 4.7 shows three types; a type with fairly-hard, screw-in plastic heads, a type with an all-rubber head and a third variant in which a metallic head is provided with plastic inserts. Types having screw-in heads are also available, which offer a choice of head shape, but they offer little advantage if the need is simply to seat work, as described. Shaped soft heads are supposed to be useful

Figure 4.7   Three types of soft-face hammer.

should panel beating be required but if there is a need to deform metal, there is no substitute for a hard-faced hammer and a good solid block to support the work.

Clamps

There is a frequent need for clamping things together for trial assembly, for brazing or soldering together or for drilling or machining two or more components simultaneously. For soldering or brazing, it is worthwhile making up some clamps especially for the task since they will inevitably be heated during these operations, perhaps to high temperatures, and are sure to become contaminated by flux, and will ultimately corrode.

The most useful form of clamp for other purposes is that known as a toolmaker's clamp, examples of which are illustrated in Figure 4.8. Each clamp comprises two similar jaws, united by opposing screws, one pulling the front of the jaws together, and the other pushing the rear ends apart. It is possible to maintain the jaws parallel in use, and the relatively large, flat jaws

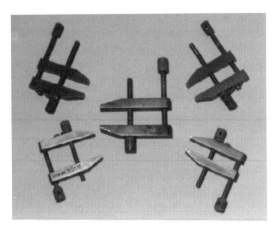

Figure 4.8 Some small toolmakers' clamps. The largest shown is a 2½ inch (65mm).

do not mark the work, even though good clamping pressure is applied.

Initial adjustment of the gape of the jaws is achieved by grasping both screw heads and winding the clamp in the appropriate direction to open or close the jaws. To assist a smooth adjustment during this operation, the head of the front screw is sometimes grooved to receive a cranked retaining strip which is secured by screws to one jaw. This ensures that the front screw takes this jaw with it when being unscrewed from the other jaw.

Figure 4.9 A toolmaker's clamp on a length of steel bar. The rear screw needs to be tightened to bring the jaws to parallelism.

Final adjustment to parallelism on the work must be by adjusting both screws, but is best achieved by adjusting the front screw until the jaw gape is too large at the front, as shown in Figure 4.9. This allows the clamp and work to be positioned correctly, after which the rear screw can be tightened to apply pressure and achieve parallelism, this being tested by rotating the clamp slightly to check whether it is gripping only at the 'toe' or 'heel' or all along the jaw.

A useful exercise is to make up a pair of clamps and Figure 4.10 shows a drawing of the parts and gives a range of suggested sizes. The retaining strip for the front screw is by no means essential and the clamps can be made as shown in Figure 4.10 without any real disadvantage, except for an occasional tendency for the jaw to jam during initial adjustment.

The material to use for the clamps is bright mild steel (BMS). The jaws can be shaped by hand and after completion should be polished with fine grade emery cloth and afterwards case hardened, as described in Chapter 3. After hardening and quenching, the bright finish can be restored by cleaning off the scale and re-polishing with emery cloth.

The screws are a straightforward turning and threading exercise. If you don't have a set of knurls, simply cross drill the screw ends so that a tommy bar can be used for tightening and releasing. This might be done anyway, following commercial practice.

Scribers, squares and protractors

A preliminary to almost all work, even the simplest of components, is the marking-out stage. A description of this process is contained in Chapter 5, but the subjects not covered there include a consideration of the marking-out tool (the scriber) itself, and the means for establishing lines on the work at the relevant angles, this

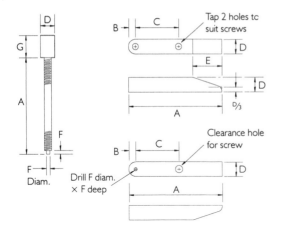

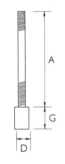

Dimensions					
A	1½ (40)	2¼ (58)	3 (75)	3¾ (95)	4½ (115)
B	⅛ (3)	³⁄₁₆ (5)	¼ (6)	⁵⁄₁₆ (8)	⅜ (9.5)
C	¾ (19)	1⅛ (30)	1½ (40)	1⅞ (50)	2¼ (57.5)
D	¼ (6)	⅜ (10)	½ (12.5)	⅝ (16)	¾ (19)
E	¼ (6)	¾ (19)	1 (25)	1¼ (32)	1½ (40)
F	¹⁄₁₆ (1.5)	³⁄₃₂ (2.5)	⅛ (3)	⁵⁄₃₂ (4)	³⁄₁₆ (5)
G	⅜ (10)	⁹⁄₁₆ (15)	¾ (19)	¹⁵⁄₁₆ (24)	1⅛ (30)
	5BA or	2BA or	0BA or		
Thread	⅛	³⁄₁₆	¼	⁵⁄₁₆	⅜
	BSW	BSW	BSW	BSF	BSF

Figure 4.10 Drawing for toolmakers' clamps up to 4½ inch (115mm).

function being performed by squares or protractors.

A scriber has a hardened, sharp point which is used to scratch the work to establish a reference line or position. For accurate marking out,

the scriber must have a fine point and this is usually ground on a finely tapering end. Figure 4.11 shows scribers of two basic sorts, an engineers' or machinists' type comprising a rod with centralised, knurled grip, and a pocket scriber which comprises a body having a collet fitting which accepts the scriber point either way round. A clip fitted to the body creates a pocketable scriber. Provided that the point is fine, there is no functional difference between the two types.

Although usually just called a square, the correct term for the tool for establishing the 90-degrees, or square, condition, is an engineer's try square. A square comprises a stock into which is set a thinner blade, centrally at one end. The completed square usually has a ground finish and both edges of the blade are nominally at 90 degrees to the stock.

Squares are available in different sizes, specified by the length of the blade, in inches. Figure 4.12 shows a 6-inch and a 3-inch. Most needs are satisfied by a 6-inch and something smaller, say a 2-inch or 3-inch, but occasional access to something larger can be helpful. Failing the availability of a larger square, a straight length of stock rectangular bar can be clamped to the work, and its squareness achieved by using a

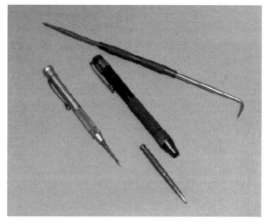

Figure 4.11 Scribers.

Figure 4.12 3 inch and 6 inch try squares.

Figure 4.13 The components of a combination set, and the protractor mounted on to the rule.

smaller square. A 6-inch square is just about adequate to set up a 24in. (600mm) bar and this method will serve for the odd occasions when something larger is required.

A longer reach is frequently provided by a combination square or combination set, the foundation for which is usually a 10-inch or 12-inch (300mm) rigid rule. The rule is grooved on one side so that sliding heads can be located and locked on it. The available heads are of three types. A square head has faces which lie at 45 and 90 degrees to the edge of the rule and thus forms a 12-inch square or a similarly sized, 45-degree bevel. A protractor head allows any angle to be set against an angular scale which is typically graduated every degree.

The third type of head is known as a centre head. It comprises a two-faced stock, the angle between the faces being bisected by one edge of the rule. If the stock of the centre head is placed on a circular workpiece, the rule edge lies on a diameter and a line scribed along the rule must therefore pass through the centre of the work. The intersection of two diametral lines thus defines the actual centre. The components of a combination set are shown in Figure 4.13 and the centre-marking operation in Figure 4.14.

Figure 4.14 Centre marking using a centre square.

Figure 4.15 A simple protractor.

A good quality combination set might cost about £100, of which £50 or so is accounted for by the protractor head. If less-frequent measurement of angles and bevels is envisaged, and such expenditure is not justified, a simple protractor of the type illustrated in Figure 4.15 will be satisfactory. This comprises a slim rule held in a pivot on a plate that allows setting to 1-degree engravings on an angular scale. One advantage of this type is that the rule can be used for depth measurements when set to the 90-degree position.

Small vices and pin chucks

For the instrument making kind of work which much of model engineering comprises, there is a frequent need to hold small items, either tools or workpieces. Sooner or later, holding devices for these will be required, several of which are appropriate to hand use.

The most frequently needed devices are pin chucks and pin vices, intended principally for holding small-diameter, circular items. Two types of pin chuck are shown in Figure 4.16.

On the left, a standard type having a knurled body and 4-jaw split collet, intended for holding small drills, up to about .040 in. (1mm) in diameter. The body is bored through, however, and the chuck will accept long rods, should these need to be held. The upper item in Figure 4.16 is a double-ended pin chuck having a differently sized collet at each end. This accepts items from virtually zero diameter up to about ⅛in. (3mm) which is sufficiently large for it to be used as a handle for needle files.

Figure 4.17 shows a pin vice. This comprises a small, springy vice jaw unit mounted to a hollow stem, the jaws being cut with vee-shaped grooves to allow small rods to be gripped coaxially with the stem. The otherwise plain jaws may naturally be used for holding small, rectangular work and the device will thus function as a simple, hand-held vice.

A hand vice is usually more substantial than the pin vice of Figure 4.17 and may have jaws

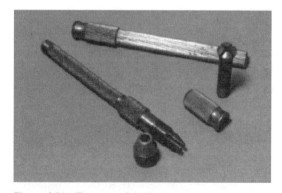

Figure 4.16 Two types of pin chuck.

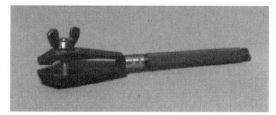

Figure 4.17 A pin vice.

up to 1¼in. (30mm) wide, opening to ⅝in. (16mm) or so. Both types of hand-held vice are useful when working with small parts, but an adaptation which can be held in the bench vice is also convenient for some work, since it allows both hands to be used for filing, allowing more precise control to be exercised.

Punches

The most important of the punches is probably the centre punch. The 'business' ends of three from my stock are shown in Figure 4.18. A centre punch has a hardened and tempered point, usually ground to a 60-degree included angle. Centre punches are used for forming dimples in the work in positions which define the centres of holes. The purpose of the dimple is to allow the drill to find the location and commence drilling in the correct position.

A drill does not strictly have a point but terminates instead in a vee-shaped line on the end of the web, or central core of the drill. The larger the drill, the larger is the core, and the larger the centre-punched dimple needs to be. This means a wider angle to the punch point and a harder blow from the hammer. Unfortunately, a wider point and harder blow are not conducive to accurate positioning, so it is useful to have two punches available, one for spotting the location and a second for widening and deepening the dimple. Too sharp a point cannot be used since it lacks strength and quickly breaks down in use.

Small punches such as those shown can easily be made up using silver steel rod, the pointed end being hardened, and afterwards tempered to dark purple, as described in Chapter 3.

An alternative to the basic punch is the automatic centre punch shown in Figure 4.19. This comprises a short, hardened point mounted into a holder which incorporates a spring-

Figure 4.18 A close-up and shadowgraph of three centre punch points.

loaded latch. Pressing the point onto the work through the holder, compresses the spring progressively until the latch releases, suddenly applying the pressure directly to the point. A knurled cap, or ring, on the holder allows the latch release pressure to be adjusted, thus allowing a variable blow to be applied to the point. Initial spotting of the hole position, using only light pressure, is thus possible, followed by setting the punch for a heavier blow and going round the hole positions for a second time.

Centre punch points do become blunted in use and they should be examined under a watchmakers' glass from time to time since a damaged or worn point makes it difficult to position the dimple accurately.

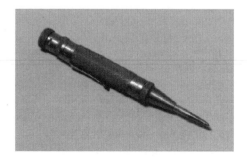

Figure 4.19 An automatic centre punch.

Parallel-shank punches, known as pin punches, will be needed for inserting and removing taper pins or any parallel fixing pins. These can be purchased from commercial sources but can also readily be made up from silver steel, again being hardened and then tempered to dark purple, as described in Chapter 3.

Silver soldering and brazing equipment

Gas blowtorches

There is, strictly, a difference between brazing and silver soldering, but the processes are so similar that they are frequently just described under the one heading of brazing. Both processes require the work to be heated to 600–800°C which is a 'comfortable' red heat and it is essential to have appropriate and adequate equipment available.

The first need is a proper source of heat and if brazing, as distinct from welding, is required, the most convenient source is a propane blowtorch. These torches are usually designed on a modular basis so that one buys a handle (incorporating the shut-off valve), a neck tube and a burner to acquire the components of a basic torch, afterwards adding other burners and neck tubes to suit the size of the job. A handle, two neck tubes and a selection of burners is shown in Figure 4.20.

To complete the blowtorch outfit, a large bottle of propane is required. This must connect to the handle through a regulator and a hose suitable for use with propane. A regulator is employed to deliver a constant gas pressure to the handle-and-burner assembly but also incorporates a detector to cut off the gas supply should the hose fail. An alternative to a regulator is to use a hose failure valve which does not regulate the pressure but does guard against hose failure. The outlet of a small propane cyl-

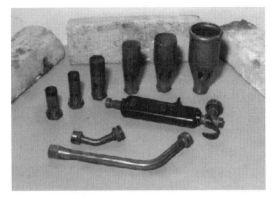

Figure 4.20 A propane blowtorch outfit, showing handle, neck tubes and burners.

inder fitted with a hose failure valve is shown in Figure 4.21.

When obtaining a torch outfit of the type shown in Figures 4.20 and 4.21, it should be borne in mind that a range of different burners is required to cope with differently sized jobs, since a large structure needs a larger input of heat to achieve and maintain the same temperature, due to its ability to radiate a greater quantity of heat. A larger quantity of heat is obtained by burning more gas, and the burners, although given simple reference numbers, are classified according to the gas consumption, in ounces or grammes per hour, when working at the standard pressure. The heat radiated from

Figure 4.21 A hose failure valve fitted to a propane cylinder.

the work increases with its size, as noted above, so it is necessary to stand farther away from the larger jobs and a longer neck tube is required for use with larger burners.

The amount of heat which is radiated from burner and work sets a practical limit on the size of burner which can be used, since the operator must be close enough to the work to apply the silver solder or brazing alloy, and this limits the maximum length of neck tube. The limit is represented by a gas consumption of about 70 ounces (2kg) per hour, but this is certainly adequate to carry out the silver soldering on a model boiler having a 5-inch (127mm) diameter barrel and being about 18 or 20 inches long overall (around 500mm). Larger work can also be managed by a burner of this capacity provided that the work is well packed around with heat retaining material. If more heat than this is required, it is best to limit the burner size but to use two torches by calling in the services of an assistant.

Two particular problems occur with bottled gas blowtorches, one associated with the bottled gas and the other with the basic torch design. When the gas burns, it consumes oxygen, and a good supply of this is vital to the combustion process. Conventional burners, of the type shown in Figure 4.20, mix gas and air within the burner, which is provided with a group of air inlet slots adjacent to the gas jet. The burner must always be able to draw in enough air through these slots to support combustion.

If the torch is directed into a closed space, the products of combustion blow back towards the rear of the burner. This prevents air being drawn in, and the flame thus extinguishes itself. One way to prevent this is to redesign the burner so that the air is drawn in away from the burner head and one manufacturer makes burners which operate on this principle. The separate neck tubes and burners shown in Figure 4.20 are replaced by an assembly which

Figure 4.22 A Sievert *Cyclone* burner and its handle.

combines both functions, the air holes being positioned at the handle end of the assembly, so removing them from the vicinity of any blown-back combustion gases and placing them in a location having a freely available air supply. A handle and burner of this type are shown in Figure 4.22.

Brazing hearth

Bearing in mind the above description of the amount of heat likely to be radiated during the brazing of a large structure such as a copper locomotive boiler, it is clear that a safe area, free from combustible material, must be available in which to carry out the process. This is provided by a specially built brazing hearth. This should ideally be located outside the area containing the machine tools but in any event must be positioned where there is adequate ventilation since poisonous fumes can be produced by the brazing processes.

A common method of making a hearth is to use one end of a circular, 50-gallon (225 litre) steel drum. If the drum is cut off 12 or 15 inches (300–375mm) from one end an open-topped cylinder is produced. If part of the cylinder wall is cut away to provide access, the remaining up-standing wall around the base prevents the spread of flame and creates a safe brazing hearth. A set of legs will be needed to create a free-standing unit, and it is usually recommended that a central hole should be provided so that one end of a long workpiece such as a boiler can protrude through from below, thus maintaining the work low down in the hearth.

If you are likely to have to deal with fairly long items, the circular shape may not be convenient. Figure 4.23 shows a rectangular hearth constructed from commercial slotted angle with the top and back formed by two parts of a shelf unit from the same manufacturer. This has a central hole in the base which is ordinarily covered by a loose plate of mild steel.

In order to minimise on the heating time (and save on gas) large work must be packed into a 'cocoon' of insulating material. Traditionally, the material used was coke, but a cleaner alternative is firebrick, or some other type of refractory material. Secondhand bricks can sometimes be obtained when boiler or flue linings are renewed. Whole bricks, as

Figure 4.24 A low-cost alternative to a small brazing hearth.

illustrated in Figures 4.23 and 4.24, can be used to support the work, or form a barrier behind it, while broken brick, or small pellets of refractory material, may be used inside hollow work, or for packing around the work generally.

For small work, a large hearth of the type illustrated (about 30in. × 14in. or 750mm × 350mm in plan) may not be required. Something smaller can be created by utilising a few firebricks in a biscuit tin, as shown in Figure 4.24. Two layers of firebrick provide the basic insulation below the work, but this is supplemented by a piece of commercial insulation, known as *Sindanyo*. In addition to being a good insulator, it provides a flat surface on which the work can rest. Since only small blowtorches are used with this hearth, two or three firebricks behind the work are adequate to provide insulation and prevent spread of the flame.

For silver soldering of very small work, such as might be undertaken by a jeweller, very small flames are ordinarily used and a small block of *Sindanyo*, or similar, will serve most needs, provided that care is taken.

Figure 4.23 My brazing hearth showing its central hole and some firebricks obtained when a boiler flue was being rebuilt.

5

Measuring and marking out

Measuring equipment

The humble rule

Much ordinary measuring is done with a rule, especially when marking out or checking the size of work in the flat, as for sheet metal work. But there are rules and rules, and what will do for making a garden fence will not do at all for most operations which might be encountered in model engineering.

First of all there is the question of accuracy. If you are making a scale model of about one-twelfth full size, one inch on the prototype will be represented by .083in. (2.12mm) on the model. While you may not think that it is important to work to high accuracy, if there are two items on the model which should be the same size, it is often noticeable if they are not. In any event, things need to fit where they are supposed to, and if they do not, that also is very noticeable.

You will need at least two rules – one for accurate measurements and the other for deciding whether you have picked up a piece of 1 × ¼ or ¾ × ¼ BMS (or whatever). This latter can

be regarded as the 'hack' rule and it may be left lying about the workshop in a casual way, or thrown into a drawer when not in use. This rule can, therefore, be the older of the two. Its corners may have become rounded as a consequence of much use (or abuse) and it may therefore not be suitable for use when marking out, or when real accuracy is required.

The accurate rule should be the best you can afford and it should be stored and used carefully so that it remains in good condition. It must have finely engraved scales so that the lines are significantly smaller than the smallest division on the rule. If you model to imperial standards, you will need a rule engraved down to ¹⁄₆₄in. (.0156in.). The most useful type is perhaps a double-sided rule engraved with ⅛, ¹⁄₁₆, ¹⁄₃₂ and ¹⁄₆₄ inch divisions on its four edges, but an alternative is the type which is engraved in ¹⁄₃₂in. increments along one edge, and ¹⁄₆₄in. on the other. Both types allow measurements of ¹⁄₆₄in. increments along the full length of the rule.

An alternative imperial rule is that engraved in ¹⁄₁₀₀in. divisions, this type being useful in instances where the scale dimensions have been 'rounded' to decimal fractions rather than the more usual imperial (vulgar) fractions such as 1/64in. The .010in. markings are difficult to see without use of a watchmaker's glass, but since estimation of sizes is frequently performed using a glass, this is not necessarily a serious disadvantage.

For modellers using metric standards, 0.5mm engravings provide slightly larger increments than the imperial 1/64in. (.0197in. compared with .0156in.) so that 0.25mm divisions may seem more desirable. However, rules engraved to this level of precision are not readily available and present the same problems of visibility as those imperial rules engraved with 1/100in. divisions.

When purchasing a rule, fine engravings and good visibility are the prime requirements, hence the popularity of satin-chrome finish rules and black-filled engraving. The other considerations are length, width and thickness. Length is obvious and most needs are satisfied by 6in. or 12in. (150 or 300mm) rules but for work on the lathe (for length measurements) there is real advantage in something shorter and a 3- or 4-inch (75 or 100mm) rule is probably more useful than a 6-inch unless you particularly wish for one which fits conveniently into a top pocket. A dual, imperial and metric, 4-inch rule is illustrated in Figure 5.1 together with a number of other rules.

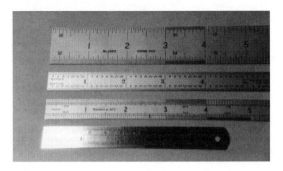

Figure 5.1 A selection of imperial rules and a 4 inch (100mm) having metric and imperial markings.

Once length is determined, the choice lies between rigid or flexible types. Personal preference dictates the decision, the two types being thin and narrow, and therefore flexible, or wide and thick, and hence rigid. Flexible rules are generally 1/2in. or 12.5mm wide, while the rigid type is usually 1in. or 25mm wide, although some rules 3/4in. (19mm) wide are available. Both flexible and rigid types are shown in Figure 5.1.

If you intend to make a large-scale model steam locomotive, you will need a longer rule; either 36in. or 1 metre long, according to your preferred measuring system, for marking out the longer items such as the frame plates, water tanks, boiler barrel and running plates.

Calipers

A rule is not suited to making really accurate measurements since it is engraved only in 1/64in. or 1/2mm increments. A rule is also not ideally adapted to measuring the diameter of a circular rod, or the bore of a circular hole. Nevertheless, before the widespread introduction of more accurate measuring devices, rules were generally the only measuring instruments available, and parts were made to roughly the right size and then finished by hand to achieve the required fits and clearances.

For measuring thicknesses, and the sizes of holes and recesses, simple devices were developed in ancient times for use by carpenters and stonemasons, and these naturally came to be used by metalworkers.

These instruments are essentially 'comparators' rather than 'absolute' measuring devices, since they are not engraved with scales and their settings must be checked using a rule. They are known simply as calipers, two basic types being used, intended for outside or inside measurements and consequently known as

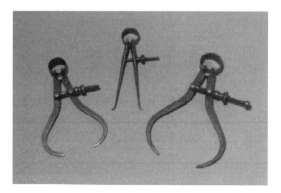

Figure 5.2 Inside and outside calipers.

inside and outside calipers. A group of calipers is shown in Figure 5.2.

A caliper essentially comprises two stiff arms with hardened tips which are either turned out or turned in to create an inside or an outside caliper. The joint between the arms may be formed by overlapping two (flat) arms and riveting them together, with substantial washers on each side, to form a stiff, but movable, joint. This is the firm-joint caliper. Alternatively, two arms are united by a circular spring and roller and fitted with a screwed stud and adjusting nut. This is the spring caliper.

For making a measurement, the caliper is adjusted to just touch either the outside or the inside faces of the work, after which its setting is measured using a rule.

In setting the caliper, the correct sense of touch needs to be used to ensure that the caliper is not set to such a size that it is 'sprung' when fitted to the work. A description of the method of setting an inside caliper into a bore or recess is given below.

Figure 5.3 shows the setting of an outside caliper being checked against a rule, one leg resting on the end of the rule and the position of the other leg (the setting) being read off on the rule.

If the setting of an inside caliper is to be measured, the rule and the caliper must abut a square block in order that the setting can be measured accurately. A simple offcut of steel bar makes a suitable block, and the operation of checking the setting is illustrated in Figure 5.4.

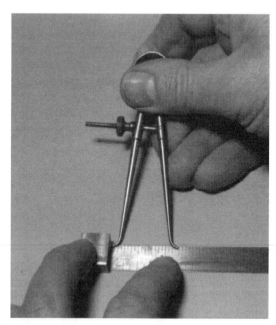

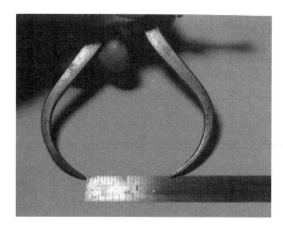

Figure 5.3 Measuring the setting of an outside caliper using a rule.

Figure 5.4 Measuring the setting of an inside caliper using a rule. Note the use of a square block of steel to provide an abutment.

Adjustment of a spring caliper is obvious, but adjustment of the firm-joint type, for small movements, is best made by tapping either the back of the joint, or one of the arms, on a wooden block. Tapping the back of the joint causes it to spring apart by a small amount, while holding one arm and tapping it on a block closes up the caliper.

Jenny calipers and dividers

An alternative form of caliper is the odd legs or Jenny caliper. Two pairs of these are shown in Figure 5.5. This instrument is basically an adjustable caliper with one side holding (or shaped into) a scriber point, and the other provided with a turned in or recessed foot which abuts the edge of the rule, for setting, or the edge of the work, in use. During setting, the plain foot rests against the end of a rule and the scriber point locates in an engraving on the rule, thus making setting to a standard, engraved value simple and straightforward. Once set, the caliper is used to scribe a line parallel to a reference edge, and provided that the caliper is always held squarely during setting and use, accurate marking off results.

Related to the caliper is the divider, two sizes of which are illustrated in Figure 5.6. Dividers

Figure 5.5 Jenny calipers, or odd legs, once a common item for apprentices to make.

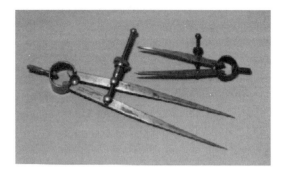

Figure 5.6 Two pairs of spring dividers, the larger is a 6 inch (150mm).

are used for measuring the distance between two points or parallel lines on a flat surface. They are also used for transferring measurements from a rule to a flat surface and are naturally used for scribing arcs or circles. Most dividers are of the spring type, but firm-joint types are still sometimes seen.

The vernier caliper

An improvement in the measuring precision of a linear device such as a rule can be obtained by using a vernier scale. This device, which is extremely simple in concept, allows an indication of sizes between the smallest divisions on the scale to be obtained directly. Figure 5.7 shows a portion of a rule engraved in $\frac{1}{10}$in. divisions. Adjacent to this is a short scale, engraved with 10 divisions similar to those on the rule. However, this scale is engraved so that it has 10 divisions in the same length occupied by 9 on the rule itself, as shown in Figure 5.7a. Thus, each division on the auxiliary, or vernier scale is $\frac{9}{10}$ of the size of those on the rule. With the scale positioned as shown in Figure 5.7a, only two lines on the vernier scale correspond with engraved lines on the rule, these being the scale's '0' and '10' lines. This is the position of the scale whenever the measured value is a

(a)

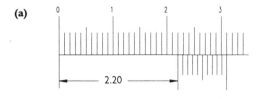

(b)

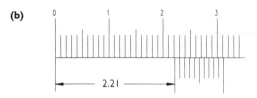

Figure 5.7 Two settings on a vernier scale, .01 inches (0.25mm) apart.

whole number of inches and tenths of an inch, the position of Figure 5.7a corresponding with 2.20in.

If the scale moves relative to the '2.2' mark on the rule, one tenth of the distance towards the '2.3' engraving, its own engraved '1' mark exactly coincides with the '2.3' line on the rule, as shown in Figure 5.7b, since the distance between the lines on the vernier scale is $9/10$ of that on the rule. The position taken up by the scale is thus 2.2in. plus $1/10$ of the distance towards 2.3in. or 2.21in. Thus, by use of the vernier scale, a rule engraved in $1/10$in. increments can yield measurements which can be read to within $1/100$in.

A simple rule fitted with a vernier scale is not a practical proposition and the device most commonly found is that illustrated in Figure 5.8. This is generally simply known as a vernier although it is correctly called a vernier caliper. This comprises a rule which incorporates a fixed jaw and a sliding jaw, integral with a frame which slides on the rule. The sliding jaw frame is engraved with a vernier scale and its position relative to the fixed jaw may be assessed accurately. Both jaws are hardened, ground and lapped so that they lie parallel to each other and at right angles to the rule.

The sliding jaw frame connects to a clamping frame through a short screw which engages with a knurled nut running in a slot in the clamping frame. Since the clamping frame can be locked to the rule by a screw, the knurled nut allows fine adjustment of the position of the sliding jaw and its frame along the rule. A second locking screw is fitted to the sliding frame so that it, too, may be locked to the rule.

The imperial scale on this vernier is divided into tenths of an inch, and each tenth into four divisions, each of which represents a quarter of one tenth, or .025in. The vernier scale is graduated so that the distance occupied by 49 divisions on the rule is divided into 25 parts. Thus, each vernier division represents $24/1000$in. and since each corresponding division on the main scale represents $25/1000$in., the distance between any two vernier and main scale divisions is $1/1000$in., or .001in. This type of vernier caliper is used to indicate a measurement within $\pm$.001in., generally called 'one thou'.

When using the vernier caliper, the knurled clamping screws are first loosened and the sliding head and clamp adjusted approximately to the correct position. The clamping frame is then locked to the rule using the screw and the measurement taken by bringing the sliding jaw into contact with the work to be measured, making fine adjustment of the sliding jaw and frame with the knurled nut. For accuracy, the

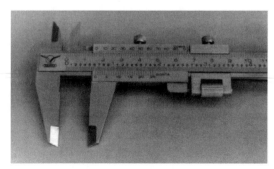

Figure 5.8 A vernier caliper.

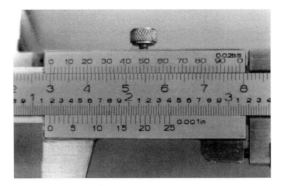

Figure 5.9 Close-up of the vernier caliper set to 1.185/1.186 inches.

sliding jaw needs to be brought 'nicely' onto the work, so that its final position is just touching the work with neither slop nor excessive tightness. There is a need to develop a certain touch in using the vernier. This sense of touch is important, too, in respect to other measuring methods, and it sometimes pays to repeat the measurement to be sure that consistent results are being obtained.

Vernier calipers are used for measuring external diameters and sizes. In some types, including the one shown in Figure 5.8, an additional pair of jaws is fitted on the back of the fixed and moving jaws and these can be used for measuring internal diameters and sizes. A rod is also fixed to the moving jaws and can be used for depth measurement.

To illustrate the use of the vernier scale, Figure 5.9 shows a close-up of rule and scale. The setting portrayed is between 1.1in. and 1.2in. this distance on the rule being divided into four increments of .025in. The position shown is thus slightly greater than 1.175in. The vernier scale is engraved with 25 divisions, each representing .001in. (one thou) and it is either the 10 or 11 engraving on the vernier which is aligned with an engraving on the main rule. The setting is 10 or 11 thou greater than 1.175in. or somewhere between 1.185in. and 1.186in.

On the metric scale which is also engraved

on this vernier, the setting is between 30.01mm and 30.012mm which may be taken as 30.011mm.

Micrometers

An alternative way in which measurements can be taken, not reliant on direct measurement along a rule, is by use of a screw thread. The instrument has essentially the same parts as a vernier, that is, one fixed and one movable 'jaw', but this latter is attached to, or in one with, a screw thread which allows the distance between the fixed and moving elements to be measured indirectly. The instrument is known formally as a micrometer, but is frequently referred to as a mike.

The components of a typical instrument are shown in Figure 5.10. It consists of a steel frame carrying a fixed anvil, A, on one horn, and opposite to it, an adjustable spindle, B, to which is fixed a thimble, C. The spindle passes through a boring in a fixed sleeve, D, which is effectively in one with the frame, although it may be held by a stiff, frictional grip, and be adjustable.

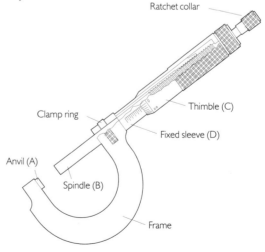

Figure 5.10 Principal parts of a micrometer.

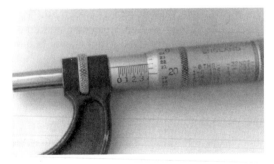

Figure 5.11 0–1 in. micrometer set to 0.397 in.

Both the fixed sleeve and the spindle carry mating screw threads so that the distance between the anvil and the spindle end may be varied by rotating the thimble. On an imperial micrometer, the thread used for the sleeve and spindle has 40 turns per inch (tpi) so that each turn of the thimble moves the spindle through .025 in. The thimble is accordingly engraved with 25 divisions, as shown in Figure 5.11, each of which thus represents .001 in. or one thou.

The actual measurement set on the micrometer is obtained by reading the scales on both the sleeve and the thimble. The sleeve is engraved with lines at .025 in. increments which correspond with whole numbers of turns of the thimble and spindle, from the zero position. Every fourth line is identified as 1, 2 etc. up to 9, corresponding with 0.1 in., 0.2 in. up to 0.9 in. Obtaining the measurement requires three positions to be noted and the setting of Figure 5.11 is interpreted as follows:

Highest sleeve engraving visible =
    3 → .300 in.

Number of .025 in. increments above this =
    3 → .075 in.

Reading on the thimble scale =
    22 → .022 in.

TOTAL MEASUREMENT = .397 in.

This procedure may seem laborious, but in practice one soon becomes proficient in interpreting the readings without effort.

A metric micrometer is perhaps easier to interpret since the spindle thread has a pitch of 0.5 mm and the thimble is engraved with 50 divisions, each representing .01 mm, as shown in Figure 5.12. Again, to save counting turns, the sleeve is engraved every 0.5 mm; whole mm above the index line, and half mm below, the whole mm lines being identified every 5 mm. To read the measurement it is necessary to read the sleeve and the thimble, and the setting of Figure 5.12 is read as follows:

Highest sleeve engraving visible = 6.5 mm

Reading on thimble scale = 0.47 mm

ACTUAL MEASUREMENT = 6.97 mm

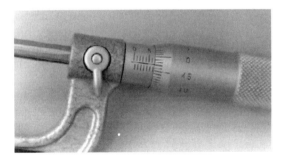

Figure 5.12 0–25 mm micrometer set to 6.97 mm.

Reading this micrometer is more straightforward than the imperial type.

Two further points need consideration in relation to the micrometer. First of all, the sleeve is not usually firmly fixed to the frame but is held by a stiff friction grip. It is therefore adjustable and can be rotated to bring its index line to correspond with the '0' engraving on the sleeve when the spindle end is closed against the anvil. To allow this adjustment, the sleeve is provided with a shallow drilled hole and the micrometer is supplied with a 'C' spanner which allows the sleeve to be rotated. These features are shown in Figure 5.13A.

A

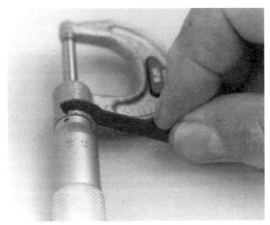

B

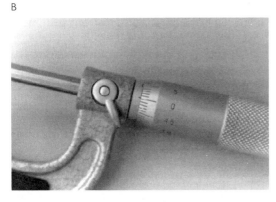

Figure 5.13 Close-up of a metric micrometer at what should be its zero position, with the 'C' Spanner which is used to adjust the position of the sleeve.

From Figure 5.13B it will be seen that the index line on the sleeve does not quite correspond with zero on the thimble, and a small adjustment of the sleeve is required.

A second point to consider is how firmly the spindle should be screwed into the anvil to define the zero. Clearly, a sense of touch is required, since the pressure needs to be just so much – neither more nor less. To remove the need to develop this required sensitivity, most micrometers are fitted with a ratchet on the thimble end which slips when a certain turning force (torque) is applied to it. Using the ratchet

thus ensures that a standard torque is always applied, both when setting the zero and when making measurements.

However, most craftsmen prefer to develop the required sense of touch for themselves and the most common way in which the micrometer is held is shown in Figure 5.14. The ratchet is contained within the small-diameter thimble end, but is studiously not in use since the third and fourth fingers are holding the micrometer frame, and the remaining digits cannot reach the ratchet. This method is the most convenient when the work must also be held, since the other hand is freed for this duty.

If several items need to be made to the same dimension, the micrometer may be used as a gauge by being set to the required size and machining each piece until it will (just) pass between anvil and spindle. To assist its use in this manner, the micrometer may be provided with a spindle lock. One type of actuation for this is the small lever visible on the frame of the metric micrometer in Figure 5.13. The imperial micrometer of Figure 5.11 is provided with a knurled ring set into the frame for the same purpose.

The micrometers illustrated and described above, cover only the limited ranges from zero to 1 inch or zero to 25mm. Reducing the meas-

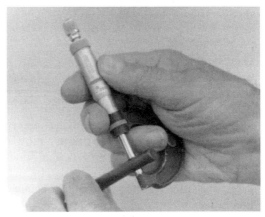

Figure 5.14 Holding a micrometer.

Figure 5.15 Three micrometers capable of measuring from zero to 3 inches, with setting bars 1 inch and two inches long.

uring capability to this fairly short range places a lesser requirement for accuracy on the screw, since it must be accurate to within, say, .0001in. in one inch (.0025mm in 25mm), rather than this amount in 2, 3 or more inches. Micrometers are consequently usually made to cover just a 1in. or 25mm range, standard micrometer heads being attached to different frames to provide 0–1, 1–2, 2–3, or even 11–12 inch (10–25, 25–50, 50–75 or even 250–275mm) micrometers.

Figure 5.15 shows three micrometers covering the range from zero to three inches, together with setting rods 1 and 2 inches long that allow the two larger instruments to be correctly zeroed.

Gauging the size

It is naturally possible to reverse the process of using a caliper, first setting it to a required size by use of a rule and then removing material from the part until it fits the caliper, allowing you to make a part having a particular dimension without measuring it directly. In these cases, the caliper is being used as a gauge to determine when the correct size has been reached. Using an outside caliper in this fashion allows the thickness, width or diameter of the work to be brought to the required dimension, whereas an inside caliper is used when a hole or recess of a specific size is required.

This is a useful technique to adopt when diameters are being turned on a lathe, for example. In these operations, the work is reduced in diameter (or a hole increased in diameter) in discrete steps, thus progressively approaching the required size. If an outside diameter is being turned, a caliper can be set a little larger than the final size. It is then possible to continue machining until the caliper 'goes' over the outside diameter, which indicates that the size is approaching that required. This allows the approximate dimension to be achieved without making any measurements directly on the work.

The same method can be adopted for circular bores, by setting an inside caliper just smaller than the size required, continuing to increase the bore progressively, without measurement of the size, until the caliper will enter, after which, the actual size of the bore needs to be measured after each cut, until the size is reached.

Any item which is being progressively brought to size, by whatever means, can be sized using a gauge, but the technique is especially valuable if several items need to be made to the same overall dimension, but the size is larger than the capacity of your really accurate measuring instrument, which might only be a 0–1in. or 0–25mm micrometer.

When building a model locomotive, the frame plates are held the correct distance apart by spacers, or stays, which all need to be the same length if the frame plates are to be parallel. There are usually several stays, and the attachments to the buffer beam at each end, and in some locomotives, the cylinder block is also mounted between the frame plates. There are, therefore, several items which need to be brought to the same length when making up the frame's components.

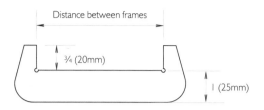

Figure 5.16 A gauge for measuring locomotive frame stays.

If a gauge of the type illustrated in Figure 5.16 is made up at the outset, it may be used to test each item as it is machined, and each reduced in width or length until it just enters the gauge without noticeable clearance. All items then conform to the standard dimension defined by the gauge.

The gauge itself can easily be made by hand, and its gauging dimension checked adequately using a rule, thereby providing an accurate gauge for making a group of similarly dimensioned components.

If rectangular recesses or slots are being produced, a simple gauge for the size can be made up by hand and used for testing as the enlargement proceeds. For example, if a standard gauge is available which measures .500in. (precisely), it may be used to determine (gauge) the width of a slot which is being machined, since, if the gauge will enter the slot, its width must be just greater than .500in.

If small gaps need to be gauged, the usual tool is a feeler gauge. These are manufactured in standard sets ranging from .0015in. (.04mm) to .025in. (25 thou) (0.6mm), the usual form being that in which ten blades are mounted together into a folded metal 'handle', or plastic moulding, as shown in Figure 5.17. The blade thicknesses (in inches) provided in the usual imperial set are:

.0015, .002, .003, .004, .006, .008, .010, .012, .015 and .025

The individual blades are etched with their thicknesses. Two or three can be used together to make up a gauge of the desired size. Similar metric sets are also available.

The usual way to employ feeler gauges is by employing a 'go' or 'no go' technique. If a gap of .015in. is required, and the gap is settable, the correct setting is such that a gauge of .014in. thickness should enter freely, or 'go', whereas a gauge of .016in. should not. By implication, this means that the gap is set to .015in.

Plug gauges

The technique for gauging the width of a slot, can also be used when machining a bore. This is achieved by employing a precisely sized cylindrical gauge, known as a plug gauge, which is made to the required size.

Industrially, plug gauges are available in closely spaced standard sizes, enabling a wide range of sizes to be gauged. As these gauges have a ground and polished finish they are consequently expensive and not usually available to the amateur, but it is quite easy to make them up to suit particular jobs.

The important characteristics are size and surface finish, so the home-made gauge needs first of all to be finish turned with a correctly ground and sharp tool. Secondly, it must be accurately sized but this can usually be arranged if a micrometer or vernier caliper is

Figure 5.17 Standard set of imperial feeler gauges.

A

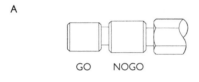

GO   NOGO

B

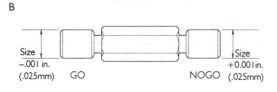

Size
−.001 in.
(.025mm)   GO

Size
+0.001 in.
NOGO   (.025mm)

Figure 5.18 'Go' and 'no go' plug gauges.

available, and is not difficult if approached carefully.

A single gauge may suffice if a hole is being bored progressively to size, but an existing hole may be measured if there are two plug gauges having a known small difference in diameter. If the smaller of the two enters the bore freely, but the larger one does not, the bored dimension lies roughly midway between the sizes of the two gauges.

The 'go' and 'no go' gauges can often be made conveniently on a single plug, as shown in Figure 5.18A, although this does limit the use of the 'no go' gauge to bores which are deep enough to accommodate the 'go' gauge also. The alternative is to arrange a plug at both ends of the gauge, as shown in Figure 5.18B, which naturally avoids this problem.

The gauges should ideally be hardened and polished, but to avoid spoiling the dimensional accuracy, it is best to aim for as good a turned finish as possible, afterwards hardening the gauge and then polishing. The simplest way to do this is to use bright mild steel (BMS) to make the gauge, afterwards case hardening it, as described in Chapter 3, and then polishing it to a fine surface finish by using emery cloth and paper to remove the scale.

As noted above, a single gauge often serves

the purpose, since, if a hole is being bored progressively, it is only necessary to reach the point at which a correctly sized gauge 'goes' into the bore. However, to assist this type of operation, it is useful to make two gauges – a correctly sized one and another which is .005in. or .010in. (0.1mm or 0.25mm) smaller than the size required, thus allowing the approach to the desired size to be detected.

Measuring practice

Using a rule

The method of using a rule may seem obvious, and so it is if the task is just to determine whether an item is ¾in. or 1in. (19mm or 25mm) wide. However, the rule is not ideally suited to measuring accurately the width of an item, and the task is different if the need is to determine whether two scribed lines are 6.75mm apart. Firstly, a metric rule having 0.25mm engravings may not be available, and secondly, even if it is, you may not be able to discern the precise size without the use of an eyeglass. It therefore becomes necessary to estimate the position because it falls between two engraved divisions, and it is also necessary to use a magnifying aid.

If the rule is engraved in ¹⁄₆₄in. divisions, it is possible, with a glass, to estimate the distance between two lightly scribed lines with an accuracy that is better than this i.e. it is possible to decide whether a scribed line actually coincides with the rule's engraving, or is displaced, and if so, by how much. This means that an estimate of actual position can be made to better than ¹⁄₁₂₈in., or better than .008in. A metric rule with ½mm divisions, allows estimation of the position of scribed lines to better than 0.25mm. This is illustrated by Figure 5.19 which shows two views which approximate to those provided by a 4½-inch watchmaker's glass.

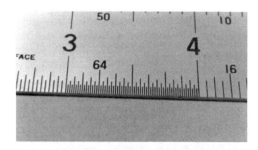

Figure 5.19 Close-up views of the two sides of a rule which has imperial and metric engravings, equivalent to the views provided by a 4½ inch (115mm) watchmaker's glass.

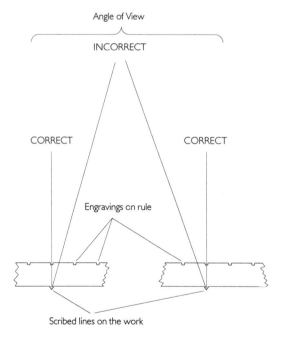

Figure 5.20 The possibility of parallax error should be remembered when checking or setting measurements on the work.

There is, however, one general point which should be made concerning the reading of any kind of scale, and that is the phenomenon known as parallax error. This arises if the position of the eye, when making the reading, is not exactly over the engraving on the rule. The situation which can arise is shown in Figure 5.20. An angled view of the scribed line and the engraving on the rule introduces an apparent error and the total error may be significant if it is in different directions at the two ends.

The thickness of the rule affects the size of the error, given a particular point of view, and a thin rule is therefore beneficial in this respect. However, in some circumstances, it may be possible to stand a thicker rule on its edge, thus avoiding the problem, but care must be taken to ensure that it is not bent since an error is introduced by the curvature.

Checking of marked-out positions must be performed as a matter of course. This requires careful use of a watchmaker's glass in order to allow the divisions on the rule to be seen adequately, in addition to care in positioning the eye to avoid the error which parallax may introduce. Used carefully, the rule is an excellent means to check marked-out work. It is slightly less well suited to the measurement of the size 'over' something and relatively poor when used to measure diameters. It is also not good for measuring bores and there is naturally a limit to the smallest division which can be engraved on the rule.

The use of calipers to measure inside and outside dimensions, in association with a rule, is described above, and the technique to adopt when setting the caliper is described below.

Measurement of depth

The ordinary rule is quite readily utilised for the measurement of the depth of a bore or recess in the work, its major drawback being that its width restricts its use to relatively large

Figure 5.21   A depth rule is simply a narrow rule which can be clamped into a shaped stock.

bores. A narrower-than-normal rule is beneficial and is usually produced in the form shown in Figure 5.21, in which a ¼in. (6mm) rule is provided with a sliding stock which can be clamped to it. The instrument shown has its rule engraved with both metric and imperial graduations thereby satisfying the needs of both systems.

The stock has a waisted shape which provides both long and short ground faces which lie at right angles to the rule and it can also be used in 'tight' situations as a substitute for a square.

An alternative to the depth rule, although not generally so useful due to its larger 'stock', is the type of protractor shown in Figure 5.22. This uses a narrow, engraved rule mounted to a screw clamp which allows it to be locked to any angle set on the scale. If the rule is set to the 90-degree position, it may naturally be used to provide a depth measurement. However, the 'regular' depth rule is generally superior for it slides in a machined groove in the stock and is always maintained in the 90-degree position, even when not clamped. The depth rule facilitates the actual measurement in that the stock can be held on the work's reference surface and the rule slid down until it contacts the bottom of the bore or recess. The rule can then be clamped to the stock and the tool removed from the work and the actual depth read on the rule.

Since the protractor does not automatically hold the rule in the 90-degree condition, there is always some uncertainty when trying to make a measurement in this way. The protractor's rule may nevertheless be useful in confirming that a bore or recess is under or over the depth required, by being set to the measurement and then clamped at 90 degrees. The depth can then be tested by probing the hole and confirming whether or not the rule reaches the bottom before the stock contacts the reference surface, or vice versa.

More accurate measurement of depth may be achieved by utilising an adaptation of the micrometer, but with the sleeve engraved in reverse i.e. showing zero at what would conventionally be the 1-inch position, the spindle then extending downwards into the bore or recess to show the increasing depth. Figure 5.23 shows a depth micrometer which measures depths from 0 to 1in.

The micrometer comprises a standard 0–1in. head but with the sleeve graduations increasing as the thimble is screwed towards the stock. A spindle is inserted through the top

Figure 5.22   This protractor uses a similar rule which can be clamped at an angle to the engraved plate. If the rule is set loosely to the 90-degree position, it might also serve as a depth rule, if used with care.

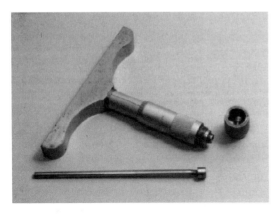

Figure 5.23 The components of a 0 to 1 inch depth micrometer.

micrometer shown in Figures 5.24 and 5.25 is properly described as an outside micrometer.

The use of the micrometer, and the 'standard closure' that is provided by the ratchet normally fitted to the end of the thimble, is described above, and the necessity to develop the correct sense of touch when using a vernier caliper is also mentioned since a standard technique is important in achieving repeatability.

Figure 5.24 illustrates a potential problem when using a micrometer to measure circular work. If the work is small, as shown at A, the micrometer naturally measures the true diam-

of the thimble and is clamped into place by a screwed cap.

The spindle has a screwed bush at its upper end which can be set so that the spindle is precisely the correct length. The micrometer is zeroed by standing the stock on a flat surface and screwing the spindle down until it, too, touches the surface, when the sleeve should be set to show zero.

Depth micrometers are usually provided with alternative spindles, typically allowing them to measure from 0 to 1in., from 1in. to 2 in. and from 2in. to 3in. Such an instrument is described as a 0 to 3in. depth micrometer. The equivalent metric instrument measures from 0 to 75mm. A fitted case is normally supplied to house all of the items – the one shown lacks these refinements, having been rescued from a secondhand tool store.

Outside measurements

Measuring the external dimension of a piece of stock material, or of the work, is carried out simply by use of a micrometer or vernier caliper. To distinguish it from instruments designed for measuring bores, the type of

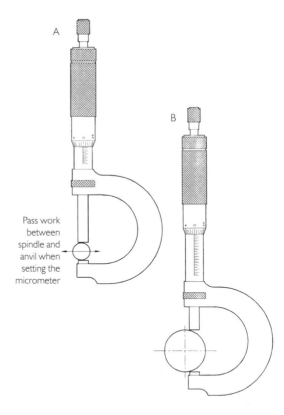

A

B

Pass work between spindle and anvil when setting the micrometer

Figure 5.24 There is a possible inaccuracy when measuring round material using a micrometer, if the material is not passed through the gap between spindle and anvil when the setting of the micrometer is adjusted.

eter if the work is roughly centred over the anvil of the micrometer. However, a micrometer applied casually to a larger bar can give a false reading if it is not applied precisely to a diameter, and it is essential to swing the bar through the gap between spindle and anvil, after closing up the micrometer on the work, to confirm that the setting does correspond to the true diameter.

One further point is also worth emphasising. The making of outside measurements is only straightforward provided that the work has a uniform shape. Unfortunately, this is not true of some circular-section cutters, notably drills, reamers and end mills. Consideration of the measurement of a drill diameter serves to illustrate the problem.

A drill is machined with two flutes that allow the cut material to be expelled from the drilled hole. The cutting edges of the drill are actually on the end and not on its diameter, but the outermost part of the drill is naturally machined to the correct size. This sized part of the drill is, however, confined to a small 'land' behind the front edge of the flute, and it is between the two lands that the measurement of diameter must be made.

Figure 5.25A shows an imperial micrometer measuring a ⅜in. diameter drill. The micrometer is not correctly positioned on the lips of the flutes, however, and is reading 0.351in. or about 24 thou too small. In Figure 5.25B, the drill is correctly positioned between the anvil and spindle of the micrometer so that the maximum diameter is being read, and the micrometer is now showing 0.375in. This type of measurement can be a problem with all-fluted cutters and care needs to be exercised, particularly with drills, since they are manufactured in very small size increments.

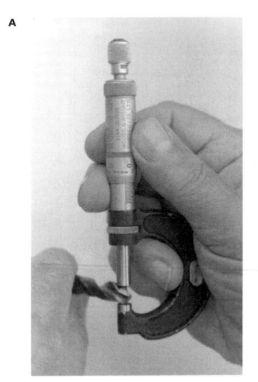

A

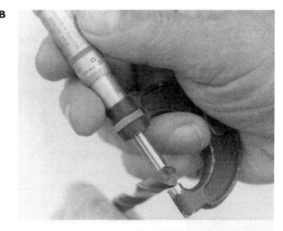

B

Figure 5.25   Since the end of a drill is not truly circular, an error in measurement can be made if the micrometer does not touch the lands just behind the edges of the flutes when the measurement is made. For similar reasons, end mills, slot drills and reamers are also difficult to measure correctly.

Measuring the bore of a hole

The vernier caliper shown in Figure 5.8 is arranged with two sets of jaws so that it may be used for both inside and outside measurements. The necessity to 'feel' the jaws onto the work when making measurements has already been referred to, and this is naturally also required when making use of the inside jaws. Equally important is the need to position the jaws on a true diameter of a circular bore and the technique of Figure 5.26 should be adopted to ensure that the correct position is found.

In a circular bore, an incorrect position results in the caliper jaws being set too close together. A rotation of the caliper about one jaw will reveal if an incorrect position has been selected. If so, the caliper jaws must be opened slightly and the check repeated until the correct position is found. This corresponds with the 'no free play' condition when the caliper is entered into the bore.

In a rectangular bore, incorrect positioning of the line of the caliper jaws may result in too large a setting of the jaws, as shown by the central sketch in Figure 5.26. Rotation of the caliper reveals this condition since movement is possible only on one direction and the jaws need to be brought closer together and the check repeated, until the caliper can just be rotated through the correctly aligned position.

When making the rotations, it is essential to keep one jaw in contact with the bore and rotate the caliper about it.

A suitable firm-joint or spring caliper may also be set to the diameter of a bore by adopting the same technique. If more accuracy is required than can be achieved by comparing the setting of the caliper with an ordinary rule, a micrometer or vernier caliper may be used to measure the caliper setting. Figure 5.27 illustrates this operation when using a micrometer.

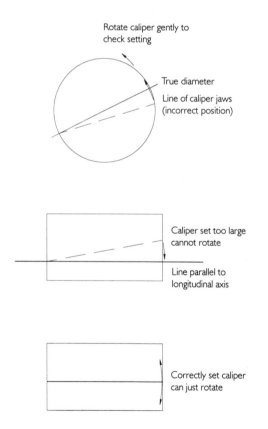

Figure 5.26 When using calipers, or the inside jaws of a vernier caliper, to measure a bore, care must be taken to ensure that the jaws, or tips of the caliper feet, are correctly aligned.

Figure 5.27 The setting of a caliper can be measured using a micrometer, carefully closing the micrometer until the caliper will not swing between spindle and anvil, under its own weight.

The caliper has been set carefully to the diameter of the bore and its setting is in turn being measured by the micrometer. One arm of the caliper is supported in contact with the micrometer spindle and the other arm is allowed to rotate past the end of the anvil. As the spindle is advanced successively towards the anvil, the point is reached at which the caliper arm will not rotate past the spindle under its own weight. At this point, the micrometer has been set too small and needs to be opened slightly to produce just the right feel to the movement of the caliper between spindle and anvil.

Two touch-sensitive settings need to be made to assess a bore diameter by this means, but a little practice should produce consistent results, and good repeatability will reveal when an appropriate level of proficiency has been achieved.

If a direct measuring method is required for bores, an inside micrometer is the correct instrument to use. However, these are expensive, and their utilisation may not be sufficient to justify the expense. A compromise, falling somewhere between the caliper and an inside micrometer in ease of use, is the telescopic gauge, three of which are illustrated in Figure 5.28.

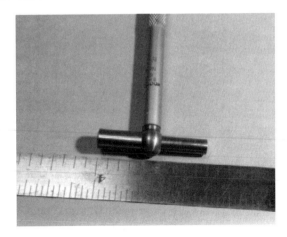

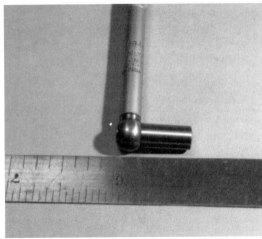

Figure 5.29 Maximum and minimum settings of the middle gauge of the set of three.

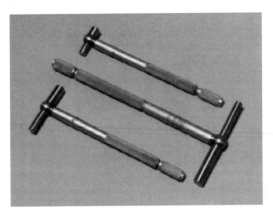

Figure 5.28 A set of telescopic gauges for measuring bores between 0.5 and 2.125 inches (12.5mm to 54mm).

Each gauge is made in the form of a tee-shaped assembly having two bars set at right angles to a plated and knurled handle. One side of the head of the tee is fixed to the handle while the other is spring-loaded within the first but lockable by turning a knurled ring on the end of the handle. Figure 5.29 shows the full range of adjustment of the middle size of the set of three, which, overall, cover the range of bores from 0.5in. to 2.125in. (12.5mm to 54mm).

To measure a bore, the gauge is allowed to expand (gently!) into the bore under the action of the internal spring. The movable plunger is then clamped using the locking ring and the fit in the bore tested by rocking and rotating the gauge and gently easing it along the hole. When the fit is correct, the gauge is carefully removed from the hole and its setting measured using a micrometer or caliper.

Being somewhat larger in diameter than the arms of a caliper, a telescopic gauge is more readily established on a true diameter and its polished and domed ends make the setting of the micrometer more straightforward than when assessing the setting of a caliper.

Various forms of small-hole gauges extend the principle of the telescopic gauge in the direction of decreasing hole size and an expanding gauge operating on a slightly different principle is shown in Figure 5.30. There are also other types of small-hole gauges which utilise a tapered mandrel which is entered into the hole, the depth of penetration indicating the diameter.

The possibility of using plug gauges to judge the diameter of a hole should not be forgotten if the size of a hole needs to be measured.

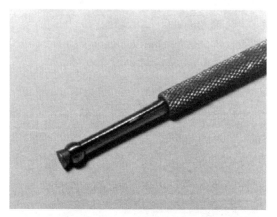

Figure 5.30   A gauge for small holes which is adjustable by drawing a shaft with a tapered end into the bore of a ball-ended split rod.

Ground steel rod (either silver steel or ground mild steel) might be used, but the shanks of reamers may provide accurate gauges for standard sizes. The shanks of drills may be used, but these are usually finished to a slightly smaller diameter than the drill itself, and they should be measured carefully before use as gauges. Nevertheless, the availability of an outside micrometer makes possible the measurement of inside dimensions, albeit by indirect methods, but that all-important sense of touch does need to be developed to allow adequate accuracy to be achieved.

It should also not be forgotten that it is not always necessary to make an actual measurement, since, if a bush is required to fit a shaft, or vice versa, the one can be made until it is a good running fit on the other, as judged by its feel. Naturally, it is helpful if one of the items is a standard size, but exact dimensions may not be important.

It is thus possible to machine the bore of an engine cylinder until it has reached the correct diameter as measured by spanning a rule across one end. This will mean that it is correct to the nearest $\frac{1}{64}$in. or $\frac{1}{2}$mm, say. When the piston is machined, it can be turned to provide the required fit in the bored cylinder and will then match the bore for which it has been made, although neither item has been measured accurately. Once again, however, that important sense of touch comes into play.

Marking out

Introduction

Except in the case of a very simple component, or when dimensions simply do not matter, the workpiece needs to be marked to show the positions of its outlines and to define important points within it. This is generally true whether the work is to be finished by hand, or is to be machined. Marking out consists simply of using a sharp-pointed tool to scribe the outline of the work which needs to be cut out, or to define the positions of holes or the centres of arcs or curves, according to the requirements of the job.

Scribing may be performed in a number of ways, but if hand methods are used, is normally performed using a scriber and straightedge, for straight lines, or dividers for arcs or circles. Dividers are also ordinarily used if several equal pitches need to be marked off.

Hand methods of marking out are usually based on the creation of two edges (or surfaces) at right angles to each other, from which measurements may be set off to define relevant points on, or within, the outline.

An alternative method of marking out uses a flat surface as a reference from which to set off measurements, the work being supported in some way on the surface. If the work can be rotated precisely through 90 degrees, it is possible to scribe lines on the work in two directions at right angles, which satisfies the basic need. The method does not necessarily require a true edge or surface to be created in advance of marking out, although it can be helpful. The method uses a flat reference known as a surface plate, the use of which is described below.

Visibility

Scribing the work removes a small groove of material, producing a bright line. Removal of material means that the surface is permanently marked and it is therefore necessary to scribe only lightly. The visibility of the line can be improved by coating the material with an appropriate colour, darkening the surface of light materials (bright steels, brass and aluminium alloys) and lightening the darker ones such as cast iron or black mild steel.

Felt-tipped pens are a useful source of colour, although their effect is not long-lasting. Commercially produced coatings are available, usually called marking-out fluid or layout fluid (blue and green are the popular colours) which comprise a solution equivalent to a rapid-drying paint. A quick brushing over of the work quickly produces a reasonably hard coating to improve the visibility of scribed lines although it is important not to apply too thick a coat or there is a likelihood that it will flake off. Once the work is completed, the remaining coating can generally be removed using a simple solvent such as methylated spirit.

For use on cast iron, a white coating is the most useful and this can be provided by one of the quick-drying correction fluids used by typists, but it is again necessary to apply a thin coating otherwise it flakes off. Rubbing over the casting with blackboard chalk and tapping

Figure 5.31 The effect of coating the work with a contrasting fluid before marking out.

off the excess is also a useful way to improve the contrast on cast iron.

No particular surface preparation is necessary on inherently clean materials, but castings usually benefit from a cleaning up with a wire brush to remove any sand adhering to the surface and may even require attention from a file in some instances. To illustrate the benefit of using a marking fluid, Figure 5.31 shows an aluminium casting and an iron casting which have been partially coated and then scribed.

Checking measurements

All marked-off lines on the work must be checked for correct dimensioning before cutting out or centre punching for holes. This means using the rule directly in its measuring role, but with sufficient precision to determine the actual dimensions marked, to the relevant degree of accuracy. In model work, this usually means working to the smallest division on the rule (or perhaps better) which will require an accuracy of $\pm\frac{1}{64}$in. or $\frac{1}{2}$mm, depending upon the engravings on the rule.

By using a watchmaker's glass (loupe) of about 3in. (75mm) focal length there should be little difficulty in estimating positions more accurately than the smallest divisions on the rule. For an imperial rule, this readily allows measurements to be checked to one-third of $\frac{1}{64}$in. or about 5 thou (.005in.). Figure 5.19 shows typical views provided by a watchmaker's glass.

Centre punching for holes

For positions which define the locations of drilled holes, it is convenient to check the measurements at the time of centre punching a hole position. It is essential to use a watchmaker's glass during this operation since it is the only way in which the centre punch point can be accurately placed on the intersection of the lines. Since the loupe is used during the dimensional check, it is convenient to associate the punching operation with the check. If the scribed lines do not exactly define the required position, it is possible to compensate for this when placing the punch, checking the dimensions once more after only lightly punching the dimple. If there is still an error, it is usually possible to 'push' the dimple in the required direction, using the punch at an angle, progressively checking with the rule until the correct position is achieved, and then punching vertically to ensure a symmetrical displacement of the material.

Hole positions in mating or matching components

If sufficiently accurate marking out and drilling can be performed, it is possible to mark and drill two items independently for the common fixing holes which will be used for their attachment to one another. This is the method used industrially.

In modelling, interchangeability is frequently not essential. It is sufficient that we have two clock plates with matching holes, they do not need to be identical with another pair of plates. In locomotive work, it is sufficient that there are two cylinder, steam chest, piston-and-rod and cover assemblies which go together, even if individual components from the right- and left-hand sides are not interchangeable.

If several holes are required in two or more items so that they may be bolted or riveted together, it is preferable if only one item is marked out, centred and drilled. The two (or more) parts can then be clamped together and one used as a jig to drill the other(s).

It is often desirable to perform all the drilling for a set of items at one operation, for example if two plates are being drilled and reamed for bearings and fixings which must be in line, on assembly. The vital point in these cases is to ensure that both (or all) parts are clamped together sufficiently tightly for them to drill as one, otherwise burrs form at the exit and entrance of the holes in the individual items. These burrs force the items apart and this naturally spoils the location accuracy of the holes – the holes may not be in line when the parts regain their natural shape.

If a large number of nominally identical parts is required it is usually expedient to make a drilling jig which has all of the required holes carefully marked and drilled. The jig can then be used to drill, or centre, all of the holes in each component. It is usually convenient to drill all of the jig holes to the same small size, so that rapid spotting of the holes is possible.

It is unlikely that a jig will receive much repeated use in the amateur's workshop so it can be made from a suitable offcut of mild steel. Only if extensive use is envisaged will it be necessary to fit hardened bushes into the jig holes. The jig should, of course, incorporate some form of positive location for the part to be drilled so that a simple placement and clamping operation positions the work correctly.

If two identical (or more-or-less identical) items are required in sheet material, the 'truing' of the edges should be performed with the items clamped or bolted together firmly. They should then be retained in this condition throughout marking, drilling and profiling so that one operation at each stage will produce the pair.

Marking out using hand tools

Use of rule and scriber

Having established two reference edges at right angles to each other, the most common way of

marking out work in sheet form is simply by use of a square, a rule and a scriber. The method has the advantage of simplicity, but unless it is carried out carefully, and the marking afterwards checked with equal care, significant errors can be introduced.

The general requirement is to measure and mark a dimension, either from a prepared edge, or from an already scribed line. Two approaches are possible, as shown in Figure 5.32. The rule can be laid on the work with the engraving for a required measurement placed over a reference line, or aligned with an edge, as shown at A, and the scribed line made on the end of rule. The alternative, shown at B, is to use only the edge of the rule. Personal preference will dictate the choice.

Both methods have their disadvantages. If a line is scribed on the end of the rule, this usually means that the line is displaced due to the thickness of the scriber point, the resultant line being

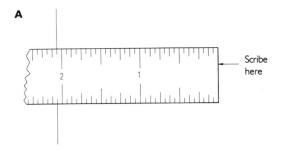

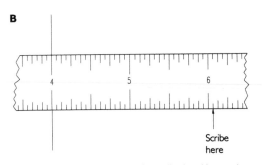

Figure 5.32 Two ways in which a scribed position can be marked off using a rule.

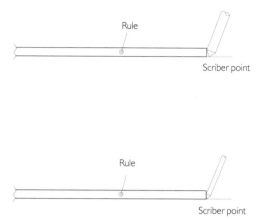

Rule

Scriber point

Rule

Scriber point

Figure 5.33 The effect of the size of the scriber point on the marked position, when using the end of the rule.

Figure 5.34 The square from a combination set provides a ready means to mark a position relative to a prepared edge.

too far from the reference, as shown in Figure 5.33. So, a scriber with the finest possible point is needed in order that the error is small.

In practice, two considerations apply; the error may be small enough for it not to be critical in any event, or the error may be significant but can be accommodated (removed) by offsetting the rule in the first instance. This is because, once a standard technique has been adopted, the error introduced by the scriber point tends to be very closely the same for each marking operation. So, provided that you know what the error is, you can readily compensate by displacing the rule slightly prior to scribing the line.

Scribing against the edge of the rule can also introduce an error due to the parallax effect, which is illustrated in Figure 5.20. This results from the eye not being immediately above the engraving on the rule. In this respect, it helps if a thin rule is used and this is one reason they are often preferred to the rigid type.

Use of a depth rule

Measurements relative to a prepared edge can most readily be set off by using the simple depth rule, if one is available, or a combination square. In both cases, the fact that the rule can be adjusted in the stock and then clamped into position, allows accurate marking to be undertaken. This operation is illustrated in Figure 5.34.

If a line is required parallel to the long edge of work which has only been prepared along one edge, the process of marking off can be repeated at two points and a straightedge and scriber used to mark a line parallel to the reference edge. Although it should make no difference, it is useful to scribe at three points as a check on general straightness and accuracy – it is only too easy to introduce an error.

Jenny calipers (odd legs)

If the distances are small, lines parallel to a prepared edge can be scribed using a pair of odd legs or Jenny calipers. For setting, the plain foot rests against the end of the rule and the scriber point locates in an engraving on the rule. Setting to a standard, engraved value is simple and straightforward. Once set, the caliper may be used to scribe a line parallel to a reference edge, and the only point to note is that the caliper must always be held squarely during setting and use.

Marking a centre line

It is frequently necessary to mark the centre line of a component. For narrower work, this can readily be accomplished by using Jenny calipers since, to define the centre, it is only necessary to scribe two lines equidistant from opposite edges. The centre then lies midway between the two lines. It is naturally helpful if the two lines are close together, but there is no necessity to measure off the distance very accurately, since an estimate will do quite adequately.

If the work is too wide to use Jenny calipers, other methods, such as the use of a depth rule from both edges, can be used to scribe two lines, nominally half the width of the material from the edges, to produce the same result, any errors in the actual measurements not being significant as long as they remain the same for each scribed line.

Dividers

For some marking operations, the rule may ultimately be abandoned in favour of something more straightforward. This is particularly the case when a number of equal pitches are required, which is frequently for a series of fixing holes. Each position must be centre punched as part of the marking operation. Dividers have two hardened steel legs, united by a spring at the top and provided with an adjusting screw which allows the distance between the pointed ends of the legs to be adjusted. Both points are ground sharply to act as scribers and they may consequently be used for marking out a series of equal pitches, locating the first hole in the series as dictated by the job, centre punching this and then using the dividers to mark the next, punching that, and so on.

Dividers can be set to a measured pitch by standing one point in an engraved line on a rule and 'feeling' the second point into another engraved mark while adjusting the setting. Once set, the dividers may be used as shown in Figure 5.35, stepping off the pitches and centre punching progressively along the required line. One advantage of this is that all pitches will be equal (given care in punching the centres) but an increasing error can accumulate if the setting of the dividers is not correct, and it pays to mark in one or two absolute measurements on a long line of equal pitches, using a rule, starting the stepping off once again at several

Figure 5.35 Equal pitches are scribed by 'walking' the dividers progressively along a scribed line, and centre punching the positions progressively.

Figure 5.36 Arcs are scribed relative to the marked and centre punched centre, using dividers. Note the sparing application of marking fluid.

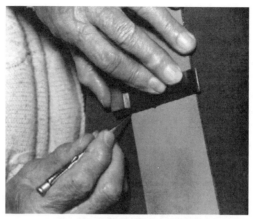

Figure 5.37 The first stage in marking out is to define a reference position from which other features may be marked.

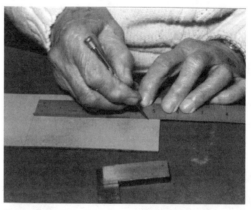

Figure 5.38 The scriber is being positioned relative to the line already scribed on the work, at the edge of the rule.

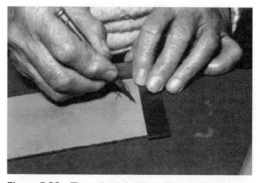

Figure 5.39 The scriber is held in position and a square brought up to touch it so that the position can be marked.

intermediate points in order to minimise the errors. An alternative is to start at the centre and work outwards.

Using the dividers, it is not necessary to use a standard measurement taken from a rule, but the dividers can be set to provide an equal number of pitches within an overall dimension, setting them by eye and then 'walking' them carefully along the desired line by way of trial, to determine whether the required pitch has been set, making successive adjustments until the setting is correct.

Dividers are also used (naturally!) for scribing arcs in the work. For this operation, the centre of the arc or circle is marked and centre punched, and the arc then scribed, as shown in Figure 5.36.

A practical example

Figures 5.37 to 5.43 show examples of the methods described. A ⅛in. (3mm) mild steel plate for a model locomotive frame is being marked out. This plate, and its twin, have been

105

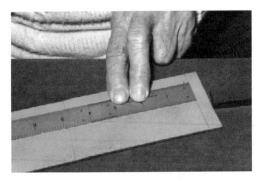

Figure 5.40 The measurement is checked using the rule.

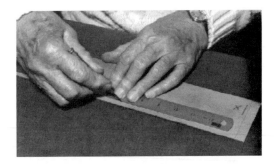

Figure 5.43 . . . so that the rule can be positioned for scribing the position of the top edge of the frame.

Figure 5.41 A short line is scribed at the required distance from the prepared bottom edge . . .

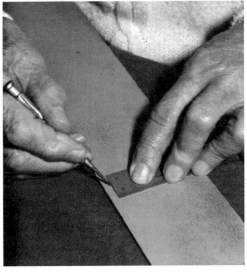

Figure 5.42 . . . and a second one further along . . .

bolted together through the nominal positions of the axle centres and the lower edges machined to create a reference edge. The plates have been separated so that a single one may be marked out, since it is easier to handle the one, rather than the pair with nuts and bolts protruding from both sides. Since the plate is in black mild steel, it has been given a very light coat of grey cellulose primer.

The first operation, Figure 5.37, is to scribe a reference line through the leading axle position, right across the plate, nominally bisecting the already drilled hole.

From this position, the distance to the front of the frame is marked off, as shown in Figure 5.38, and the scriber held at the marked point while the rule is put to one side and the square brought up to the scriber so that the front end can be marked, as shown in Figure 5.39.

The marked distance is checked with the rule, Figure 5.40. Notice how the rule which is not engraved to the desired precision along its full length is again not helpful to this task. Although the modeller doing the work thought the measurement was good, parallax error caused by the photographer's viewpoint makes it appear that the distance marked may be in error.

Following the check on the marked lines, the top edge of the frame is marked by scribing two short lines against the end of the rule. Notice the angle at which the scriber is approaching the end of the rule, in order to minimise any error, in the views of this operation in Figures 5.41 and 5.42.

With two lines marked to define the top edge, the scriber point is placed in one of the lines and the rule brought up to touch it. The rule is then pivoted about the scriber point until it is aligned with the second mark also, and the line marked. A longer rule would be more helpful here, since the 12in. (300mm) rule used spans only a short part of the frame, as Figure 5.43 shows.

The remainder of the marking-out process follows the same pattern, with the hole positions being centre punched prior to the pair of frame plates being bolted together once again, with their lower edges in alignment, for hole drilling, sawing and shaping to be performed.

Marking out using a surface plate

Introduction

Although the rule-square-and-scriber method of marking out is generally satisfactory and will serve if carefully performed, there are potential sources of error. Most of these can be removed by adopting an alternative method which uses a flat reference surface from which to mark off the required dimensions.

The reference surface is provided by a specially made surface plate which usually has a finely machined and hand-scraped surface which is as near flat as possible and is carried on, or integral with, a substantial cast base. Usually, the plate is made as a one-piece iron casting, but granite surface plates are now quite common commercially. How flat the surface is depends on how much you want to pay and how large a surface is required. A typical 12in.

by 18in. (300mm×450mm) cast iron plate may be flat to within .0004in. (0.01mm) and cost between £250 and £300. A good surface plate is thus a highly valuable asset and if you have one it should be reserved for the precision tasks of marking out and measuring, and definitely not for supporting work for hammering, straightening, banging or centre punching.

For amateur use, a less-costly alternative is a sheet of plate glass. This is ordinarily about ¼in. (6mm) thick and is today usually made by the 'float' process which more or less guarantees flatness, at least to the sort of tolerances which generally apply. Suitably supported on a piece of non-warping material such as chipboard or blockboard, a sheet of glass is a perfectly adequate substitute for its cast iron counterpart, the only disadvantage being its fragility should something be dropped on it.

An expensive adjunct to the surface plate is a height gauge. This is effectively a vernier caliper mounted to a substantial base and provided with a scriber which may be set precisely above the base by use of the vernier.

The ability to set the scriber of a height gauge precisely above the base indicates the way in which the surface plate may be used for marking out. If the work has a true surface or edge which may be stood on the surface plate, the height gauge may be used to scribe lines parallel with that edge. The vernier-setting capability of the height gauge confers accuracy on the process and therefore represents the best marking-out practice.

A height gauge is, however, expensive, being a precision instrument with ground and polished base, which precisely matches the depth of the scribing edge, an offset naturally existing between the underside of the base and the zero position on the vernier. A 12-inch (300mm) height gauge might cost between £100 and £200.

A less-expensive alternative to the height gauge is a scribing block, or surface gauge, illustrated in Figure 5.44. This comprises a steel base, ground flat on the underside, which houses an L-shaped arm, to one end of which is attached a cylindrical pillar. Fitted to the pillar is a two-part clamp which houses a double-ended scriber, usually bent at one end and straight at the other. The cylindrical pillar is adjustable in the end of the L-shaped arm and can take any position from the horizontal to the vertical. For marking out, its normal position is, however, roughly vertical, as shown in Figure 5.44.

The L-shaped arm is hinged in the base and fitted with an adjusting screw so that fine angular adjustment of the position of the pillar is possible, thereby raising or lowering the scriber fixed to the pillar.

Used on a surface plate, the scribing block is capable of scribing lines parallel to the refer-ence surface, and provided that the scriber point can be set to the correct height, and one edge of the work established parallel to the surface, accurate marking out is possible.

For most purposes, an ordinary rule provides adequate accuracy for marking out and may be used for setting the scriber point. For accuracy, the rule must have its zero in contact with the surface plate and it must be vertical (in both directions). A simple way to ensure correct setting up of the rule is shown in Figure 5.45. Here, the rule is simply clamped to an angle plate standing on the surface plate. This automatically achieves squareness in one direction and a square can be used to check square-ness in the other. Use of a watchmakers' glass allows accurate setting of the scriber point. Figure 5.45 shows a 12-inch rule clamped to an angle plate.

Figure 5.44 A scribing block, or surface gauge.

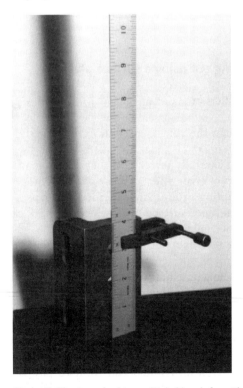

Figure 5.45 An angle plate used to hold a rule for setting the scribing block scriber point.

Marking sheet materials

If a flat surface and a scribing block are available, together with some means of setting the scriber point to the required height, the marking-out operation may be carried out with respect to a prepared straight edge on the work which is in contact with the surface plate. To bring the item to a convenient height, it can be placed on parallel packing of known height, as shown in Figure 5.46. This shows a ground parallel in use as packing, but anything will do, provided that it is parallel and of some convenient (and known) height.

Figure 5.46 shows a small brass plate being marked out using a scribing block. The plate has been filed to prepare two straight edges

Figure 5.46 Marking out a small brass plate on a surface plate.

which are square with one another. One of the prepared edges is standing on the ground steel parallel, of known height, to bring it to a more convenient working position. Lines can be scribed on the plate parallel to this edge, and the plate afterwards turned through 90 degrees to stand on the other prepared edge for the remaining lines to be scribed. The marked-out plate is illustrated in Chapter 7.

There is no doubt that this method of marking out is by far the most satisfactory and accurate. The point of the scriber can be set very accurately to the required height against the rule by observing with a watchmaker's glass, and the scribed lines are always parallel to the reference edge, conferring real accuracy on all of the marked positions. The best type of rule for use as the reference is clearly one in which the whole length of the rule is engraved in the smallest divisions with which one needs to work, either ¹⁄₆₄in. or 0.5mm. The rule illustrated in Figure 5.45 has the sole benefit that it photographs well, since it has a satin chrome finish. It was bought for the metric scale which is engraved on the reverse side.

Marking castings

By their nature, castings are three-dimensional and therefore do not lend themselves to marking out using a square and rule since they frequently do not incorporate sufficiently large reference surfaces or edges adjacent to the surfaces to be marked. This is illustrated by the cylinder block casting shown in Figure 5.47. This casting has been prepared by machining the underside of the base to produce a reference surface, having first determined how much should be removed to bring the base flange to the design thickness.

Having established one true surface, a second has been created by machining the front of the block, again bearing in mind the amount to

Figure 5.47 A part-machined cylinder block casting being marked out on a surface plate.

and then decide how much needs to be removed from the first surface to be machined. As a preliminary to the machining, it is usually necessary to flatten at least one face (filing is usually sufficient) to produce a sufficiently accurate base from which to set off the first cut. A case in point is shown by the two steam chests in Figure 5.48. Since machining at the first stage was simply to bring them to thickness, no preliminary marking out was performed, once the machining allowance was determined by use of a rule.

be removed to bring the casting to the correct overall length by ultimately machining the rear end. With two adjacent faces, at right angles to each other, machined flat, the casting can be set up as shown in the figure for marking out for hole centres, overall length and height, and so on.

As surfaces on the casting are machined progressively, further marking out can follow and the process becomes one of marking and machining alternately.

Some machining operations on castings may not actually require preliminary marking out, it being sufficient to 'run a rule over' the work

Figure 5.48 A pair of steam chest castings which have been machined on both sides without benefit of marking out.

6

Basic handwork

Introduction

In the amateur's workshop there is relatively more handwork carried out than would be the case in industry. Much of this occurs because the amateur is not equipped to deal with larger items which, in industry, would be profiled by machining. These larger parts include such items as locomotive frames and bodywork, the larger clock plates, long rods or levers which need to be profiled, sheet materials for boilers and so on. Anything large enough to take it beyond the capacity of the amateur's machines is worked by hand, even cutting a small piece out of a large sheet.

Conversely, small items are also frequently cut out and profiled by hand, perhaps due to the difficulty of holding such items for machining, or simply because the machining capability is again not available. For these reasons, undertaking the basic shaping by hand is often the only option which is available, making it essential to acquire reasonable skills in the basic handwork of sawing and filing.

The normal progression of work is naturally from selection of the basic material, through marking out, to sawing, filing and final cleaning up or polishing, and the processes can, therefore, be dealt with in this order. Before these processes are performed on some types of stock steel, it is highly desirable to carry out some initial heat treatment, however, and this can be considered first of all.

Initial preparation of materials

Heat treatment of steel sections

Chapter 3 provides a description of the more common materials and includes a description of the way in which internal stresses may be built up in the material during its manufacture or preliminary cutting. Provided that the material is not 'worked' in any way, internal stresses are not usually evident so that a bar of bright mild steel (BMS) although internally stressed, will not distort further as long as it remains unmachined.

Removal of material, particularly from rectangular-section BMS, removes part of the structure that is essential to the stability of the bar's shape, thus allowing it to distort. This is a nuisance to say the least, but it can be avoided by 'normalising' the material before any operations are carried out. Normalising equalises the internal stresses and allows material to be removed without distortion of the remainder.

In addition to normalising BMS, it is frequently useful to carry out the same process on sheet materials which have been cut off by guillotining. This process usually causes some bending of the cut strip since the guillotine blade normally acts like a pair of scissors, cutting progressively from one end to the other and bending the sheared strip downwards, away from the fixed blade. The result is a bow in the material along its length, and perhaps, also, a degree of twist.

The problem frequently occurs when sheared plates are provided for locomotive frames. Since the frames are to be paired, it is often recommended that they should be used so that the bowed shapes complement one another, both bowing outwards or inwards, whichever produces the straightest result. This can be tried by bolting the two plates together through the axle centres, for initial marking out, drilling and shaping. Unfortunately, this doesn't always produce frames which are straight, and the technique cannot in any event be used if you need a single piece of plate for a frame stay, or whatever, so the best way is to normalise the plates before commencing work.

The process is very easy to perform, simply requiring the material to be heated up and then left to cool slowly. A fuller description of the normalising process is given in Chapter 3.

Preliminary preparation of sheet materials

Some preliminary preparation of sheet materials is generally required prior to marking out the work – this applies also to castings. This preparation is usually to bring at least one edge (for sheet materials) or one surface (for castings) to a straight and/or flat condition. Since castings are distinctly different from sheet materials, they are considered separately below.

The first point of note concerning sheet materials is that they are ordinarily cut into the required sizes using a guillotine. While in theory, the guillotining process provides a clean cut, in practice there is normally some distortion of the material along the cut edge. This renders it useless for modelling since the guillotined edge usually has the appearance shown in Figure 6.1, one surface having a rounded profile and the other showing a 'pulled' or extruded appearance. In model work we are representing a scale appearance, and any gross distortion of the edge of the sheet is generally not tolerable, except perhaps for 'hidden' work such as model boilermaking, when much of the structure will be hidden by cleading, and even if not, some out-of-scale appearance may be acceptable. In general, therefore, the edges of guillotined sheets must first be cut back to obtain a good edge which is then machined or filed to create a square and straight edge prior to marking out.

If sheet work has a reasonable length in both dimensions, it is usually convenient to prepare

Cut away this material before use

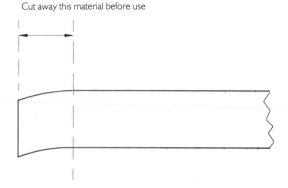

Figure 6.1 Deformation on the edge of a guillotined sheet.

two adjacent edges, by machining or sawing and filing, to establish them at 90 degrees to each other with both edges straight and square to the surface of the sheet. The major dimensions of the work can then be specified with respect to the two adjacent prepared edges, which are used as references. If the two edges are at 90 degrees, a square placed on one edge allows a line to be scribed parallel to the other edge, and measurements along such scribed lines will define the required positions.

If sheet work is large, or long and narrow, a sufficiently large square may not be available to span the job and alternative means must be adopted. Large work can be accommodated by clamping a straight edge (or suitable length of stock rectangular bar) to the work, using as large a square as possible to establish the square condition.

For long and narrow work, such as locomotive frame plates, it is usually best to work solely from one of the long edges, establishing lines parallel to this using one of the methods described in Chapter 5, and scribing lines at right angles to the edge by use of a square. The short ends can then be brought to the finished size and shape after marking out.

Saws and sawing

Introduction

The basic metal-cutting saw is the familiar hacksaw fitted with a replaceable blade. Saw frames are produced in a range of different styles, three of which are shown in Figure 6.2. Which type to choose really depends on personal choice, but if you have to make do with only one, make sure it is adjustable to accommodate blades of different lengths. If a single saw must suffice, the traditional type with a handle in line with the blade is probably most useful.

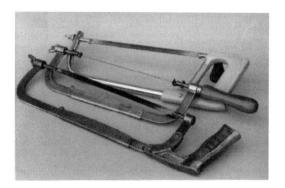

Figure 6.2 A selection of hacksaw frames.

The replaceable blades for these saws are punched with holes which engage pins in square-section fittings at each end of the saw frame. Blades are manufactured in a range of different lengths and types, some being all carbon steel (the cheapest), others all high-speed steel, some with hardened teeth on a relatively flexible (and soft) backing, some even with tipped teeth. Lengths vary from 9in. to 12in. (230mm to 300mm) and tooth pitches from around 14 to 32 teeth per inch (25mm).

Since there is such a choice of blade available, even the simple hacksaw needs to be set up and used in the correct manner and since this fundamental tool also comes in other forms, a general introduction is in order.

Choice of tooth pitch

First of all, a consideration of the relationship between the saw blade and the material being cut is necessary. Figure 6.3 shows two views of a saw blade in the process of cutting the work. At A, the pitch of the saw teeth allows three points to be in contact with the work, whereas at B, the tooth pitch is so large that the material being cut can be accommodated between two points. In this case, the saw can fall down into the work, causing it to jam against the leading

A

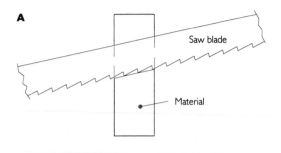

Saw blade

Material

B

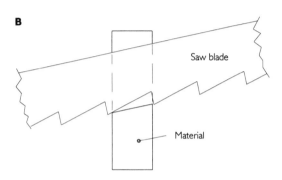

Saw blade

Material

Figure 6.3 The relationship between the saw teeth and the work.

edge of the material. The resulting 'collision' may be sufficient to break off a tooth, and thus ruin the blade, but even if this does not occur, sawing is very hard work and the jam-ups may well bend the material if it is thin and/or soft. It is essential that the pitch of the saw teeth is chosen to suit the thickness of material being cut. It is usually recommended that three teeth should be in contact with the cut material, and the pitch of the teeth (and the method of approach to the work) must be chosen with this in mind.

Material thickness is not the sole consideration, however. When a material is cut, the form of the chippings, or swarf, is dependent on the way the material cuts. Brittle materials tend to chip, whereas greater ductility produces long 'shavings' of material which may clog up the

spaces between the teeth if these are small. The quantity of material cut on each stroke is also significant since even small chippings may clog the saw if they are produced in large quantities. A tooth pitch to suit the material is also required, the usual recommendations being:

- for cast iron   14 teeth per inch
- for mild steel   18 teeth per inch
- for brass and copper 20 to 24 teeth per inch

Harder steels can be cut with a saw having a finer tooth pitch than that used for mild steel.

The above recommendations are not universally applicable however, since the work may need a fine-pitch blade in order to maintain three teeth in contact with the cut surface, and material thickness must be considered, rather than material type, in making the selection of tooth pitch.

GOOD   BAD

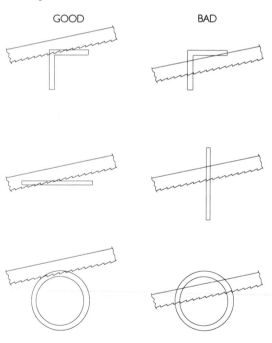

Figure 6.4 Good and bad ways in which a saw blade can approach the work.

The approach to the work must also be varied to maximise tooth contacts and Figure 6.4 shows both good and bad approaches to three types of work. The solution for angles is obvious; they should always be approached so that the cut face presents its long dimension to the saw. The problems posed by thin materials when supported vertically in the vice are well illustrated, and this suggests that cutting sheet materials is normally better served by holding them down on to the bench surface, rather than in the vice, when the only good approach is from a kneeling position below the vice, which is hardly conducive to comfort!

Tubes, particularly if thin-walled, present particular difficulties. The saw may jam on the thin material in any event, but additionally, there is always one position in which the saw jams on both sides at once. If the tube is small and in a soft condition, this usually means a bent tube, but whether soft or not, sawing does not proceed readily beyond this position. The remedy is to saw around the tube rather than across it, but if the tube is suitable, a properly designed tube cutter is usually a better alternative.

Sawing action

The sawing action may appear self-evident, after all, it just needs to be pushed backwards and forwards. And so it does. But how can you ensure that the cut material is expelled from the saw and from the slot (called the kerf) which it cuts? At what speed should the saw move and how can blade life be maximised? In answering these questions the sawing action is seen to be not so simply defined after all.

First, the question of blade life. To minimise wear, all teeth should do their fair share of cutting. This means using long strokes to ensure even wear throughout the length. A saw blade ideally needs a backing for the teeth which is slightly narrower than the teeth themselves so that the body of the blade slides freely through the kerf (slot) which the teeth produce. To simplify the manufacturing process, this effect is produced either by 'setting' alternate teeth outwards in opposite directions or by forming a wavy edge to the line of teeth so that some project to the left, and some to the right, progressively down the blade.

Teeth naturally wear where they do most cutting and if the saw is habitually used by making short strokes using just the centre of the blade, this naturally wears the centre but not the ends. Once the blade has worn near its centre, a stroke longer than usual causes unworn teeth at the ends to jam in the kerf and the cutting process is interrupted. This process repeats itself frequently and sawing becomes a frustrating business. The prime requirement is, therefore, for full strokes from the saw.

The next questions are the speed of sawing and the pressure to be applied. Firstly, the saw must be allowed to cut for itself and sufficient pressure should be applied only to prevent the saw from sliding over the work on the forward stroke (saws of this type should always cut going away from the operator). Concentration should be applied to guiding the saw along the line rather than forcing it to cut by the application of pressure. Secondly, the saw should cut at about 60 feet (18 metres) per minute. Thus, if the blade is 12in. (300mm) long, a full stroke should occupy one second – sawing is not a rapid process and should not be hurried.

Finally, there is the question of clearing the teeth, and the kerf, during sawing. If the work is thin, this usually presents no problem, but for thicker work the sawing action should be such that the 'driving' hand is lowered slightly towards the end of the stroke. This lifts the front of the blade within the work and allows room for the swarf to be expelled.

Fine-pitch saws

Replaceable blades for the types of hacksaw frame shown in Figure 6.2 are available with tooth pitches up to 32 teeth per inch. For very thin materials, or intricate work, something much finer is required, preferably mounted into a lighter frame, two types of which are shown in Figure 6.5. The first of these is the familiar junior hacksaw with its shaped 'springy' frame which can be squeezed up by hand to remove and replace the blade. Only a single type of standard blade is available for the junior frame – this is 6in. (150mm) long and comes in a standard pitch of 32 teeth per inch and is ideal for the small, light jobs around the workshop.

The second type of saw frame in Figure 6.5 is a jewellers' piercing saw. This is the metalworkers' equivalent of the woodworker's fretsaw, but intended for very light work. It employs replaceable, fine-pitch blades which are only a little deeper than they are wide and they are capable of sawing very small radius curves or even being turned sharply through 90 degrees during the sawing process.

The piercing saw shown is adjustable and blade tension is achieved by sliding the handle-and-clamp along the frame and then locking it by use of the wing nut. The blade should not be adjusted 'bow-string tight' but just sufficiently to take up any slack. One advantage of the adjustable frame is that blades with broken-off ends can still be utilised. Less costly, non-adjustable frames are also available in which blade tension is adjusted by 'springing' the frame.

The method of use of the piercing saw is quite different from that employed with other types of saw. The very fine blades, which range from 32 to 80 teeth per inch, are not strong enough to support the weight of the saw and the force of cutting if the saw is used in the conventional horizontal position. The saw is consequently used with the blade vertical and the work supported on a table having a vee-shaped notch cut in it, as shown in Figures 6.6 and 6.7. The blade carries the stress of cutting along its length and is inserted into the frame so that it cuts going downwards, the table supporting the work to resist the force of cutting. No saw should be forced into the cut, but this is especially true of the piercing saw due to the extreme fineness of the blades. An illustration of work which can be produced is shown in Figure 6.8.

Piercing saw blades are manufactured with tooth pitches from about 32 to 80 teeth per inch, in 12 grades referenced from No. 6 (the coarsest) to No. 6/0, the blade thickness decreasing as the tooth pitch decreases. A No. 6 blade has 13 teeth per cm (33 per inch) and is 0.5mm, or .020in. wide, whereas a No. 6/0 has 29 teeth per cm and is 0.22mm, or .09in. wide. In the centre of the range, the Nos. 1 and 1/0 have 21 and 22 teeth per cm (54 and 56 per inch) and a blade width of about .013in. (0.3mm).

When making abrupt changes of direction with a piercing saw, the saw must be kept on the move but with absolutely no pressure applied in the cutting direction. Instead, the saw is rotated slowly to widen the kerf until it has changed position to face in the required direction.

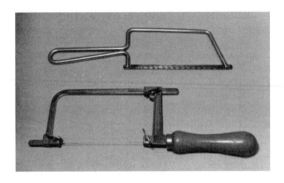

Figure 6.5 Junior hacksaw and piercing saw frames.

Figure 6.6 A piercing saw on the notched table with which it is used.

Figure 6.7 A piercing saw in use.

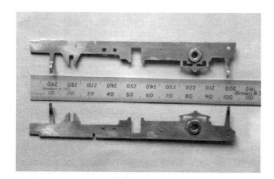

Figure 6.8 Locomotive frames for a 4mm scale model, cut from 1mm brass using a piercing saw.

Pierced work

The name of the piercing saw comes from the fact that it can produce 'pierced' work. A hole may be drilled in the work to allow the blade to be passed through and afterwards clamped into the saw frame to allow the cut to commence within the work.

Pierced work is frequently required in more substantial material than that for which the piercing saw is designed. The hacksaw can be adapted for this work by fitting to it a saw blade of circular cross-section which can cut in any direction. The most commonly available type is the *Abrafile*, a trade name now used universally for this type of saw.

Figure 6.9 shows one end fastening of the file in a hacksaw frame. An adaptor clip fits to the blade mounting pin and allows a button on the blade end to be retained in a slotted angle. The file can then be inserted into the clips and tensioned in the normal way.

The files are around $\frac{1}{16}$in. (1.5mm) in diameter and may be used for any intricate work in substantial material. Files are available in a range of tooth pitches, usually described as coarse, medium or fine, and are produced in different lengths, although the most common is the 9in. (230mm). Like the normal hacksaw blade, the files should always be inserted into

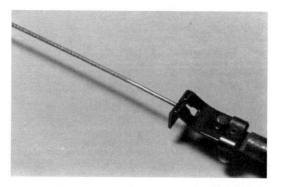

Figure 6.9 An adapter clip for fitting an *Abrafile* blade into a standard hacksaw frame.

the frame to cut when going forwards, the direction of the cut being determined by sliding the file through the fingers (gently!) to decide which way the teeth face.

The files are hard, but relatively brittle, and can easily be broken if too much pressure is applied, or if they jam in the kerf. The sawing action needs to be gentle and great attention must be given to keeping the saw absolutely on the same alignment during the whole stroke if blade breakages are to be avoided.

Cutting large sheets

The hacksaw frames shown in Figure 6.2 utilise square-section bolts with inserted pins as the means of holding the blade into the frame. The bolts may be inserted in four distinct positions, allowing the blade to lie in the plane of the frame, or at right angles to it. Long work may, therefore, be sawn conveniently as shown in Figure 6.10, provided that the cutting line is within reach of the frame depth.

If a saw with the blade turned over in this way cannot reach the position of the cut, a special holder, known as a pad saw handle, may be used. This attaches to one end of the blade, leaving the other free to enter a sheet of material at any position. Since a standard blade is

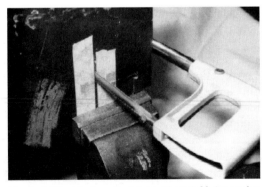

Figure 6.10 A hacksaw blade turned through 90 degrees for cutting parallel to an edge on the work.

not designed to be pushed into the cut (it is not stiff enough) the blade must be inserted so that it cuts when pulling. A regular hacksaw frame is used with the points of the blade teeth pointing away from the handle and is sufficiently robust (or should be) so that a 'push' on the handle reaches the blade as a 'pull' from the front. Cutting with a pulling action with a pad saw is very awkward, but it is sometimes the only alternative which is available.

To enable the saw blade to enter the sheet for commencing the cut, a slot is created by drilling a line of closely spaced holes adjacent to the cutting line and joining them by filing away the material between them so that the saw blade will enter to commence the cut. If holes are drilled all along the cut, the amount of work which the saw must do is reduced, and this helps to reduce the awkward work of sawing with the cut taken on the pull stroke.

If a line of holes is drilled all along the cutting line, it may well be possible to complete the cut by using a cold chisel to cut through the remaining material. This is carried out by supporting the work on a substantial block and hammering the chisel down into the webs of material remaining between the holes. For soft materials such as brass and aluminium alloys, a chisel having a narrow angle at its tip may be used, but for steel sheet, a larger tip angle is required (to give adequate strength) and the distortion that this produces should be borne in mind when drilling the initial holes. Do not despise the humble chisel – it is a very effective tool and may well solve the problem of cutting large sheets of thin materials.

Alternative cutters, known as nibblers, are available in both hand- and power-operated form. These remove a narrow strip of material, up to ⅛in. (3mm) wide, depending upon the design, by utilising a 'pecker' or 'beak' which is drawn up between the two jaws of an anvil. Beak and anvils thus act as two guillotines, removing the narrow strip of material.

Provided that the cutting edges are not worn, there is little distortion of the parent material. The cutter is capable of cutting curves and changing direction quite sharply, and is very versatile in use. Types sold as attachments for DIY electric drills will usually cut steel up to 18 swg (.048in. or 1.25mm) and light alloys up to ¹⁄₁₆in. or 1.5mm.

Reducing noise

Noise is sometimes a problem when cutting large sheets since the small area gripped by the vice allows most of the sheet to vibrate. This can sometimes be overcome by clamping the sheet to the apron of the bench, if it has one, or by hanging a heavy cloth such as a towel over the unsupported part. It is frequently useful, however, to clamp a heavy bar to the work, parallel with, and adjacent to, the cutting line. The bar then provides sufficient mass to damp any vibration and the work and bar may be quickly repositioned in the vice as the cut proceeds.

Files and filing

Introduction

The ability to file surfaces flat, or nearly so, is an undoubted asset, and some of the 'old boys' brought up in the days when hand fitting was more widely practised than it is today, can produce reasonably flat surfaces, seemingly with little effort. In those days, files were used as the intermediate stage in a chiselling, filing and scraping process which produced both flat and curved surfaces which matched one another or matched a standard reference, such as a flat surface plate.

Figure 6.11 The tang end of a double-cut flat file.

Files were also at one time used for filing hexagons, squares and approximately circular shapes out of forged or cast (but roughly shaped) billets. A great deal of expertise in the use of the file was needed and this was the first skill which the apprentice was expected to acquire.

Files are available in a range of grades and shapes to suit the finish required and the size and shape of the work. The normal grades are known as coarse, bastard, second-cut, smooth and dead smooth, these terms describing the size of the teeth on the file.

File teeth are cut by raising ridges across the body. The ridges lie at about 25 or 30 degrees to the long axis of the file and if only a single set of ridges is cut, each forms a long cutting edge across the file. Most files are double-cut, however, two lines of ridges lying at 50 or 60 degrees to one another. This is easily discerned at the handle, or tang, end of the file, as shown in Figure 6.11.

The double-cutting of the file produces individual teeth on the body. This allows swarf to clear the teeth more readily and produces a cleaner cutting action which allows more material to be removed. Most files are of the double-cut type.

Standard file types

The commonest file shapes (cross-sections) are flat (rectangular), square, three-square (triangular), round and half-round. Most files are tapered in their length, being smaller in the point than in the heel (tang end), but rectangular-section files known as hand files normally have the same cross-section throughout. A rectangular cross-section file with a taper in its length is described as a flat file. A variant of the flat file, known as a warding file, is thinner in relation to its width and more suitable for working in narrow slots. These are sometimes known as pillar files. A three-square file not having equal sides, but having teeth cut on two sides having a very shallow included angle, is known as a knife shape.

Square files have teeth cut on all four faces and are intended for cutting into corners or cutting square shapes. A flat file also generally has teeth on all four edges but a hand file is normally provided with one 'safe' or uncut edge so that it may be used in a corner without significantly removing material from the side. With the exception of the knife shape, other shapes have teeth cut all round, or on all surfaces. The more common file shapes are shown in Figure 6.12.

Needle files

The files described above are available in lengths from 4in. to 12in. (100mm to 300mm) in increments of 2in. (50mm). A 12-inch hand file is typically 1¼in. (32mm) wide and 9/32in. (7.5mm) thick and a 6-inch (150mm) might be 5/8in. (16mm) wide and about 5/32in. (4mm) thick. A 4-inch (100mm) file is somewhat smaller than this, but still not suitable for extremely fine work.

A range of miniature files is needed, which, because of the form of some shapes, are described as needle files. These are available having the same or similar shapes to those described above. Each file is a one-piece forging incorporating both the file blade and a small-diameter handle. The overall length is usually about 6in. (150mm) divided 50:50 between handle and blade. Although different grades are available these are often not specified, so it pays to browse around a little when purchasing these files.

On some needle files the handle is plain and the file may be held in a long pin vice, which makes for a more comfortable grip than the small-diameter forged handle. On other files the handle is provided with knurled rings, either to improve the grip directly or to permit a push-on handle to be fitted.

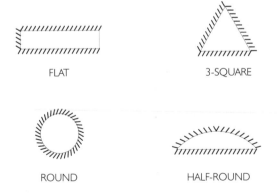

FLAT   3-SQUARE

ROUND   HALF-ROUND

Figure 6.12 Common file shapes.

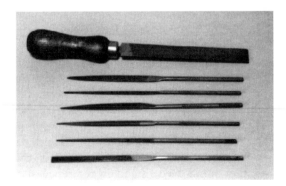

Figure 6.13 A group of needle files compared with a 4 inch (100mm) smooth file.

With the exception of the hand shape, needle files taper to a fine point. This is illustrated in Figure 6.13 which also shows a 4-inch (100mm), dead smooth file. Needle files are commonly referred to as Swiss files, whether originating in that country or not.

Specialised files

Related to the needle files, there are various other types of small file, described as precision files, escapement files or rifflers. Riffler files comprise a square-section forging, about 6in. (150mm) long, with file teeth cut on both ends, each of which is forged into a curved shape. Different sections are available, such as round, half-round, square, three-square and flat, which, with the short, curved ends allow access with a file to otherwise difficult places.

Precision files are more finely machined and cut than their general counterparts, and are available in a very wide range of cuts at the finer end of the scale, from about 23 to 300 teeth per inch. Being this fine, they are not available in the larger sizes, but something up to an 8-inch, having 250 teeth per inch can be obtained.

Escapement files are a variant of the standard type of needle file which have square handles rather than the more usual round type.

Some specialised shapes of file are also available in the smaller sizes (up to 6in. or 150mm). A crossing file is radiused (like a half-round) on both sides, usually a different radius on the two sides, making it suitable for working into corners. Also useful for this task is the barrette shape which is something like a half-round, but having teeth cut only on the flat face.

Using a file

Before use, a file must always be fitted with a handle. Plastic handles are nowadays available which incorporate a device for gripping the tapered tang. The traditional handle, and still the cheapest, is a wooden turning fitted with a metal ferrule. An appropriate size should be chosen from the available 3-, 4- or 5-inch types. As supplied, these are bored centrally with a pilot hole and are fitted to the tapered tang by being wedged into position.

The simplest way to do this is to heat up the tang (it doesn't need to be red hot) insert it into the handle and allow a tapered hole to develop as the tang burns into the handle under pressure. If the tang burns its way into the handle almost, but not quite, far enough, the two can be separated to allow tang and socket to cool. Afterwards, the two can be jammed firmly together by fitting the handle and then dropping the assembly, handle end down, onto a solid support such as the bench top, the vice or the floor.

Once a file is fitted with a handle, it is ready for use, but it needs to be applied to the work in the correct way. On a double-cut file, the cutting edges form lines of teeth on the faces. If the file is pushed straight across the work, succeeding teeth tend to follow the grooves cut by those that lead and the result is a scratched and poor quality finish. The correct filing action is obtained when the file moves both forwards and sideways, as shown by Figure 6.14. At A, the file moves forwards and rightwards, and at B, forwards and leftwards. In practice, it is best to alternate the directions shown at A and B, making a few strokes leftwards followed by a few strokes rightwards.

It is naturally necessary that the file should always traverse the work in the same plane and the action must be practised to acquire the skill to apply the correct amount of pressure to the driving and leading hands during each stroke, to keep the file level. The natural tendency is to produce a 'crowned' surface and this trait should be corrected until a flat surface can be produced.

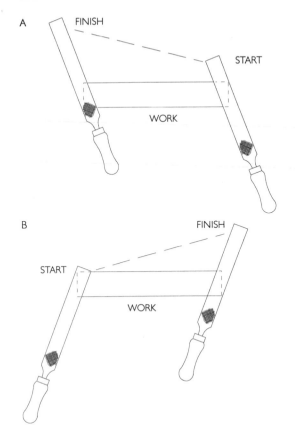

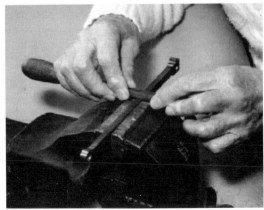

Figure 6.15 Draw filing.

To use it for draw filing, a file is grasped in both hands, around the point and the handle. It is then stroked backwards and forwards along the work, as shown in Figure 6.15. If the file is double-cut, it removes material going in both directions and cuts more effectively than might be expected. Since the usual aim in filing is to produce a fine finish, draw filing is usually performed using a smooth or dead smooth file, as the finishing filing stage.

Figure 6.14 The correct filing action.

Like sawing, filing should be slow and leisurely.

Draw filing

Since the cuts in the working faces are generally made at 25 to 30 degrees to the axis of the file, cutting can take place even if the file is moved at right angles to the long axis, because the teeth are partly facing the edge of the file. If the file cuts in this manner, the teeth marks in the work are less obvious and the finish on the work is finer. The technique of using the file in this manner is known as draw filing.

Clogging of file teeth

When working with soft materials such as aluminium and copper, the chippings can become embedded in the file teeth. This reduces the cutting efficiency and also produces scratches as chippings become trapped between the teeth and the work and get pushed over the surface. Should a build-up of particles occur, they must be removed, for which a file card is used.

A file card comprises a substantial fabric backing into which stiff steel tines are inserted. The material is ordinarily used for carding wool prior to spinning, but the sharp tines illustrated in Figure 6.16, will also remove most

Figure 6.16 A file card.

embedded chippings from a file. To prevent the tines being pushed out of the fabric backing, the piece of card should be mounted on a wooden backing, but my sample continues to lack this refinement.

Any chips which adhere stubbornly to a file after the use of the card can be removed using a sharp point to pick them out, or by filing an offcut of sheet brass.

One way to prevent the build-up of swarf in the file is to fill up the teeth with some readily removed substance. If a block of French chalk is rubbed up and down the file it tends to pack into the teeth and thus prevents the swarf or chippings becoming embedded. Filling up the teeth spaces in this way does also affect the cutting action of the file since the teeth project less and therefore take a finer cut.

Filing straight and flat work

General

One of the most common tasks is that of cleaning up a sawn edge and filing down to a scribed line. If the sawn edge to be filed is broadly parallel to the scribed line, and the work is wider than the file itself, there is no need to force the file, it can be allowed to lie naturally on the

work as the file is worked progressively along the edge. The ends of the edge do need to be approached with care, however, since the file tends to 'fall off' the end and produce a rounded, rather than flat, surface.

Narrow work which the file overlaps needs a slightly different technique and the file may need to be twisted slightly to correct any lack of parallelism with the scribed line, during each cutting stroke.

If the sawn line is a series of humps and hollows, work must start by attending to the high spots. These may be quite localised, and the file needs to be held very firmly to prevent it lying on the sloping side of a high spot thereby removing material from the adjacent hollow. At this stage, it is not desirable to use the normal forwards-and-sideways motion until the edge has been roughly brought to straightness.

Filing ordinarily starts with a coarse file and progresses towards a fine file. Coarse and fine are relative terms however, since what will do for removing $\frac{1}{16}$in. (1.5mm) from the edge of a $\frac{1}{8}$in. (3mm) mild steel plate is not at all appropriate for cleaning up the edge of a sawn-out component in 20 swg (1mm) brass.

It does help in producing a straight surface if a wide file is used rather than a narrow one, and a selection of long (and wide) files down to at least a second-cut, will be found to be useful. One or two 6-inch (150mm) dead smooth files should, however, meet the need for the finest cut.

Testing the work

As filing progresses, it is necessary to check that the surface produced is at right angles to (or the correct angle to) the already prepared surfaces, and is straight, or curved appropriately in its length. Curved work is dealt with below so the initial consideration can be of straight and square work.

Squareness of the filed surface to an existing face should be checked as filing proceeds. Squareness is checked by using an engineer's try square. The check is performed by placing the stock of the square firmly against the reference face, as shown in Figure 6.17, and resting the blade on the edge or surface being prepared. Holding the work and square up to the light, reveals any out of squareness and shows which way the filing action needs to be adjusted to achieve a square edge.

The stock of the square should be undercut at the root of the blade, as shown in Figure 6.17, so that any burr at the filed edge does not prevent the stock from lying in full contact with the reference face. In this respect, it is usually best to use as the reference the face with the marked

line scribed on it. This should be nearest to the operator when filing (so that the line is visible). The major burr occurs on the far side since it is there that the file tends to 'turn over' the edge.

Even though the filed edge is at right angles to the reference surface, it may not actually be straight in its length. Straightness should properly be judged using a straightedge, but if one is not available, the blade of a try square will serve for most purposes, or even a rigid 12-inch (300mm) rule. It goes without saying that the straightedge, or its substitute, must be longer than the work.

To test straightness, the reference edge is placed along the filed edge and it and the work held up to the light. It is then necessary to note or mark the high spots and file these down as part of the process of using finer files progressively to achieve a straight and square surface or edge.

As the line (and squareness and straightness) are approached, finer files are used, and the burr consequently reduces in size. Even so, as completion is approached the remaining burr should be removed by draw-filing, first of all, along the edge which is being worked on, and then along the surfaces at right angles to it.

A single operation along the two surfaces doesn't always remove all of the burr which remains but can simply 'turn it over' onto the adjacent surface, so the process has to be repeated once or twice until the burrs are completely removed. A burr is also produced on the edge nearest to the operator, but this naturally cannot be removed until the scribed line can be dispensed with.

Flattening and testing large surfaces

The filing and testing of large surfaces can be accomplished by adopting the technique used for edges of sheet materials, as described above, provided that there is a suitable reference

Figure 6.17 Testing squareness using a small try square.

surface against which the stock of the square can be placed. Without such a reference, it becomes difficult to detect any twist in the filed surface since this does not become evident simply by checking with a straightedge that the work is flat along particular lines.

If the surface of a casting is being prepared as a preliminary to machining, there may be no reference available from which to make an assessment of the surface being worked on, so the surface must be judged against some other reference. For flat surfaces, this is ordinarily a surface plate, but the alternative sheet of plate glass can be used, or even the table of the drilling machine.

Assuming that a casting is being filed to produce a flat surface on which the item may be clamped to a machine, the first stage is to file

off any surface bumps or moulding 'flash' to produce a smooth and clean surface. This surface can be tested by standing it on something which is nominally flat, and checking for any instability, or rock. This test will reveal any significant high spots, which can be attended to, progressing at least to a second-cut file as the work becomes flatter.

Even when the work will stand on a flat surface without rock, this does not guarantee flatness, since the situations shown in Figure 6.18 may apply. Three points of contact between the work and the reference surface provide a stable condition although pressing on an unsupported corner should reveal its 'un-flat' state.

If the high spots on the work are more numerous, and do lie in the same plane, the work may stand on its four corners, and although it is not flat may appear to be so, due to its stable condition on the surface plate. The actual contact areas need to be highlighted in some way. The method to use depends on the accuracy required, and that is naturally dependent on the quality (flatness) of the reference surface and the required result.

Figure 6.19 shows a feeler gauge in use to test the surface of a casting. Gaps between the underside of the casting are detected by probing with a thin feeler and the magnitude of the

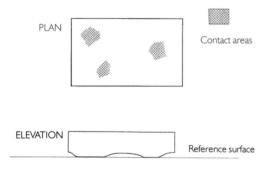

Figure 6.18 Illustrating how an apparently flat surface may have high spots.

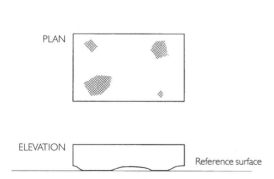

Figure 6.19 Using a thin feeler gauge to locate high spots and hollows on the underside of a casting.

125

gap can be assessed by determining the largest feeler gauge which will enter the gap.

The above supposes that there is not complete contact between the casting and surface plate all round the edge, thus allowing the feeler blade to enter. If there is complete peripheral contact, the method cannot be used, in which case, a marking medium must be employed. This requires a thin coating of a greasy marking medium to be applied evenly to the surface plate before standing the work on it. High spots on the work then pick up the marker colour, revealing the areas from which material needs to be removed.

A thin coating of marking medium (generally called marking blue, or engineers' blue, since this is usually the colour used) is necessary, particularly as flatness is approached, otherwise the medium bridges the gaps and shows all over the surface, giving a false indication.

The method is usually employed when a large surface is to be brought to flatness by hand methods, the process traditionally progressing from chiselling (chipping), to filing and finally, to hand scraping. This produces a fine and relatively decorative finish. Since a scraper is capable of removing very fine shavings from the work, the accuracy achievable is also very high, and with care and patience it is possible to match two surfaces fairly exactly.

If it is simply necessary to make the surface of a casting (or whatever) sufficiently flat to avoid distortion when it is bolted to a machine table, a straightforward filing operation is usually sufficient.

Filing curves

Introduction

The filing of curves is generally more difficult than the creation of straight and square sur-faces and unless great skill with a file has been achieved, some form of aid is a necessity.

The creation of external (convex) curves is undertaken using flat files. Internal (concave) curves must be created using round or half-round files. Using these is similar to using flat files; the cut is taken by using both forward and sideways motions, when this is possible, but it is often necessary to impart a rotary motion, in addition, as this helps to eject the swarf and create a smoother finish.

The sort of curves likely to be needed are blend radii between adjacent straight edges, radiused ends to levers and arms and compound curves such as are found at the bases of domes, chimneys and safety valve covers on locomotives. The first two of these may possibly be produced by machining, concave blend radii being created by drilling and lever end radii being produced by milling, if a rotary table and milling facilities are available.

Compound curves must generally be produced by hand.

Radiused lever ends are the easiest of the curved shapes to produce and so this technique is described first.

Using filing buttons

Small levers frequently take the form shown in Figure 6.20, having a tapered form and being radiused at both ends. They are usually drilled or reamed at the ends for the attachment of an operating or control rod, the end radius being drawn from the centre of the end hole.

The presence of the hole makes possible the use of a pair of guides which allow the radius to be produced easily. These guides are known as filing buttons and may take the form shown in Figure 6.21. Each button is simply a turned cylinder, the correct diameter for the lever end, having a through bore which fits a suitable bolt or plain pin.

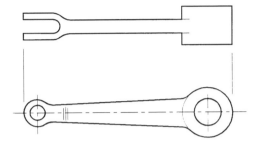

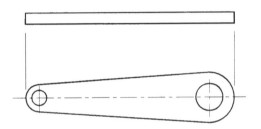

Figure 6.20 Typical small levers with radiused ends.

If the buttons, mounted on a bolt, are fitted one on each side of the lever end, and the bolt is nutted and lock-nutted as shown in Figure 6.21, the lever end may be filed to the correct radius by filing along the lever and around the end so that the buttons act as rollers to guide the file at the correct radius. If the buttons are made from mild steel, the file will mark them, as will be seen from the illustrations, but this is really not a problem and hardening is not essential.

The alternative to the rather 'posh' form of button shown here is simply to utilise two thick washers rotating on a pin which fits the lever's end hole directly. Provided that the file is worked squarely around the end of the lever, there is little tendency for the washers to wind themselves off the pin and the plain type will be found satisfactory. Either type of button is very easily made up and once a range is available the filing of end radii is straightforward as long as the lever is provided with an end hole. Two washers and a plain pin are shown in use to radius the corner of a square plate, in Figure 6.22, and Figure 6.23 shows a small lever and the buttons and pin which were used to round the end.

Figure 6.21 A pair of small filing buttons fitted to a rectangular rod for rounding the end.

Figure 6.22 The corner of a cast baseplate being rounded.

Figure 6.23 A part-made lever with the filing buttons which have been used to round one end.

Radius gauges

If convex curves without a coaxial hole are required the radius must be produced entirely by hand and some form of gauge is needed to judge the shape of the curve. If working in sheet materials, the required curve can be marked out in the usual way to provide a basic cutting line. Once the filing stage is reached however, judging the shape by eye is extremely difficult since the scribed line seems to deceive the eye into believing that the curve is good. It is usually beneficial to observe the curve from the unmarked side when its true shape will usually be discernible.

The use of a radius gauge is the best method of judging the shape. Sets of these are available commercially, each gauge being provided with two radiused edges providing both a convex and concave surface of identified radius. Sets of gauges are shown in Figure 6.24.

If commercial gauges are not available, they can very easily be made up from the plastic (polystyrene) card which is available in sheet form in various thicknesses. This is used extensively for plastic model building and small-scale model railways and is available from shops catering for these activities.

For a radius gauge, .020in. or .030in. thick (0.5mm or 0.75mm) card is satisfactory. If the radius is small, and convex, a drilled hole will suffice, but larger radii or concave curves must be marked out and cut by hand. The material takes pencil markings very well and curves can be marked easily with compasses, or dividers can be used to cut a shallow channel at the required radius. A sharp craft knife is then used to cut along straight lines up to the curve (only a

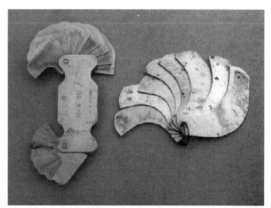

Figure 6.24 Three sets of radius gauges.

deep score mark is required, not cutting right through) and then the curve is worked round with the point of the knife. Once again, it is not necessary to cut right through, just to score the surface.

Once the plastic card has been scored it can be broken off along the line by simply bending in the direction which opens the cut. For a curved line, it is necessary to work around the curve progressively, but it will break off quite cleanly. The initial cutting does raise a burr but this can quickly be removed with a fine file. A

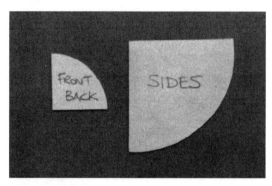

Figure 6.25 Purpose-made radius gauges cut out of polystyrene sheet.

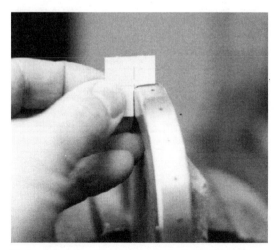

Figure 6.26 Checking the radius on the front of a smokebox casting using a purpose-made gauge.

pair of radius gauges for concave curves is shown in Figure 6.25.

Figure 6.26 shows a plastic card gauge in use for radiusing the front of a large casting. In this situation, a gauge represents the only alternative to simply judging by eye, a technique which is not generally very satisfactory unless both the eyes and the hands are well practised.

Compound curves

Compound curves are frequently found around the bottom of chimney, dome and safety valve castings for model locomotives and a chimney base will serve as an example of this type of work. Before the upper surface is shaped, it is best to finish the bottom radius on which the base will 'sit'. Since it is important that this is square, both side-to-side and front-to-back, some preliminary marking out of the casting is required. The best method to adopt for such items is to mount the casting in the lathe and bore it through to the correct size and then to re-mount it in the lathe and turn the basic diameters at top and bottom. Ways of doing this are described in Chapter 13.

The result of these operations is a partially turned casting which can be marked out with four reference lines. At front and rear, the correct height can be scribed, and at the sides an approximate position can be marked which will act as a 'witness' as the underside is filed to fit the smokebox on which it will be fitted. The marking out is most easily performed on a surface plate, using the scribing block, with the base casting upside down, standing on the turned top. The marking-out operation is shown in Figure 6.27.

Once the part-machined casting has been marked out, the underside can be filed to fit the smokebox. Front and rear marks will show when the correct centre height is reached and the side marks will confirm that side-to-side

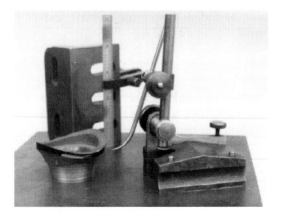

Figure 6.27 Scribing reference lines on a part-turned chimney base casting prior to filing the skirt to shape.

squareness has been (is being) achieved. For holding the casting, the method adopted for turning the outside diameters may be used, the casting being wedged on to a stub of material held in the vice.

When filing curves such as these concave shapes, a round or half-round file is naturally used. The radius required is certain not to match that of the file and it must be moved sideways and forwards during each stroke in order to remove metal evenly since it is difficult to persuade a half-round file to cut on the top of a high spot. If the opposite happens, and hollows form, the file tends to remove yet more metal from them, making matters much worse and increasing the difficulties of effecting a cure.

Once the base is filed to fit the smokebox, attention can be turned to the upper profile. A wide range of curves is required in this case, most of which are difficult to determine, but it is possible to estimate the shapes required for the front and back radii and also for those on each side. Two radius gauges can be made up, as shown in Figure 6.25. Filing can start at the front and back and then progress to the sides, removing metal where it is really proud of the desired line, and keeping away from the edges as far as possible.

A start should be made with a coarse file to get under the skin of the casting and remove metal from front, back and both sides separately but progressively. As the shape develops, metal can be removed from the 'quarters' filing around the casting in whatever way seems easiest. The aim should be to get as good a shape as possible with the coarse file before changing to something finer, continuing to keep away from the edges as far as possible.

The problem with the edges is that they easily become rounded (convex). The casting is a convenient way to make the model, but the prototype chimney skirt is almost certainly a thin steel spinning or a thin-wall casting and the filing of the model base should aim to give this appearance to the skirt. Figure 6.28 illustrates the problem. At A, the required shape is drawn, suggesting the separate skirt of the prototype, while B shows the shape which is likely to result if too much pressure is placed on the rim of the

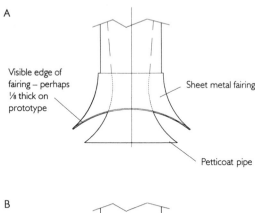

Figure 6.28 Desirable and undesirable shapes for a chimney base.

skirt when filing. The 'hump' is difficult to remove but in any event the haphazard filing of the rim edge might already have reduced it to a knife edge which is not desired until the body of the casting has attained the required shape.

The same problem can occur where the narrow top collar meets the body of the chimney and there can be additional difficulties if the chimney is an all-in-one casting since the chimney barrel will be scored should the file overrun the skirt. In these cases the barrel should be protected with masking tape or something similar.

Once the basic shape has been achieved, it should be worked down by progressively finer files and then emery cloth to give the desired finish. Figure 6.29 shows a partially completed base. The chimney barrel has been protected by a few turns of masking tape and reference lines have been drawn on the tape to define the positions at which the radius gauges should be used. The required shape has been achieved using coarse files and the surface now needs to be taken down to reduce the thickness of the base rim, and reduce the upper diameter to meet the already turned ring at the top of the base. The

Figure 6.29 A chimney base casting roughly filed to the shape required. Having achieved the shape, it needs to be smoothed and reduced to the desired size.

radius gauges for the base are illustrated in Figure 6.25.

Scrapers

Historically, the products of forge or foundry were brought to final size by hand. This meant that once parts had been made, they were assembled together, and during this process, any inaccuracies in the manufacturing processes were accounted for by 'fitting' the parts, the fitter using hand methods to correct the fit when this was required. Traditionally, the final hand-finishing process was the scraping of the fitted surfaces.

If skilfully carried out, scraping produces well-fitting parts and finely rippled surfaces which have both a decorative appearance and good oil-retaining properties. The acquisition of the levels of skill possessed by the experts is doubtless a long process, and not having scraped more than the odd big-end bearing in the days when a pre-1939 car was the only type I could afford, I am not equipped to impart that knowledge here.

Scrapers are useful for general purposes and the tool kit should at least be provided with a three-square. Craftsmen traditionally ground up scrapers from an old file, and indeed, files can be ground or forged into useful shapes, but it is easier to buy the useful three-square and flat forms readymade as they are supplied with handles and are ground to the correct shape.

Figure 6.30 shows the cutting ends of the two most useful shapes of scraper; three-square (triangular) and flat. The other common shape is the half-round, but this finds little use in my workshop.

The three-square scraper is slightly hollow-ground and designed to cut on its edges, which have a short straight portion nearest to the handle, but which otherwise curve towards the

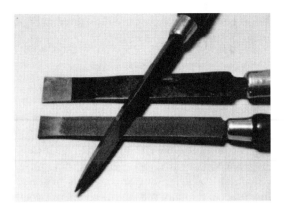

Figure 6.30 Small 3-square and flat scrapers.

end so that the scraper can be applied in such a way that only a narrow line of contact is provided with the work. The scraper can be applied to a high spot and narrow shavings removed locally until the high spot is removed.

Scraping can be used to provide the required clearances in bearings, for example, but is equally suitable to achieve the correct flatness and fit for machine slideways, by hand. Since slideways mostly present flat surfaces, a flat scraper is used. The tool is designed to cut on its end, and is normally used by making successive passes over the part of the work being scraped in two directions at right angles to one another. This equalises the removal of material and helps to improve the appearance.

A flat scraper is held by both hands, pointing away from the operator, and held at a shallow angle (about 15 degrees) so that one edge cuts the work and removes a small shaving, when the scraper is pushed away from the operator. Large flat scrapers have huge handles which are long enough to be held between the upper arm and the body. The sort of work which a model engineer might undertake needs a 5- or 6-inch (125 mm or 150 mm) blade which is fitted with something like a small file handle.

Both types of scraper may not be required, but a small three-square is essential. It is useful

for de-burring work, whether finished by hand or by machine, for scraping off paint, marking fluid, general muck, or even light rust.

Provided that it will go through them, a three-square is good for de-burring holes, but some specialised scrapers, known as de-burring tools, have been introduced during the last few years, especially for this purpose. The business end of one of these is shown in Figure 6.31. It comprises a handle, into which is inserted a hardened, cranked rod which has been ground away to form a cutting edge. Being about ⅛ in. (3 mm) in diameter, the tool can be used in holes down to about ¼ in. (6 mm), or a little less. Its shaped end allows the cutting portion to lie on the drilled surface of the hole and it removes the burr without producing an unsightly chamfer to the hole. It can naturally also be used to de-burr straight edges which have been filed or machined, although these are better de-burred by filing a small chamfer with a smooth file.

Figure 6.31 The cutting edge of a de-burring tool.

The final polish

Introduction

A polished finish is not always required, but it is generally necessary to smooth the surface left by the file, or even by machining, to produce a clean and bright finish. It is also frequently necessary to smooth the unmachined surfaces of stock material, which sometimes show surface blemishes or inclusions of dirt, as a result of the manufacturing process.

Surfaces turned in the lathe can be carefully polished while they are rotating, provided that there are no 'lumps' going round on the work which pose a special danger. Turned outside corners should be de-burred using a small, dead smooth file, in the way described in Chapter 15 and the outside surface of turned cylinders can be smoothed using the same file. Flat-machined surfaces should also be smoothed with the same file and burrs on the edges removed in the same operation.

For ordinary machined surfaces, the above operations may be sufficient, but there are cases where a proper polish is required. The process of polishing is progressive, starting with coarse, followed by fine, files, after which finer and finer grades of abrasive cloths and papers are used until the stage is reached at which very fine polishes may be applied. If carried out correctly, it should not be necessary to return, at any stage, to the use of a coarser abrasive or file, the progression always being towards a finer surface finish.

Abrasive papers and cloths

Abrasive papers and cloths are used for surface cleaning and preparation. These materials comprise abrasive particles cemented to a cloth or paper backing, various grades or 'grits' being available. The grit number relates to the mesh of the sieve used to select the abrasive, the higher numbers representing the smaller particle sizes and hence smoother finish produced. The two main types of material are the emery cloths and papers and the so-called wet-and-dry papers. For these latter, a waterproof cement is used to attach the abrasive to the backing and they can be used in the presence of

water under circumstances in which the paper would otherwise rapidly become clogged with abraded material, for example, when rubbing down painted surfaces.

One particular advantage of wet-and-dry paper is that it is available in very fine grades, up to 1200 grit. It is a natural successor to the coarser emery cloths and papers, although it is available from some sources in grades as coarse as 80 grit.

A cloth backing provides greater wear resistance, and for general use at the coarser end of the scale, an emery cloth is more satisfactory. These cloths tear very easily going 'down' the sheet, but less readily 'across', and they should be cut in this direction. Paper-backed abrasives need to be cut in both directions if tidy strips and smaller pieces are required, as is generally the case.

Backing up the abrasive

There is a great temptation to use emery cloths in the hand, but unless its use is simply to clean up stock material which will later be machined, the method should be avoided. Lack of support for the abrasive allows the cloth to 'fall over' the edges, which rapidly become rounded, and some kind of solid backing must therefore be provided.

On large surfaces, a hand-size wooden block can be used, or a woodworker's cork block, and a strip of emery cloth or paper wrapped around it. Provided that care is taken to keep the pressure firmly on the main surface and away from the edges, there should be no degradation of the corner profile. It is important to ensure that the abrasive material is stretched tightly around the block, otherwise any slack can roll itself over the edge and spoil the corner.

Support for the cloth or paper is also naturally required for work on smaller items. One convenient method is to select a file whose size suits the job and to tear off or cut a piece of material long enough to lie along the blade and fold over the end. If the file is held as for draw filing, the emery can be gripped firmly to the file and the work cleaned up without damaging the corners. This operation is shown in Figure 6.32.

An alternative to the hand-held emery strip is to glue a strip of abrasive paper to a strip of wood, to produce what are known as emery sticks. A group of these is shown on Figure 6.33. The sticks are commercially prepared wooden strips, about 1in. × ¼in. (25mm × 6mm) to which emery paper has been glued,

Figure 6.32 A strip of emery cloth can be wrapped around the blade of a file for the later stages of smoothing a surface.

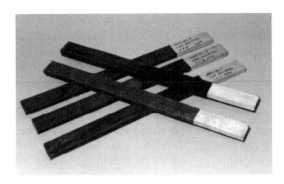

Figure 6.33 Several grades of emery sticks.

creating in effect, an emery paper file. These sticks are a much more satisfactory arrangement than the emery-around-file method shown in Figure 6.32, and can readily be made up, or the emery paper replaced, using double-sided tape. Paper is normally used, rather than cloth, since it creases sharply when folded, and the sticks can be 'wrapped up' completely making both wide and narrow edges available.

These methods of providing support for the abrasive are very convenient for the edges of small levers and rods, but are less useful when their faces require treatment, particularly if they are tapered and cannot readily be held in the vice. More flexibility is provided by turning the process upside down, laying the abrasive material on a flat surface and rubbing the work up and down on it. This must, however, be done carefully, ensuring that the cloth or paper remains flat and taut, otherwise ripples develop ahead of the work as it moves to and fro, and this tends to round the corner profile.

A useful alternative is to stick the abrasive paper down onto a board using heavy-duty, double-sided tape, preparing three or four boards with different grades of abrasive. The boards can be sufficiently large to accommodate a complete sheet of cloth or paper and will last very well, particularly the finer grades,

Figure 6.34 A sheet of emery paper or wet-and-dry can be stuck to a piece of board and used for smoothing small parts.

which do not, in any event, remove much material.

On the subject of the wearing capabilities of emery cloth, it is worth noting that the larger grit particles tend to become dislodged from the backing first and the emery thus tends to become finer as it wears out and can be downgraded to the later stages until it is finally useless.

A board to which a sheet of emery paper has been glued is shown in Figure 6.34.

Direction of working

If cleaning up is being performed as a preliminary to painting, no great fineness of finish is generally required, it being necessary to ensure that the overall surface has a good shape and provides a key for the paint. Shape is determined by the filing and machining processes and emery cloth is used to provide the key. A medium grade of emery is usually appropriate for the larger models and a good key for the paint is created by adopting a circular motion.

For work on levers and long rods which are not to be finished to a high state of polish, but do need to be clean and bright, it is best to rub along the length using progressively finer grits, until the desired appearance is achieved. For models, the finish required is usually such that no scratches are discernible with the naked eye, since the work is, after all, only a small proportion of its prototype size. How much polish is required (or desirable) is dependent on the model, the item and the builder's notion of what is needed.

If the work is required in a highly polished state, which is generally the case for clock plates and wheels, great attention needs to be paid at all stages to ensure that surfaces are not spoiled during the initial stages of marking out, sawing and filing. Great attention also needs to be given to the polishing processes. First of all, the work must always progress from coarse to fine, the aim of each stage being to remove the scratches produced by the previous (coarser) stage.

Scratches created by the previous stage can best be seen by carrying out each stage at right angles to the previous one, continuing to use finer and finer abrasives until the scratches are so small that the required degree of polish has been achieved. At this stage, very fine abrasives are being used and the direction in which they are applied is not usually significant.

Polishing progresses through the finer grades of emery cloth and paper towards the use of compounds which are applied by being loaded onto a cloth pad. These compounds include various grades of rouge, which can be purchased from modellers' suppliers who specialise in providing a service for clockmakers, but cans of liquid polishes such as *Brasso* are readily available from DIY and hardware stores, sometimes simply described as 'metal polish'.

7

Bending, folding and forming

Introduction

Preparation of the material

The bending of metal is a frequently required operation in model work. Since thin materials are normally involved, bends are made with the metal in the cold condition, rather than by forging at red heat. If the material is to be bent cold, it may require preliminary heat treatment to render it sufficiently ductile (soft) to bend easily and without fracturing. The process is known as annealing, and it may be applied to steel, copper, brass and aluminium alloys, although not always for the same reasons. Annealing is performed by heating the material (usually to red heat, although not for aluminium alloys) and allowing it to cool, although the manner of cooling can vary, depending upon the material and the aim of the process.

Since copper and brass are ductile metals, they will usually bend readily in the 'as bought' state, although they actually become harder as they are worked, a process known as work hardening, and do need to be annealed at intervals through the forming process if significant,

repeated shaping needs to be performed. This is usually when these metals are being flanged or formed into complex shapes.

The most common form in which steel is purchased is that known as bright mild steel (BMS). This is drawn down into standard bar forms (rounds, squares, hexagons and flats) using a cold process which produces a smooth, bright surface finish to the bar. The bar has, however, been pulled and squeezed into shape and the process leaves the material with an elongated grain structure which is quite brittle.

The result is that a bend across its width, formed cold, frequently produces a fracture. This can be avoided if the steel is normalised before bending. The need for normalising, and a more detailed description of the process, are given in Chapter 3.

The bending operation

Before considering the bending process in detail, it is useful to consider what happens to the bent material, as this helps to avoid likely problems. In making a bend, the work is

136

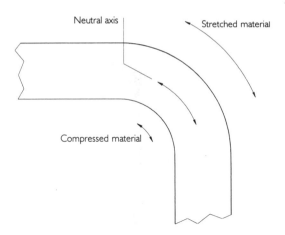

Figure 7.1 Stretching and compression of the material in a bend.

Figure 7.2 Deformation which occurs on a bend, due to stretching and compression. This effect is particularly noticeable on narrow strips.

distorted, as shown by Figure 7.1, the material being compressed on the inside of the bend and stretched on the outside. The effect of this on a narrow strip is illustrated in Figure 7.2. Compressed material on the inside of the bend is expelled outwards and stretching on the outside tends to draw material in from the edges. The result is that the material is significantly distorted on the edges, as well as on the outside of the bend, and this produces a distinctive 'turned-up' appearance. It is important that the technique adopted for forming the bend tends to mitigate these problems.

The distortion which occurs on the edges means that they need to be 'dressed' (filed or machined) after the bend is formed. A narrow strip which is to be bent needs to be wider than the required size in order to allow the final shaping to be performed after bending.

Local distortion also means that holes close to the angle cannot be drilled until the bend has been formed, otherwise the hole distorts with the material (more strictly, the material around the hole distorts) and the hole is 'pulled' towards the bend on the outside. The general procedure must therefore be: bend, shape the edges and then mark off for, and drill adjacent holes.

Cold bending

Use of a bending bar

In the absence of a specialised bending or folding machine, the operation is most conveniently performed by 'pushing' the metal around a former which approximates the shape required, nothing more elaborate than a stout bar being required. Unless the material is very thin, and very ductile, 'dead sharp' internal bend radii are not ordinarily achievable, and the former needs to be provided with a suitable radiused edge, around which the material can be folded.

For general use, a simple length of BMS rectangular bar can be retained, especially for the purpose, and a 12in. (300mm) length of ¾in. or 1in. (20mm or 25mm) square bar, having its longitudinal corners radiused, can be kept specially for the purpose. Ordinarily, an internal bend radius equal to the material thickness can be applied for general use, unless there is a specific need to represent a particular radius from scale considerations. A short bar of the sort shown in Figure 7.3 serves adequately for short work. This example has lived in the workshop for many years and has been used both for bending and as a holder for riveting dollies, hence its chequered appearance.

Figure 7.3 A much-used bending bar produced from a length of ¾in. (19mm) square steel.

This particular bar is about 9in. (230mm) long and has one long edge filed to a radius of about ⅟₃₂in. (0.8mm). It suits the making of bends in materials up to 20 or 22 swg (up to about 1mm thick). At one end, one edge is filed to a radius of ⅛in. (3mm) for a length of 1in. or so. This allows narrow materials about ⅛in. thick to be bent satisfactorily and was originally prepared many years ago for bending mild steel strip for model locomotive guard irons.

In use, the bar and material to be bent are clamped into a vice with the upper surface of the bar just proud of the top of the vice jaws, and the bending line, already scribed on the work, carefully aligned with the bar's upper surface. The angle which the body of the work makes with the bar's upper surface must be checked, if this is possible, although this naturally depends on the shape and size of the work, principally the length of the upstanding part. Figure 7.4 shows the essentially different types of work.

When making the bend, it is not usually adequate simply to pull on the upstanding part of the material, even assuming it is long, as shown in Figure 7.4A. It is essential to push the material closely around the bar, and to do this requires pressure to be exerted on the material just above the bending bar's corner. This pressure can best be applied by using a block of soft material, such as aluminium alloy, holding it firmly against the work and hammering it down to press the material around the radius on the bending bar, as shown in Figure 7.5.

A

B

Figure 7.4 A brass plate positioned in the vice with the bending bar. If the plate has a large upstand, a small square can be used to check for squareness.

Figure 7.5 An aluminium alloy block being used with a hammer to form a bend over the edge of the bending bar.

The aluminium alloy block serves two essential purposes: it allows a large hammer to be used without fear that the work will be marked, and it spreads the load so that the full width of material moves together around the bend.

It might be thought that the same result could be achieved by using a soft-faced hammer to apply the force directly. Unfortunately, this simple technique doesn't work, and for several good reasons. Firstly, if the upstanding material is long and the blow is applied as shown in Figure 7.6, the inertia of the upper part of the upstanding material tends to make it want to remain undisturbed. The consequence is that the end lags behind the remainder of the material as the bend is formed, and a distortion of the work results.

Secondly, the localised load, even from a soft-faced hammer, turns the material over unevenly, and can produce an uneven bend. By applying the hammer blow through a block, the localised load which the hammer applies is distributed over a larger area, consequently reducing the actual force applied locally and smoothing out the unevenness.

The third problem caused by the individual hammer blows is that the bending force is not necessarily applied to the material immediately above the edge of the bending bar, and its shape cannot therefore be imposed on the work. A block of soft metal can be positioned and angled so that the force of hammer blow is applied exactly where required.

If the upstanding portion of the work is short and fairly narrow, it may be possible to tap it over the former with a hammer since a squarely aimed blow may effectively contact all of the material. For a short upstanding part on wide work, some means of applying the bending force over a wide area is again beneficial, and the use of a block some two or three inches (50mm to 75mm) long through which to apply the force is helpful. In these cases, it is preferable to bend progressively, deflecting the material through a small angle along the whole bend, and then repeating the process in this manner until the required angle is formed along the complete edge.

Forming large-radius bends can naturally be by utilising an appropriately radiused corner to a bending bar or by using a circular bar as the

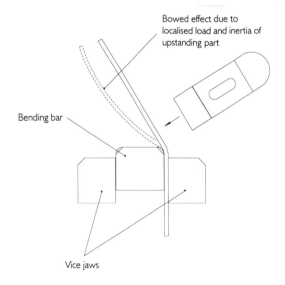

Figure 7.6 If an attempt is made to hammer a long strip over the edge of the bending bar, a bow is usually produced due to the inertia of the workpiece.

former. The difficulty comes in locating the bar, since the scribed line, which usually shows the start of the radius, must be positioned such that it is not visible. It is frequently best, therefore, to put the bend in at the approximate position on a larger-than-necessary piece of material, afterwards trimming material on both sides of the bend to the correct length. If a little extra is left on the width, this might also allow for some lack of squareness in placing the bending bar which can be taken out when the sides are trimmed to size.

If a narrow strip is being bent, it must initially be wider than that which is finally required, in order that the turned-up edges can be removed afterwards.

Using a 'proper' industrial bending or folding machine the problem of lack of visibility does not occur, since the scribed line remains visible, as shown in Figure 7.7. The work is clamped to the machine bed such that the

scribed line coincides with the front of the clamp. The bend is formed by rotating a substantial angle or blade against the work, but at such a position that it rotates at the required radius. The blade rotation forces the work around and upwards, forming the bend radius as it does so. If the correct allowance has been made, the (now) upstanding part has both the correct radius and the correct amount of upstand.

Material in the bend

It is obvious that some of the material is 'consumed' in forming the bend. It therefore becomes a question of what allowance needs to be made for this. For large-scale, production items it is convenient if material can be cut, and the bend formed, leaving both sides of the angle exactly the correct length. Most engineers'

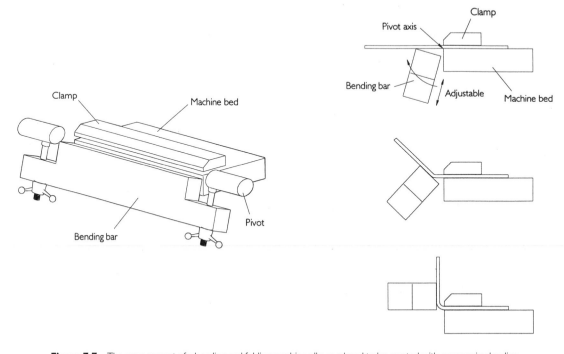

Figure 7.7 The arrangement of a bending and folding machine allows a bend to be created with any required radius.

reference books, even quite abbreviated ones intended for the workshop rather than the library, provide tables showing the allowances which need to be made in laying out the work 'in the flat', both in relation to the material in the bend and the overall size of the work, bearing in mind also the thickness of the material.

In model work, materials are usually thin and bend radii very small, often amounting to dead sharp with little concession to normal allowances. Published data are not therefore especially relevant to model work and if only a single bend is required, it is usually more convenient to form the radius or bend and then mark out the two sides individually for cutting and finishing to length. The bend allowance then takes care of itself.

If items of the form shown in Figure 7.8 need to be made, the use of a conventional bender, together with calculation and application of the correct bend allowances, permits accuracy to be achieved quite simply. Lacking a bender, other techniques will yield acceptable results, if carefully applied, but much naturally depends upon the accuracy which is desired in the completed item.

If several identical items need to be bent into the sort of shape shown in Figure 7.8, they should certainly all be marked out together if possible, and all bent up in the one working session thereby readily allowing the same tech-

nique to be applied to them all. If the span of the U shape is not too large, it is useful to cut out a former of the required width, radius the two edges and bend each piece closely around the former. In this case, the upstanding arms of the U can be marked out and cut off after bending.

Folding up a box

Small boxes frequently need to be bent up, together with flanged lids to fit them. These items can be bent using the techniques described above to form the actual bends, but since two adjacent sides will be bent up, the corners need special treatment. This is usually to cut away material at the corners to form a notch, but before this is done, the intersections of the lines defining the internal corner positions must be centre punched and the corners drilled. Following this, the corners are notched so that the bent edges will abut nicely at the corners. The purpose of the holes is to allow the bent material to expand at the edges of the bends, and if this space is not allowed, the corners of the box 'bunch up' and distort.

Folding up the box is straightforward once the technique is understood. The initial bends should be formed on two parallel sides, it usually being convenient to bend the two long sides on rectangular work. This results in a U-shaped channel, as shown in Figure 7.9.

Examination of Figure 7.9 will show that the brass sheet for the box lid has been drilled and notched at the corners, operations which are described elsewhere. The two sides have been bent using the methods illustrated in Figures 7.4 and 7.5 and the lid is ready for the remaining sides to be bent.

To bend the remaining two sides, a bending bar is needed, to fit within the width of the box, sufficiently thick to project above the top of the already folded sides. This is then clamped into

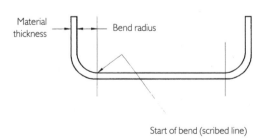

Figure 7.8 Allowances for the material needed to form the bends.

Figure 7.9 A half-folded box in thin sheet brass.

Figure 7.10 The box of Figure 7.9 fitted into the vice with a short bending bar which allows the third or fourth side to be bent.

Figure 7.11 A completed box lid.

the vice as shown in Figure 7.10, and the remaining sides bent over the bar.

The completed box lid, with the drilled-out corners in evidence, is shown in Figure 7.11. The material of the lid is .022in. (0.6mm) brass sheet and the corner relief was put in using a ⁵⁄₆₄in. (2mm) diameter drill, which has created ample clearance for material pushed outwards from the bends.

Bending a large sheet

Bending rolls (see below) might be used to bend large sheets, but they cannot produce small radii, and it is usual to bend material over a former of the required diameter in order to form such bends. Round bars immediately suggest themselves as suitable formers, and if a round bar can be clamped to the sheet material, and the 'assembly' put into the vice, the bend can be formed by pressing the material over the bar to achieve a bend of the required radius.

With a long sheet it may not be possible to clamp it centrally in the vice, in which case the sort of arrangement shown in Figure 7.12 has to be used. In this case, a sheet of brass, 11in. (280mm) wide and 24in. (600mm) long has been bent to form the outer of a water tank.

Figure 7.12 Large sheets can only be bent by arranging the bending bar and material outside the jaws of the vice. This illustration shows a pair of specially made bars which have been dowelled together and clamped to the work, support being provided by a wooden dowel held in a vice standing on the floor.

142

The corner radii are ³⁄₈in. (10mm) and the sheet is sandwiched between two rectangular steel bars which are located to one another by dowels. One bar has a cut-down ³⁄₈in. diameter steel bar screwed to the top surface and this has been used to form the bends. The brass sheet was annealed along the bend lines before bending, fitted to the pair of bars on the bench, for easier handling of an awkward assembly, and toolmakers' clamps used to hold the sheet firmly.

Since the sheet could not be accommodated within the vice it was held in the vice at one end and supported by a length of 1in. (25mm) wooden dowel at the other. The dowel was held in a vice standing on the floor of the workshop in order to provide a safe support, and the bend put in by hammering on a length of wood placed on the sheet, to force the material down into contact with the ³⁄₈in. diameter bar attached to one of the rectangular bars.

Rolling metal

Even though a bend radius is large, it may still be put in by pressing the material over a suitable piece of circular bar, but this is ultimately likely to become impossible, simply due to the non-availability of suitably large bar material. In these cases, something akin to an adjustable former is used. This takes the form of a set of bending rolls. The operating principle of these is that the material is guided through a set of rolls which includes one deflecting roller which is adjustable to alter the radius formed. This is illustrated in Figure 7.13A which shows the usual form of these rolls.

The two lower rolls can be rotated manually so that the material passes through the rolls and is bent into a curve as it passes beneath the central, deflecting roll. The vertical position of this roll is adjustable and the curve produced in the material can be varied. Repeatedly passing the

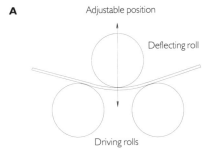

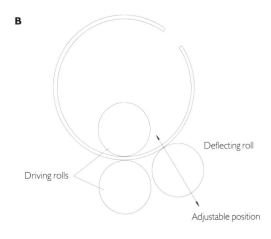

Figure 7.13 Two forms of bending roll.

material through the rolls and progressively lowering the deflecting roll, causes a progressively decreasing radius to be formed.

The disadvantage of this form of rolls is that the material cannot be curved right to the end since there is inevitably some material which lies on a straight line between the contact points on the deflecting roll and each of the lower rolls. This can be overcome by rearranging the rolls into the form shown in Figure 7.12B. Here, the deflecting roll acts directly and independently and the arrangement allows a cylinder to be rolled from flat sheet. Provided that this roll can approach the upper roll

Figure 7.14 A nicely made pair of bending rolls seen at an exhibition.

closely, the radius achieved can be very closely that of the upper roll. A set of bending rolls of this type is shown in Figure 7.14.

An example of the type of work carried out using the rolls is the rolling of the outer wrapper plate for the firebox of a model traction engine boiler. The simplest type of wrapper has straight, parallel sides and a semi-circular top.

The length of material required for the wrapper can be determined by calculating the length of the semi-circular arc and adding on the height of the two sides. The semi-circular arc is naturally one half of the circumference, which is calculated from the expression 2π (pi) multiplied by the radius, or, π multiplied by the diameter. The ratio π may well be available

these days on a scientific calculator, but if not, can be taken as 3.142. The commonly accepted approximation for π of 22 divided by 7 is actually closer to 3.143 than 3.142, but will make no difference in practical terms. In any event, it is wise to commence the rolling operation with a sheet which is slightly over the required length, allowing, say, an inch (25mm) on each side, for cutting off once the shape has been formed.

When the annealed sheet of material is available, a centre line should be marked using a felt-tipped pen and the rolling operation started by inserting the sheet into the rolls and tightening the adjusting screws until the sheet is just 'nipped' and the adjustable roll is parallel to the others. This condition is illustrated in Figure 7.15.

The rolling operation should be started from the marked centre line, progressively decreasing the bend radius until the desired curvature is achieved around the central mark but not yet extending over the full arc. As rolling and bending continue, the work must be taken out of the rolls to check the radius which has been formed, in this case by testing its fit to the boiler barrel, as shown in Figure 7.16.

Once the correct radius is being rolled, the material can be inserted into the rolls and

Figure 7.15 The initial position of a sheet of copper in the rolls. The position of the deflecting roll has been adjusted so that the work is nipped and the rolls are parallel.

Figure 7.16 Testing a part-rolled wrapper on the boiler barrel.

Figure 7.17 The almost-complete wrapper in the rolls.

passed progressively further through the rolls on each side of the centre line until an arc of the required length is achieved. Since the rolls exert only a friction grip on the material, it is necessary to push the material firmly into the rolls while turning the handle.

This operation is shown almost complete in Figure 7.17. Starting with a sheet which is slightly over-length, and then cutting to size after rolling and fitting to the flanged plates,

Figure 7.18 A commercially made firebox tubeplate.

removes the need for accurate centring of the rolled arc. This naturally allows concentration on the important aspect of ensuring a good fit without worrying about the bend's actual position on the sheet.

Naturally, if using the type of rolls shown in Figure 7.13B to form a cylinder, the length of sheet must exactly match the diameter required, and this should be calculated from 2π multiplied by the radius, as described above.

When rolling work using the rolls, it is vital to ensure that the work is inserted squarely into the rolls, and this should be checked carefully with a square before commencing. It is also important that the deflecting roll remains parallel to the driving rolls at all times, otherwise tapered work is produced since different radii are forced into the material on the two sides. Indeed, this is the way in which a tapered boiler barrel is produced.

For some rolled work, the shapes required are more subtle and complex. The flanged plate shown in Figure 7.18 is the firebox tubeplate for a small locomotive-type boiler. This plate is 4⁵⁄₁₆in. (110mm) high and 2⁵⁄₁₆in. (59mm) wide at the bottom. The top is curved at a radius of 3¼in. (83mm) and the corners have a radius of ⁹⁄₁₆in. (14mm), little larger than the diameter of the rolls shown in Figure 7.15.

Since the wrapper plate is formed from copper sheet, it must be annealed before commencing, to render it soft and ductile. Since unworked, annealed copper is extremely soft, it can be formed by manual pressure over suitable bars (for the corners) and a wooden block, for the larger radii. An alternative is to shape up a wooden block to the form required, clamp the sheet to one bottom corner of the block and then to use manual pressure or a soft-faced mallet to force the sheet down into contact with the former, applying further clamps to hold the part-formed sheet as work progresses.

Due to the inevitable spring-back which occurs whenever metal is bent, a wrapper formed in this way inevitably ends up being slightly bigger than the former, whose size might need to be adjusted to compensate.

Although a wrapper to fit the tubeplate illustrated in Figure 7.18 (and its associated backplate, which is the same size) can be bent up by hand methods, the bending rolls do offer the most precise control over the shape produced, although a little hand intervention is sometimes useful. The process must be progressive, however, and the larger radii must be put in first, followed successively by the smaller ones. This means beginning with the large top radius, then putting in the corner bends and afterwards attending to the sides.

Rolling the top radius is straightforward, the sheet being first marked to identify the centre and the rolling starting on each side of this reference mark. Since the top corner radii are very close to that of the rolls themselves, and also because they represent arcs of less than 90 degrees, the normal rolling action of the rolls is not helpful in producing these curves.

Since model boilers are generally made in copper, which is kept soft by repeatedly annealing it during bending, it can actually be bent by hand. A circular bar of the appropriate diameter can be used as a former, and if the size of the job allows, it can be clamped into the vice with the circular bar correctly positioned against the sheet and the bend put in by hand pressure. If the corner radius is similar to the diameter of the bending rolls, it may be possible to form the bend while the sheet is in the rolls, simply by pulling on the upstanding part of the wrapper. This at least solves the problem of holding the work, which can otherwise be difficult.

Although the bending can be done by hand, direct manual pressure is not satisfactory since it cannot be applied evenly across the work without something interposed to spread the load. For soft materials a piece of wooden board is satisfactory. This can be used to pull or push the material over the bar, say, through an angle of 35 or 45 degrees, before checking that the bend is progressing evenly at the two edges and the material will fold squarely over the bar.

If one edge is not folded over to the same extent as the other, gentle hammering on the required side of the board should be used to bring the work to squareness before proceeding to pull or push the bend into its final position. During the hammering, remember that soft material can be easily crushed if hammered too hard against the roll. Provided that good contact with the circular bar is achieved throughout the operation, there should be no need to re-bend already formed material, which is, in any event, harder than the remainder, due to its already having been worked. There should be little spring-back effect in the bend itself, but the sheet will need to be pulled through a slightly larger angle than is needed for the bend, since that is where any spring becomes evident.

A trial fit against the flanged plate is necessary to confirm the correct form and fit of the bend and the sheet can be marked for the second bend at the same time by using a felt-tipped marker to show the line which indicates the starting point of the radius on the other side. Getting the starting point in the correct place is the difficult part of the operation and the initial bending of the second corner should concentrate on this aspect before too tight a bend is put in. Once the radius is started in the correct place, the work almost always beds down nicely onto the bar at the correct position and the final part of the radius can be put in without difficulty.

If things do go wrong, don't despair! The copper can be rendered soft again by annealing it once more, and there is no limit to how often this may be done. Once it is soft, it can be straightened out again, bringing to near-straightness by hand and then rolling it flat in the bending rolls. This doesn't always bring the material to absolute flatness, but this is seldom absolutely necessary since it is in any case going to be rolled again into a new shape.

Flanging and forming

Perhaps the cart has already come before the horse, since the rolling of the boiler wrapper which is described above has assumed that the flanged plates are already available for use as a pattern for the shape required. The flanging process, which at its simplest just means raising a turned-up edge around the outside of a plate, is simply an extension of the straightforward bending process, except that it allows ductile materials to be formed into bends which comprise two curves at right angles to each other. That is, to be formed into the type of compound curve illustrated in Figure 7.19.

The shape shown is the top corner flange of the tubeplate of Figure 7.18, the flange extending around the two edges which are roughly at right angles. The corner is formed on a 9/16in. (14mm) outside radius and has an internal radius of about ⅛in. (3mm) where it leaves the flat surface of the main plate. To form this corner, the copper sheet is annealed and

Figure 7.19 A detailed view of the corner of a flanged plate.

clamped to a thick former (the flanging former) in the vice, and the copper turned over the radiused edge of the former by hammering with a substantial hammer. Since the copper becomes hard again due to the hammering, a process known as work hardening, the annealing must be repeated several times during the flanging to ensure that the copper remains soft and ductile.

During flanging, there is little use in attempting to turn over the flange using a soft-faced hammer, in the hope that the flange can be formed without damaging the material. It must be hammered into the desired shape, and in a corner such as that illustrated in Figure 7.19 must be compressed into the shape required. This means firm support for the work and a large enough hammer to effect the deformation.

Provided that the copper is annealed frequently enough, and the hammer blows are sufficiently heavy, there is little difficulty in turning over the flange except, of course, in a corner of the type illustrated. Material has to be forced into the corner from both sides so there is too much material in this location and it tends to bulge outwards as blows are applied to the top and the sides. The additional work which is being done on the material in the corner hardens the metal more rapidly and this naturally makes it less amenable to being squashed into the desired shape, thus aggravating the problem.

When cutting out a blank sheet for flanging, it is naturally cut oversize to allow for the material which forms the flange. There is a natural tendency to want to make this allowance just that bit larger, on the basis that the excess can afterwards be trimmed away, but in the corners a smaller allowance is actually required since metal is pushed into this region from both the top and the side. It is not helpful, to say the least, if the allowance has been made too great as this actually makes the job of forming the

Figure 7.20 A cut copper blank and the flanging former in the vice.

corner correctly that much more difficult. So, what is required is a somewhat sparing allowance, particularly in the corners.

Figure 7.20 shows a former and a pre-cut and annealed copper plate in the vice, ready for the flanging. The plate is ³⁄₃₂in. (2.4mm) thick and it will be formed using a 1 pound ball pein hammer. The objective is to turn the flange over, all round the edge, progressively. The work should receive good solid blows, and any tendency for the corner to bulge outwards must be resisted by firm hammer blows to correct it. Figure 7.21 shows the corner shape achieved at the stage at which a second annealing is required.

Figure 7.22A shows the appearance of the plate after further forming. The corner is now

A

B

Figure 7.21 Progress achieved on a flanged corner at the stage at which a second annealing is required.

Figure 7.22 A flanged corner at the stage at which a third annealing is required.

bulging noticeably, but the copper has become hard due to the hammering and a further annealing is required. Figure 7.22B shows that the flat part of the plate is bulging away from the former. This is due to the hammer blows being directed too much in a downward direction (attempting to make the flange lie flat on the former) while the material in the corner was hard, but the flat still soft. This bulge can easily be corrected when the plate is again soft, but its presence can be discouraged by backing up the copper with a stiff plate.

Figure 7.23 The final fit of the corner of a flanged plate.

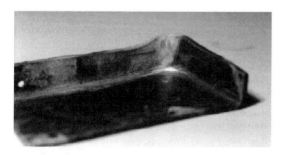

Figure 7.24 Excess material pushed into the corner during flanging.

One more annealing, and some heavy blows on the corner, produce the final form shown in Figure 7.23. There is an adequate fit for this sort of work, and the outside shape has still to be produced by filing. Figure 7.24 shows how the extra material has been forced into the corner, but this excess can be trimmed off when the plate is cut to its final shape.

Formers used for the flanging process are most conveniently cut from aluminium alloy sheet which is sufficiently solid to allow annealed ⅛in. copper to be flanged over it, yet is much easier to cut out and shape than mild steel. Wood can be used for light work, say, for the forming of thin brass sheets, around 18 or 20 swg (0.9 to 1.25mm) but is definitely not adequate for flanging thick copper around tight corners, where really solid hammer blows are needed.

Apart from the additional work involved, there is naturally no objection to steel being used for the former, but it must not be thought that effort can be economised by using only a thin plate. Considerable force is needed to complete the forming of the flange, and if the flange depth is greater than the thickness of the former, the situation of Figure 7.25 usually results.

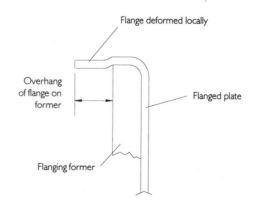

Figure 7.25 Possible effect of using a thin former for flanging.

Due to the overhang of the flange on the former, the flange is distorted locally by the hammer blows, since there is no support, with the result that the former is jammed into the plate and cannot readily be removed. The former is ideally a little thicker than the flange depth, plus any additional allowance, and the ease with which it may be removed from the flanged plate well repays the extra effort put into its production. The hammering process tends to produce an irregular flange depth, in any event, as exemplified by Figures 7.19 and 7.24.

It is sometimes suggested that a composite former can be used, comprising a wooden blank faced with a thin steel plate, but this too is often not satisfactory since the wood tends to compress and the flanged plate can easily become jammed on the former, as shown in Figure 7.25.

Although copper boiler plates are common flanged items, the same technique can also be applied to brass, which is used extensively for various fittings for model locomotives. Typical of these is the locomotive tank filler shown in Figure 7.26. Being an alloy of copper, brass can be worked in substantially the same way, and the cover shown was flanged from 26 swg (0.45mm) brass sheet over the aluminium alloy

Figure 7.27 Flanged end covers for small boilers made in thin brass.

former shown alongside the completed and fitted item.

This cover is $1\frac{3}{16}$in. by $1\frac{5}{8}$in. (31mm by 42mm) with radiused ends, and has a flange all round which is $\frac{3}{32}$in. (2.4mm) deep. The cover required annealing three or four times during the flanging operation which was carried out using a 2oz hard-faced, ball pein hammer. This left the flange surface somewhat marked, but a suitable finish for painting was achieved by cleaning up the flange using a smooth file and medium emery cloth.

The flanging former was in this instance made quite deep, due to the available material, and it was used also to hold the rough-flanged covers for sawing off the excess material from the flange which is evidenced by the saw marks visible in Figure 7.26.

Figure 7.27 shows a group of brass discs similar in size and material thickness to the cover of Figure 7.26. These are end covers for small-diameter, toy boilers which need to be polished after silver soldering to the tube forming the boiler shell. Figure 7.28 shows the flanged rim after soldering to the shell. The damaged to the rim is evident in this view but this was polished out (well, filed and polished) to produce the result shown in Figure 7.29.

Figure 7.26 A locomotive tank filler cap and the former used for its manufacture.

Figure 7.28 An end cover silver soldered to the boiler tube.

Figure 7.29 An end cover which has been cleaned up and polished.

With this type of work, in which the flange material is being compressed all around the rim, it is frequently difficult to 'persuade' the rim to turn over through a full 90 degrees all around the disc, since the compression of material which occurs in the rim increases the problem of spring-back. There is also inevitably the problem that hammering one part of the rim inwards causes an adjacent part to bulge outwards so that a point is ultimately reached at which further hammering produces no

progress. It has to be accepted that simple flanging cannot produce such sharp and evenly turned over flanges as pressing the blank between a punch and a die, which is how such items as these are produced commercially.

If you are an experienced coppersmith, the type of work described here is of the utmost simplicity. A skilled copperbeater can stretch and 'pull' annealed copper into almost any shape extremely easily by supporting it on a range of fairly standard shapes of former, known as stakes, and hitting it with a hard-faced hammer. The stakes resemble miniature anvils which can be supported in a vice or held in a 'stake hole' on a real anvil.

It is a real pleasure to watch someone 'knocking up' a model locomotive dome from a disc of copper with seemingly little effort, but it obviously requires a great deal of practice to achieve this degree of proficiency, and unless you are well beyond the stage of being shown how to beat up an ashtray from a copper disc, which always seemed to be the starting point for school metalwork classes, it is probably wise to keep it simple. Nevertheless, domes and suchlike can be formed, simply by use of a hammer and appropriate stakes.

The model locomotive dome of Figure 7.30 was formed from a length of 3-inch (75mm)

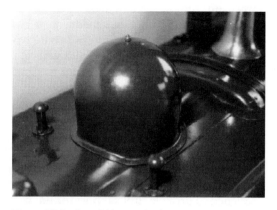

Figure 7.30 A model locomotive dome.

151

Figure 7.31 The flanging former for the bottom of a locomotive dome.

Figure 7.32 A stake for forming the top of the dome.

diameter brass tube, the base being flanged outwards using the light-alloy former of Figure 7.31, and the top being cut away to form a three-lobed tulip shape before being beaten over a wooden former to obtain the approximate shape. Final forming of the top was carried out using the large stake shown in Figure 7.32. This not-very-elegant device is a billet of cast iron mounted eccentrically to a ¾in. (19mm) diameter steel bar and turned 'freehand' on the lathe by manipulating both feed screw handles simultaneously. It was finally smoothed by hand filing but could have been brought to a much better finish with some benefit. However, the upper part of the dome was finally shaped by hammering against the stake and the edges of the 'tulip petals' silver soldered together to produce an acceptable fitting without the need to utilise a casting.

The stake of Figure 7.32 illustrates the principle of these items. They do not at all need to be shaped like the item to be formed, but are designed to provide firm support for the work so that it can be hammered into the desired

shape by striking the parts just overhanging the support, until the correct shape is achieved. Regular annealing of the brass or copper is essential to ensure softness and ductility and to prevent the material cracking or splitting during forming. As work hardening occurs, the material deforms less for a given hammer blow (force) and the sound and 'feel' of the operation changes, making it obvious that another annealing is required.

One point about annealing is worth making. Heating of the brass or copper to red heat encourages the formation of oxides on the surface, and discolouration occurs. Sometimes, quite large amounts of a black residue are created which flake off when cold or when the hot metal is quenched in cold water. Some oxides always remain on the surface, however, and if the forming process is continued without cleaning the deposits away, they can become embedded in the surface of the work, hindering or preventing proper cleaning. It is necessary to clean the material thoroughly after each annealing.

Tube bending

Preventing collapse

Steam engines (generally) require numerous small tubes of various sizes for such functions as steam feed, exhaust, cylinder drains, lubricator feeds and so on. In the scales normally used, these tubes range from ⅟₁₆in. (1.5mm) diameter up to ½in. (12.5mm) or so, for the exhaust pipes for the larger steam locomotive models. Quite tight bends are sometimes required in these pipes and it is usual to use copper tube since it is easily softened by annealing and takes both soft and hard solders extremely well.

The use of soft copper tube makes bending easy but it also brings with it some problems. If a piece of copper tube is bent, either around a former, or just in the fingers, the result is inevitably that shown in Figure 7.33 – a collapsed, unsightly and blocked tube. As mentioned in the introduction, and illustrated in Figure 7.1, material on the inside of a bend must compress, and that on the outside must stretch. If the material is tubular, there is naturally no central core to prevent the material on the inside and outside of the bend from doing precisely what it wants. Material on the outside of the bend, therefore, fails to stretch sufficiently, and 'cuts across' the corner, while that on the inside fails to compress and excess material collapses inwards.

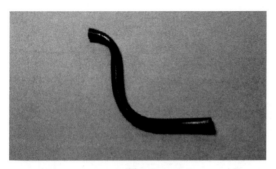

Figure 7.33 A bent and collapsed copper tube.

The solution to the problem is to 'put the middle back' into the tube during the bending. In the plumbing trade, only a strictly limited range of tube sizes is used, and standard bending springs are available which are a close fit inside the tube. Being coiled, the spring is flexible, but since it is wound from square-section wire and hardened and tempered, it cannot readily be crushed. It therefore allows the tube to be bent without it collapsing.

Unfortunately, the limited range of tube sizes used by plumbers means that the method is not widely usable by modellers, but if you need to bend a largish tube, and its actual size doesn't matter within reason, one of the plumbing sizes should be considered, especially if you can borrow a spring for an evening.

An alternative to the fitting of an internal spring is to fit a 'restrainer' to the outside, a practice which suits the smaller tubes used by modellers quite well. Sets of small springs such as those shown in Figure 7.34 are available, this set covering tubes of ⅟₁₆, ³⁄₃₂, ⅛, ⁵⁄₃₂ and ³⁄₁₆ inches in diameter. Being of a somewhat softer temper, and therefore less rigid than a plumber's spring, these small coils do not completely prevent collapse if very sharp bends are required, since they rely on preventing expansion of the tubes on the sides of the bend rather than actively preventing collapse on the inside. So, it is helpful if the bends can be as easy as possible.

Failing the availability of bending springs, the tube must be filled with a substance which is rigid enough to prevent collapse, but which can readily be introduced and afterwards removed. The material which is used is a low melting-point metallic filler, one commercial variety being known as *Cerrobend*. This is a metal having a melting temperature lower than that of boiling water, so that sticks or beads of the material can be placed in a tube (with one end plugged) and the filler melted into place by placing the tube in a saucepan of boiling water.

parallel to the vice jaw, thus permitting the rod to be placed quickly in the jig.

The rod is initially cut oversize in its length so the position of the first bend is not critical. A line drawn on the square bar with a felt-tip marker serves to indicate the initial position.

For the second bend in the rod, the two dowels in the square provide a positive location for the rod and also set the position of the bend.

Remember the proverb and bend the steel while it is still red hot. Although there is some softening effect below red heat, it is a case of the hotter the better. It is possible to discern when the strength is returning to the steel, just as the work hardening of copper or brass becomes obvious during cold forming, and it is best to reheat and continue, rather than persist when the material is too cold.

Forging

If you are equipped with an anvil and the means to bring the work to red heat, there is no reason why items cannot be forged into shape. However, forging is not much practised in model work, due principally to the small cross-sections of material which are ordinarily used, and the consequent difficulty in maintaining them at a sufficient degree of 'redness' to allow the forging to proceed. It is a useful technique however, particularly for awkwardly shaped items which might otherwise need to be made from quite large cross-sections.

An example of a small part which might be forged is the latch for the locomotive reversing lever shown in Figure 7.38. This requires thickness in two directions at right angles which ordinarily would require much cutting away of material to create the shape required.

Although small, this item is an ideal candidate for forging since it can be cut out of a thin strip suitable to form the spade shape at the

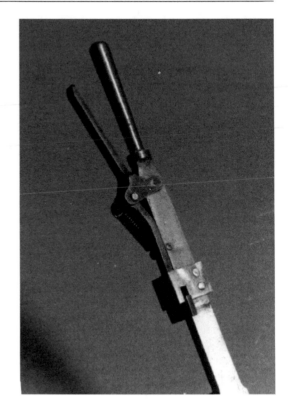

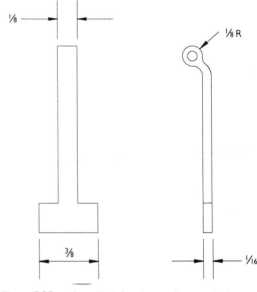

Figure 7.38 A forged latch for a locomotive reversing lever.

bottom and then the top forged to produce the flat at right angles to the bottom spade. No large anvil is required for such a small item since a small, hardened steel block properly supported in the vice is all that is required. If a support for the blowtorch can be arranged near the vice, the work can be held in the flame with a small pair of pliers and immediately transferred to the block when the end is nicely red.

The amount of metal to leave when cutting the top needs to be nicely judged, something of about the same volume being required as is contained in the desired shape. With a job such as this, you may find that the bar tends to fold over on itself rather than be flattened, but this does not actually matter since the folded-over parts weld together when hammered at red heat and the result is, therefore, the same as a simple flattening of the work. But, for the welding to be successful, the steel must be clean, and it must also be well into the red hot range of temperatures. So, 'strike while the iron is hot' and make each hammer blow actually do

some work on the material. It is no use tapping at it tentatively.

Forging can equally be used for larger sections of material such as are required for locomotive coupling and connecting rods, which generally have small cross-sections in the centre portion but large bosses at each end.

Such items can be made by 'bumping up' the ends of a small bar to form the bosses at the ends, or a large bar, suitable for the bosses, can be forged down to form the slimmer cross-section in the centre.

In 5-inch gauge (1$\frac{1}{16}$in. to the foot scale) coupling and connecting rods are relatively large items and forging becomes akin to 'real' blacksmith's work. A proper hearth and adequate source of heat are required, and the work must be packed round with firebrick or ceramic blocks and allowed to 'soak' adequately to attain the necessary temperature. A large, well-supported anvil is required, together with large hammers, and the component must naturally be struck while it is still red hot.

is most likely to be satisfied by a socket-headed screw. These are manufactured in high-strength (high tensile) steel alloys in all of the common threads and in a range of headed and headless types which equate with the normal types of screws, bolts and grub screws. These screws and bolts are provided with a hexagon-shaped socket in the head which takes a key or wrench which allows them to be easily inserted or removed. The three head types shown in Figure 8.2 can be obtained readily, being described as socket-head cap screws, grub screws and countersunk screws.

A common method of using cap screws is to counterbore the top of the clearance hole to allow the head to be recessed below the surface. The head depth must suit the thickness of the

Figure 8.2 Three types of Allen screw.

item to be secured, but if reasonably thick items need to be fastened by screws, the result of counterboring and recessing produces a neat arrangement, as Figure 8.3 shows.

These screws, known colloquially as Allen screws, are intended as a high-strength engineering fastener. They are consequently available in a very wide range of sizes, modellers' suppliers usually stocking BSF, BSW and BA sizes, in various lengths, from ⅜in. diameter down to 8BA. They are also available in a wide range of metric sizes from M20 to M1.4. They

Figure 8.3 Recessed Allen screws on a rear toolpost.

are generally available having a black finish and are ideal for the assembly of tools or when their high strength is needed, as for big-end bolts for IC engines, for example.

The hexagon wrenches are universally known as Allen keys.

Providing location for bolted fastenings

One important point about nut and bolt fastenings is that they are designed to clamp items together, not to locate them in some particular alignment. The hole drilled to accept a bolt or screw is consequently called a clearance hole. This is deliberately made larger than the screw in order to allow for tolerances in screw size and to permit slight positional errors in the placing of fixing holes in the components to be joined. The fixings themselves cannot be relied upon for location, only for clamping.

If positive location of the screw-clamped parts is a requirement, this must be arranged by separate means. For some components this is easily done by slight modification to the design, but for others, specially fitted pins or specially made bolts are required. Two examples are shown in Figure 8.4. The items shown comprise the upper and lower halves of a two-tool turret designed to hold commercial parting-off blades on a lathe. To provide location, the lower half is

Figure 8.4 The two parts of the tool clamp on a rear toolpost showing the locating register and alignment pins.

provided with a precisely turned spigot which locates in a boring in the toolpost body.

To prevent the upper and lower halves of the clamp from becoming disoriented, two ground silver steel pins are pressed into the lower half. These locate in accurately sized (reamed) holes in the upper half, ensuring accurate location of the parts.

This use of locating pins, or dowels, is very common, both as a means to ensure that no misalignment occurs when the clamping bolts are slackened off (as here) or to ensure that parts are fitted together only in the correct orientation. This latter use is simply arranged by placing dowels asymmetrically so that incorrect assembly is not possible.

Hardened and ground dowels are available commercially and can be fitted into accurately reamed holes in the items to be joined. Alternatively, silver steel pins may be used, since this material is accurately ground to size. It can, of course, be hardened also but this is not always necessary unless frequent removal and replacement of the items is envisaged. Indeed, it may be desirable to fit a hardened steel bush to mate with the dowel if very frequent movement or dismantling is envisaged.

The lower part of the toolpost turret, with its accurate register for location, exemplifies the usual approach to location of circular work. Provided that there is room, a register or spigot locating in a bore is frequently the best way to ensure accurate location. The rear cover for a steam locomotive cylinder is shown in Figure 8.5 and will be seen to have a shallow spigot turned on the inside which locates in the cylinder bore. This ensures concentricity between the bore and the central piston rod hole when the cover is assembled to the cylinder block, provided that both the central bore and the register, or spigot, are machined at the same setting in the lathe.

Figure 8.5 Cylinder covers are usually provided with a shallow register which locates the cover in the bore. This is essential for covers which are bored for the piston rod.

Bolts and fitted bolts

If items having other than minimum thicknesses are to be fastened using nuts and bolts, bolts should strictly be employed, rather than screws, the difference being that a bolt has an unthreaded portion of shank below the head. This portion is usually machined at the nominal diameter of the thread and is consequently liable to be closer to its nominal size than the threaded portion. A bolt is, therefore, capable

screws, together with two thicknesses of nut, known as full and half or standard and thin nuts.

Thin nuts are provided for use as lock nuts and are frequently described as such. For locking, they may be used in pairs, one to provide the fastening and the second locked tightly against the first to prevent slackening. Where there is room, it is usual to fit a full nut for fastening and to use a half nut for locking.

Liquid thread locking

Paint and varnish have been used industrially for the locking of small threads for many years. A thread-locking varnish has been the most common material used, this usually being painted onto the protruding thread of the screw after tightening the nut. This is perfectly adequate as a locking method since the varnish fills the screw thread (and any clearances between the nut and the screw at the outer end) and therefore prevents the nut slackening until adequate force is applied to it. The method is cheap, causes no surface damage and does not add to the number of items to be assembled, yet allows removal when required.

Specially prepared varnishes are not necessary since ordinary paints can be used, enamel or gloss being perhaps best since these have more 'body' than emulsion or undercoat. Rather than applying the paint or varnish externally, it is perhaps best to place a drop on the screw thread before its insertion into the tapped hole or before placement of the nut. Clearances between the two threads are then filled with paint and removal requires application of adequate torque to break the paint film.

Current techniques for metal joining have made available various grades of locking fluids having more precisely defined characteristics than a drop of whatever old paint happens to be available. *Loctite* is the most readily available range which offers medium- and high-

strength versions (*Loctite 242 Nutlock* and *Loctite 270 Studlock*, respectively) for placement on the threads before engagement, or *Loctite 290*, which is a penetrating adhesive for application to threads after assembly.

These fluids are members of a group known as anaerobic compounds, the group including several other fluids which form permanent or semi-permanent bonds between metals. All of these fluids are described in Chapter 9.

Shortening screws

One frequently required operation is the shortening of screws. The requirement arises because of the non-availability of short screws in the smaller sizes, these often being produced in only a limited selection of lengths. Shortening a screw is, of course, not a problem, but creating a nicely square end which will readily engage with a nut does frequently cause aggravation. The way out of the difficulty is to utilise a device which was once one of the first items made by the instrument maker's apprentice. This is the nut plate.

At its simplest, the nut plate is a piece of mild steel plate having in one end a few tapped holes to suit the range of threads which you normally use. The holes can be quite closely spaced and the plate should be no thicker than the shortest screw you think you will need. For the smaller sizes, such as the BA threads, this might be ⅛in. (3mm).

For shortening, a screw is inserted into a tapped hole and screwed in sufficiently to leave the unwanted thread clear of the plate. The plate should be sufficiently long that it can be held in the vice, with the end with the tapped holes upstanding sufficiently to use a hacksaw to cut off the excess thread and to use a file to smooth the ends. If the holes are tapped squarely in the ends of the plate, the end of the screw is automatically made square when it is filed off flush with the plate's surface. Removing the screw from the tapped hole removes most of the burrs that remain on the end, and a few strokes from a smooth file quickly produce a small chamfer.

The nut plate may be as elaborate or as simple as you please, from a piece of mild steel plate, as described above, to a stepped plate which allows specific screw lengths to be produced every time. It might be hardened, if you wish, but this is by no means essential, and the choice therefore rests with the individual. My plate, made from a strip of black mild steel, is shown in Figure 8.10.

Figure 8.10 A simple nut plate.

Rivets and riveting

General introduction

Until the development of reliable, high-strength welded assemblies, larger engineering structures were normally assembled by riveting together cut and shaped plates and suitable 'doubling' or strengthening pieces. The resultant structures possessed a degree of flexibility (which is beneficial) and enabled large assemblies to be created from relatively small plates. Hot riveting was the method frequently employed, red hot round-headed rivets being pushed into pre-drilled holes and their plain ends formed into a round head while still red hot. The round-headed shape provides a good clamping area, and forming the rivet while still red hot ensures that it can be expanded to fill the holes in the plates, and the head can be formed fully. The rivet also increases the pressure on the plates as it contracts on cooling.

In modelling, hot riveting is not necessary, but it is a requirement that the rivet fills the holes in the plates and therefore provides lateral location of the parts being joined. A good head also needs to be provided on the formed rivet.

In modelling, this is usually desired principally on the grounds of appearance, but a good head shape, matching the commercially produced head, also allows the required strength to be achieved, if this is necessary.

Rivets are produced in many forms – flat head, mushroom head, tubular, semi-tubular, bifurcated etc. – in a variety of materials and finishes. Those of interest to the modeller are likely to be countersunk or round-headed (usually called snaphead rivets) in steel, iron, brass, aluminium or copper.

Rivet spacing

In an engineering sense, a riveted joint needs to be correctly proportioned if it is to function correctly and realise the full strength potential of the materials forming the joint. When considering modelling, the basic engineering criteria seldom apply since the aim is to produce a miniature of some device which closely resembles the device itself, usually called the prototype. Rivet spacing and head size are therefore frequently determined by the prototype itself and the choice of size is removed. In

ing of the plates and butt strap(s) without distortion of either so that the silver solder can do its job of uniting the items to provide proper strength to the joint.

Unless you are particularly skilled in the design of pressure vessels, it pays always to follow a particular boiler design meticulously, particularly regarding the arrangements specified for any butt joints.

Basic procedure

One of the problems posed by the riveting operation is possible damage to the plates being joined. This is caused by the hammering which shapes the rivet head. This can have the effect of squashing the plates and hence stretching them. The effect is more noticeable if the plates are copper, soft brass or aluminium since these materials are very ductile and can readily be squeezed into alternative shapes.

If the plates to be joined are substantial, the effect may not be noticeable. This is usually also the case where the length to be riveted is short and the rivets are not placed too close to the edges of the plates. If the items to be joined are long, with closely spaced rivets, there may be significant stretching and considerable increase in length, especially for an item such as the locomotive running board shown in Figure 8.13.

In the case of this assembly, the running board itself is a robust 16 swg mild steel sheet, but it will be riveted to a brass angle since this was the only correctly-sized material available. The mixture of materials is unfortunate since the soft angle will distort easily during riveting.

Two points arise from the above. First, the rivets used should not be significantly harder than the plates to be joined. They are normally of the same material as the plates, or the softer of the two, if the metals are dissimilar. Second, for a long length such as the running board of

Figure 8.13 A model locomotive running board.

Figure 8.13, only one item should be drilled fully prior to starting the riveting operation.

If both are fully drilled, holes in the two items rapidly become displaced as rivets are successively inserted and their heads formed. It is essential to drill only the one component, say the running board itself initially, and then to drill the angle progressively as riveting takes place.

It is helpful to start at the centre of a long length and also beneficial to insert some rivets here and there, say every three or four inches (75mm or 100mm) along the plates, as an initial operation. This establishes the two parts in correct alignment and the effects of distortion are minimised.

Forward planning of this sort is required, since the brass angle will curve into an arc as the rivets are inserted, as shown in Figure 8.14, and without some holes drilled, and held, at least by temporary screws, progressive distortion will render the task of positioning the angle for riveting more and more difficult.

Distortion of the angle can, of course, be minimised by using soft rivets and by avoiding any excessive hammering beyond that which is absolutely required to form the head of the rivet. In this respect it is probably best if the commercial head is placed against the angle but since this brings the home-made head uppermost in this instance, it may be preferable to place the commercial head at the top and to

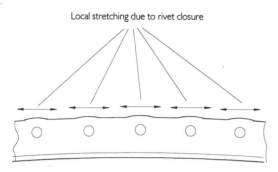

Local stretching due to rivet closure

Figure 8.14 Distortion during riveting.

deal with the distortion as riveting progresses.

For all riveting, it is important that the rivet is properly supported during the forming operation. This is especially important when riveting soft materials otherwise the force of the hammer blows may be taken by the plates and severe distortion will result. The riveting of soft copper, as in boilermaking, is frequently a problem in this respect and it is often preferable to use screws (but not brass or steel) for holding the plates together for silver soldering.

Riveting dolly

The operation of inserting and forming the head of a rivet is referred to as 'rivet driving', 'heading up' or 'closing up', this latter description describing the action of the rivet in bringing the plates together. With few exceptions, one side of the formed and closed-up rivet will comprise a round or snaphead. On occasions, this can simply be the commercial rivet head since, if the rivet is to be countersunk on one side it is the countersink which can most simply be formed. This can be done using only a suitable hammer, but a correctly shaped support for the commercial head will be required in order to allow rivet closure and preserve the head shape.

This support is usually called a dolly and traditionally takes the form shown in Figure 8.15.

A sturdy bar is drilled at some convenient point to allow the dolly to be inserted. This is a short length of hardened and tempered silver steel machined to fit the drilling in the bar and having a machined recess in the upper end into which the rivet snaphead is a good fit.

This form of support is not necessarily the best for the single-handed worker however, since the edge of the hard dolly will mark the work if it is not supported absolutely squarely in relation to the dolly during the closing up (hammering) process. It is frequently better to dispense with the dolly and to machine the recess for the rivet head directly in the bar itself. This allows the plates which are to be joined to lie directly on the bar's surface and there is little likelihood that they will be marked during riveting.

From descriptions in many books it would seem that the traditional method of creating the recess for the rivet head is to drill an appro-

Figure 8.15 The traditional form of riveting dolly.

priately sized 'starter' hole using an ordinary twist drill and then to hammer a suitable hardened steel ball into the drilling to create the required shape of recess. This doesn't seem very scientific even assuming that the correct size of ball is available, and it inevitably results in the ball flying across the workshop to be lost in some far corner.

A much better method is to make a cutter which cuts the correct profile directly, and avoids the need for hardened balls and hammering altogether. The most suitable form of cutter may be made from silver steel, turned to a suitable diameter and radiused on the end to suit the rivet head. This end radius, as machined, will resemble the rivet head (a spherical radius) and to create a cutting edge, roughly half of the end is machined or filed away to create a round-ended D-bit (see Chapter 15). The cutter should be hardened and tempered (pale straw) after which it may be used to cut the snaphead recess as a follow-up to drilling an initial hole using a twist drill.

The dimensions of the recess can be obtained from Figure 8.16 which shows the proportions of the common snaphead rivets. In the USA, the snaphead rivet is known as the button-head rivet.

Figure 8.17 Dimples for snaphead rivets, produced by drilling, in the end of a 1in. (25mm) square bar.

The purists will perhaps frown, but it is possible to grind the end of a twist drill to form the recess for the snaphead and those shown in Figure 8.17 were cut by this means. A set of three recesses is shown, two close to the corners of the bar and one placed equidistant from the sides and the end. This allows rivet heads to be supported close to obstructions on the workpiece (or other rivets) and allows the bar's top surface to be used in different situations to support the work.

When forming the recess, it is best to test the shape and depth by using a commercial head but care must be exercised in selecting the standard against which the recess will be judged since some rivets in a batch always have slightly malformed heads, so select the best out of half-a-dozen or so.

Closing up the rivet

The essential stages in closing up a rivet can be summarised as follows:

1. the drawing-together of the plates
2. the squashing of the rivet so that it expands into the holes in the plates

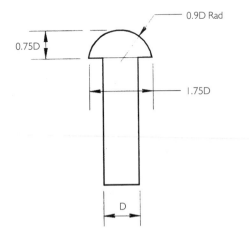

Figure 8.16 Snaphead rivet dimensions.

3. the forming of the head.

These essential steps must be performed in the above order. If the plates are not drawn together into intimate contact before the processes to swell and form the rivet are complete, the rivet will expand between the plates and the joint will not be sound.

All of the processes concerned with the closing up and forming of the rivet must be performed with the above objectives in mind and this influences the method and the design of punches used to form the head. In this respect it is essential that the rivet is properly supported during the closing operation so that hammer blows, or blows from a forming punch, truly affect the rivet and not the plates being joined. The more solid can be the support, the better the results are likely to be.

The simplest type of rivet closing is that in which the unformed end of a snaphead rivet is hammered into a countersink and afterwards filed off flush with the surface of the plate. This method is frequently used in situations in which the strength of the riveted joint demands a larger rivet (and hence rivet head) than scale appearance would demand on a model, such as the attachment of frame fixing angles, or horn-blocks, to model locomotive frames, as shown in Figure 8.18.

For such situations, the snaphead of the rivet is placed inside and the countersink formed on the outside of the frames. A plain, or ordinary, countersink is not ideal in these situations since it leaves a depression tapering away to nothing creating a very weak edge to the formed rivet. It is best to use an ordinary twist drill to form the countersink, creating a recess as shown in Figure 8.19 having a shallow, parallel-sided portion which provides increased strength for the formed head.

Before bringing the two items together, each must be de-burred so that good contact is achievable. The holes should have been drilled so that the rivets are a tight fit, and the rivets cut off, leaving enough material to fill the countersink. Cutting off must leave the rivet end reasonably smooth and burr-free since the first operation is to ensure that the head is seated correctly against the lower plate, and the two plates are in close contact. To do this, a brass or copper punch of the form shown in Figure 8.20A is used to punch the two items down on

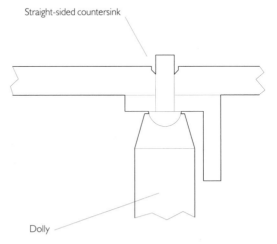

Straight-sided countersink

Dolly

Figure 8.19 The recommended shape of the countersink has a short parallel portion which gives strength to the formed rivet. The method of supporting the head of the rivet on a dolly should be noted.

Figure 8.18 The rivets holding the small angles to this buffer beam have been hammered down into countersinks and filed off flush.

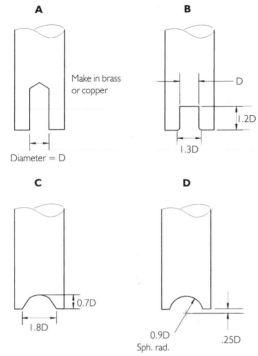

A

Make in brass or copper

Diameter = D

B

D

1.2D

1.3D

C

0.7D

1.8D

D

0.9D
Sph. rad.

.25D

Figure 8.20   A set of punches such as these is needed for forming a snaphead on the end of a rivet.

to the rivet head which is supported in a dolly, as described above. This ensures that the work is ready for rivet closing; the rivet head is properly seated in the dolly and in contact with the lower plate, and the plates are closely in contact.

For forming the end of the rivet to fill the countersink, an ordinary hammer is used, the first blow being directed straight down onto the rivet with the objective of swelling it within the holes. It should be tested at this stage to check that it is indeed firm in the hole.

With the rivet head once more supported in the dolly, successive blows are now aimed at the end of the rivet to spread it into the straight-sided countersink. Once each rivet has been dealt with in this way, the ends are filed off flush, when they should be virtually undetectable, as Figure 8.18 shows.

Forming snaphead rivets

If a snaphead is required on both sides of a riveted joint, the rivet must be closed up using a series of shaped punches. The object is still to ensure good contact between the plates, to swell the rivet in the hole and then to form the head. These processes cannot readily be achieved using a single, snaphead-shaped punch and the closing up must therefore be performed in several stages. A suitable set of four punches is shown in Figure 8.20.

The recess shapes shown are designed to perform the closing functions progressively. Punch A should be made in something soft such as copper or brass since it is designed to be used directly against the upper plate to force both plates together and to press the rivet into firm contact with the lower plate.

Once the rivet is cut off to length (see below) Punch B is used to spread the top of the rivet slightly, but more importantly, to compress it so that it expands in the hole. After use of this punch, it is wise to check that the rivet is firmly held in the plates and correctly seated.

Following the initial compression, Punches C and D are used to form the head progressively into the required shape.

The above description assumes that the rivet length standing clear of the upper plate provides the correct amount of material to create a satisfactory head shape. For the snapheads of the types shown in Figure 8.16, an upstand of between 1.6 and 1.7 times the rivet diameter is generally satisfactory, although the amount required is dependent upon the relative hardness of the rivet and the materials being joined, and the fit of the rivet in the holes. It is beneficial to treat the first one or two rivets as the means of determining the length required, assuming that the drilling has been (or will be) carefully carried out so that a standard fit between rivet and holes will be achieved.

For the common rivet sizes, the following upstands should be used as a guide:

Rivet Diam.	Upstand
$\frac{1}{16}$in.	0.10in. to 0.11in.
$\frac{3}{32}$in.	0.15in. to 0.16in.
$\frac{1}{8}$in.	0.20in. to 0.21in.
$\frac{5}{32}$in.	0.25in. to 0.27in.
$\frac{3}{16}$in.	0.30in. to 0.32in.

As noted above, the cutting off of the rivet needs to leave it relatively burr-free and clean to allow Punch A of Figure 8.20 to be used to close the plates and seat the head correctly. Cutting off with side cutters is convenient but does not satisfy the basic requirement for a cleanly cut end. Cutting off by this means cannot, therefore, be performed before the rivet is inserted into the hole and if performed afterwards will frequently be found to eject the rivet from the plates, rendering the use of Punch A obligatory. With side cutters it is difficult to judge the cut correctly to leave the required amount.

The best compromise is to use a drilled plate (like the nut plate described above in relation to the shortening of screws) perhaps with a spacer washer below the head to ensure the correct length. Side cutters (or similar) may then be used to cut the rivet and a few strokes with a sharp file will leave the end sufficiently burr-free to use directly. If very many rivets are needed, a hardened plate and a sharp cold chisel will be found to expedite the rivet-cutting process.

Press fits

Introduction

An extremely common way to assemble two cylindrical items in a semi-permanent way is to utilise a press fit. In this type of fit, a boring or housing is made just too small in diameter to accept the item which is to be fitted into it. The two must be pressed together with some force in order to allow the inserted item to enter the bore. To achieve this type of fit there must be 'interference' between the two items, this just being another way of saying that the bore is slightly too small.

In pressing the items together, there must be some expansion of the housing and some contraction of the inserted item. How much distortion occurs is dependent on the relative strengths of the two parts and on the amount of interference which has been allowed for in making them. The amount of interference also affects the difficulty, or otherwise, of removing the inserted item and whether removal may be required is one factor in determining the allowance for interference.

Press fits are typically used for fitting bearings, either plain bearing bushes or ball- and roller-bearing types. For plain bearings, the bush is made or fitted so that it provides running clearance for the shaft without further boring or reaming after assembly. Commercially produced plain bushes are available having precisely sized bores and outside diameters, specially for this type of fitting, but it is clearly important that the interference fit used is just right in the sense that the bush must not be squeezed in by any significant amount when fitted.

For the fitting of ball and roller bearings, both the housing and the shaft must be sized to provide the correct interference fit with the relevant part of the bearing. As for plain bearings, only a 'light' interference is allowed. This calls for accuracy in sizing both components so that the inner and outer of the bearing are fitted sufficiently tightly to prevent their rotation in the housing, or on the shaft, but the bearing is not distorted when assembled, hence taking up its internal clearances and damaging the balls or rollers, and the tracks.

An alternative use of interference fits is the permanent pressing together of two items which will not need to be subsequently disassembled. In these cases, a greater amount of interference is allowed in order to ensure no relative movement between the two items in service. In modelling, or model engineering generally, common uses for this type of fit are for fitting locomotive wheels onto their axles, and for the 'building up' of crankshafts, in addition to the ordinary fitting of bearings, described above.

The interference fit

To allow precise control of sizes and tolerances in industry, ranges of standard fits are specified by the various national authorities. These allow for the manufacture of shafts and housings with a range of tolerances for both which provide for various fits from loose clearance, through running and location to push and press fits. In some specifications, the description 'force fit' is used rather than press fit. For the fitting of bearings to their shafts and housings and permanent fitting together of cylindrical components, push and press fits are ordinarily required.

Industrially, it is valuable to establish separate tolerances for shafts and housings since items can then be manufactured independently and yet still guarantee that a specified fit will result when the parts are fitted together. Therefore, the published tables are more concerned with tolerances of the individual components, and the guidance provided does not immediately permit the recommended amount of interference to be deduced. Since both the hole and the shaft are allowed to vary from the nominal size, the resulting interference is a variable amount, depending upon the actual sizes achieved. The amount of interference allowed is also related to the diameter of the work and this must be borne in mind when making the parts.

As noted above, bearings must be fitted using only small amounts of interference in order that distortion does not result. The lighter push fits from the British Standard designation are used in these cases since the frictional drag created by a properly lubricated bearing is low. Push fits are also used in any situation in which one component is weak, for example in the form of a thin sleeve, like the commercially produced plain bearings. For these cases, an interference of .0005in. per inch of diameter (.005mm per 10mm of diameter) is normally recommended.

For a press fit, an interference of between .001in. and .0015in. per inch of diameter is normally satisfactory for those cases in which the fitted length is not greater than twice the diameter and neither component is in the form of a thin sleeve. The figure of .001in. per inch of diameter is that normally suggested as the maximum for good quality cast iron components, but for fitting model locomotive wheels onto their axles, or similar situations in which the components are reasonably robust, most authorities recommend an interference of .002in. per inch of diameter on the basis that movement of a wheel on its axle is, to say the least, highly undesirable. The same interference is also recommended for the pressing together of built-up crankshafts.

If you are used to working in the metric system, these figures translate to .01mm and .02mm per 10mm of diameter.

The figures for interference fits given above are those recommended for ferrous parts. Such fits are less successful for non-ferrous parts due to the softer (more elastic) nature of the materials. Consequently, if these fits are required, the above allowances should be increased by a factor of 2 in order to ensure a satisfactory joint.

The pressing operation

The assembly of press-fitted items can naturally be performed in a press. This usually comprises a one-piece casting or forging, consisting of base, column and head not unlike a bench drill in concept. The head carries a sliding sleeve which moves up and down at right angles to the base, driven through a lever or arm by a rack or screw arrangement providing considerable mechanical advantage. If a press is not available, the bench vice may be used instead, but is not so convenient since the two parts of the workpiece, together with any spacers or load-spreading plates, need to be juggled into position between the jaws before closing the vice onto the assemblage while holding all squarely.

Since the vice is not all that convenient a replacement for the press, it is sometimes better to use a hammer, together with suitable punches, to force the items together. Another alternative is to use a drawing-in bolt (or draw-bolt) and nut to pull a bush or bearing into its housing. This is only applicable to situations in which both items have a through bore.

Whichever method of pressing is used, care must be taken that the force is exerted exactly at right angles to the axis of the work and in the case of fitting ball or roller bearings, that the load is taken on the correct part of the bearing. This generally requires the making up of drifts, spacers or bushes to assist the pressing operation, to ensure, for example, that the inner member of the bearing is not used to press the outer into its housing, or vice versa.

If an assembly containing several axles and their bearings needs to be put together, the assembly sequence must be chosen so that it is not necessary to fit bearing inners and outers at the same time otherwise the forces required will be transmitted through the balls or rollers, thereby damaging the races. If shafts or axles need to be driven into race inners, either a soft-faced hammer must be used or a brass or aluminium alloy bush should be used to avoid damage to the axle end.

The fitting of plain bearing bushes or sleeves is naturally more straightforward since only one pressing or pulling operation is required and in many instances a simple bolt-and-washers arrangement will suffice.

Additional security

Properly executed, interference fits in ferrous components are perfectly suitable for built-up crankshafts, for holding locomotive wheels onto their axles or other similar, high-torque situations. However, if you lack faith, or have a particularly heavy-duty application in mind, a little additional security may be desirable. One way that this can be provided is by pinning the items together by use of a radial pin passing through both. An alternative is the use of pins or screws fitted partly into both items.

An example of this method of fitting is shown in Figure 8.21 which is the driving pulley cluster for a small lathe. It comprises three parts; the steel driving disc shown, a cast centre which incorporates a drive gear, and a two-step pulley casting. The steel disc is provided with a hole in which the driving pin which passes through the bull wheel locates. The disc is held to the pulley cluster (which is fitted firmly to the cast iron centre) by four countersunk screws. To ensure that adequate torque is transmitted, the press fit of the disc on the centre is reinforced by two parallel pins pressed into holes drilled half in the disc and half in the centre piece. The pins were made a close fit for these holes and were driven (hammered) into position. They provide a positive location for the disc on the centre and reinforce the press fit of these two items which would otherwise be relatively weak because of the use of a thin disc due to lack of space.

Alternatives to the press fit

As described above, the allowance for press fits in ferrous components for ordinary duties is normally something between .001in. and .0015in. per inch of diameter. This places quite tight tolerances on the associated machining processes and means that extremely careful work is required when making both shaft and housing. Very small departures from nominal sizes can easily render an interference fit impossible to press together or, at the other extreme, so loose as to have no interference at all. Fortunately, it is possible to assemble such items as the pulley cluster of Figure 8.21 by gluing, using one of several adhesives which are now available.

The assembly shown has three components, as described, plus the small pins providing the positive location between the steel disc and the centre. Adhesives may be used for all of the joints; pulley cluster to centre, disc to centre, pins into drilled holes and even screws into pulley cluster. Since some adhesives provide adequate strength even when radial gaps are as large as .004in. or .005in. (.10mm to .13mm) the machining of the various diameters can be undertaken assuming wider tolerances, subject to there being sufficient accuracy to maintain the various items in reasonable alignment and concentric to one another. There are particular benefits with the two securing pins since the holes for these may simply be drilled and the pins pushed in after coating with adhesive. This is much easier than attempting to create short, blind holes of sufficient accuracy to allow the manufacture of pins having a carefully controlled amount of interference.

A further method of assembly which might also be suitable in particular instances is the use of soft or hard solders. Although soldering is not especially suitable in the case of the assembly shown in Figure 8.21, it is a valuable means of metal joining. Since the use of both solders and adhesives requires the introduction of the

Figure 8.21   The driving gear cluster for a small lathe, showing the axial pins which transfer the drive from the pulley.

binding material into the joint, and both methods generally require some clearance between the parts, the use of solders and adhesives is described in Chapter 9.

Semi-permanent attachment

Introduction

It hardly seems to be covered by the chapter heading of 'Metal joining', but there is a frequent need for cylindrical parts to be attached to one another so that torque can be transmitted between one component and another, and there remains the possibility of simple separation of the items. So, something having the assembled characteristics of a press fit is required, but without the need to apply great force when separating the mating components.

Wheels, arms and sector plates fitted to circular shafts immediately come to mind, where a firm fit is required, and good transfer of applied forces, but the fitting must incorporate the possibility of later removal and replacement without damage to the parts, or spoiling of the fit. In these instances, it is usual to ream or bore the hole for the shaft or axle accurately to size so that the fitted item has a push fit onto its shaft. Additional security is then provided by using screws, pins or keys to resist the anticipated turning forces.

The items used to provide the additional security may be fitted on the radius of a circular section through the two items and are then described as being 'radial'. Alternatively, they may be fitted parallel to the axis of the axle or shaft, in which case the description 'axial' is adopted.

Radial screws and pins

The simplest form of radial fitting is the simple grub screw which is used to hold a bored sleeve onto a shaft. The sleeve is drilled and threaded to accept a screw which is inserted to bear on the shaft and lock the shaft and sleeve together. To provide additional security and allow more force to be transmitted, it is usual to drill a shallow dimple in the shaft, into which the end of the screw can locate, to provide a more positive drive.

This type of fixing is not only used for such simple things as control knobs, but is frequently the method by which steam engine eccentrics are fixed to their shafts. Eccentrics need to be set to the correct position after assembly to their shafts and the process of setting is a repetitive one. Consequently, the screws are not normally tightened fully until several adjustments have been made.

It may be difficult to hold the items together sufficiently firmly to drill a dimple at the required location, without the possibility that the setting will be lost, and in such cases Allen grub screws having shaped ends are used, which provide additional security. Some types present a circular edge to the shaft, and since the screw is hardened, the end 'bites' into the shaft and creates its own positive location. This type of fastening is illustrated in Figure 8.22, together with a flat-ended type which is used if the shaft has a machined flat.

A more secure attachment can be provided by using a radial pin to secure a boss or collar to a shaft. Three different types of pin are used for this duty: a plain, parallel pin, or a taper pin, both of which are fitted into a carefully sized hole which is produced by a process known as reaming (see Chapter 10) or a hollow, hardened pin which is fitted into a plain, drilled hole.

These latter pins, known as roll pins, are naturally the easiest to fit since they require only a drilled hole, which is quickly and easily made. Roll pins are rolled (from a type of spring steel) so that they are just larger than the nominal diameter. Being springy they can be driven into a suitable hole and just as easily be driven out, yet they are quite secure when in position in a hole. The amount of turning force (torque) which can be transmitted is limited by the fact that the pin is hollow, but if the load is

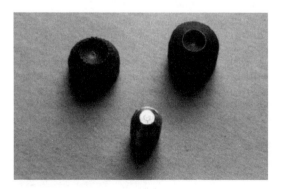

Figure 8.22 Illustration of the end shapes of Allen grub screws.

177

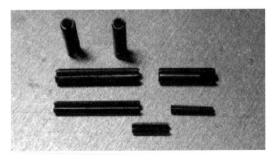

Figure 8.23 A selection of roll pins.

not severe, the roll pin is ideal. Figure 8.23 shows a range of roll pins.

If greater torque needs to be transmitted, a solid pin is required. This might be just a length of silver steel or can be a taper pin. Both are accurately ground to size and can provide a secure and solid attachment provided they are fitted to a correctly sized and well-finished hole. In the case of a taper pin, the hole needs to be finished with a specially produced reamer whose taper matches that of the pin. This requires a hole to be drilled sufficiently large for the end of the reamer to enter the hole. The reamer is then put in the drilling machine in place of the drill and the hole enlarged until the length of the hole is tapered and the pin will enter to the required depth.

The shallow taper on the pin and the hole is designed so that the pin jams itself into the hole when it is 'seated' by a sharp tap from a hammer. Any surplus pin projecting from the hole can be cut off with a hacksaw and the end of the pin rounded or flattened with a file, since the

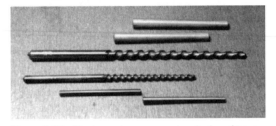

Figure 8.24 Two taper pin reamers and a selection of taper pins.

Figure 8.25 The fitting of an outside crank on a model steam locomotive. The ends of the taper pin have not yet been cut off.

pins are not especially hard. Figure 8.25 shows an external crank on a model locomotive which is held in place by a taper pin. This pin has not yet been cut off at the ends.

Two taper-pin reamers and some taper pins are illustrated in Figure 8.24.

Similar considerations apply to the fitting of parallel pins as to taper pins. The pin must be nicely parallel and the hole must be correctly sized and have a good internal finish so that the pin fits properly, without shake or sloppiness. This means finishing the hole to size by reaming after drilling an initial hole just large enough for the reamer to enter.

Obviously, it is necessary to drill both items together when fitting any sort of radial pin, and it might also be necessary to identify which way round the items should be fitted together in some instances, although a taper pin naturally only fits its tapered hole when entered from the correct side.

Axial pins and keys

The type of pin illustrated in Figure 8.21 lies parallel to the axis of the assembly and is correctly described as an axial pin, or key. For the placement of this type of key, the items are first assembled and the hole for the pin is drilled by starting on the junction between the two parts and drilling down in the usual way. If the pin is to be glued into position, it is adequate to use a simple drilled hole since a little clearance is required for the adhesive and little other attention needs to be paid apart from cleanliness.

If the pin is to be held by friction, it must be a tight fit in the hole and this must be reamed or otherwise finished to the correct size, so that the pin can be driven in. Provided that a method of withdrawal is provided, there is no reason why the pin cannot be removed, or the two fitted items pulled apart, should it be necessary to separate the components of the assembly.

Drilling the hole for the pin is straightforward if the assembly can be positioned under the spindle of the drilling machine and the two components which are to be drilled are of similar hardness. If the shaft is steel, say, and the fitted item is a soft material, such as aluminium, the drill is likely to 'favour' the aluminium and wander off the required line. It is effectively pushed away by the hard steel into the soft aluminium and the result is a disaster.

In these instances, it is better to machine the two parts separately and the usual type of key is rectangular in cross-section. This allows a keyway to be machined in both parts, and the only dimension which is critical during these

Figure 8.26   The leadscrew on my lathe is driven by a changewheel which engages a key set into the end of the leadscrew shaft.

operations is the width of the slots since the key does not need to take up the full depth which is available.

An example of a keyed fitting of this type is shown in Figure 8.26 which shows the driving end of the leadscrew on a lathe. The leadscrew can be driven by a gear fitted to the end, and is fitted with a half-moon key, known as a Woodruff key, which fits a machined recess in the end of the shaft. The gear has a corresponding square recess (a keyway) machined in it, which fits the upstanding part of the key.

Further illustrations of this arrangement are shown in Chapter 16, together with examples of straight rectangular keys which serve a similar purpose.

9

Metal joining 2 – solders and adhesives

Types of solder

Solders are described as hard or soft. Soft solders are alloys of tin and lead, usually around 60 per cent tin and 40 per cent lead, which melt at about 185°C. Both constituents are soft, as are the resultant alloys. Colloquially, the tin–lead alloys are normally described simply as solder, whereas hard solders are referred to as silver solders.

Hard solders are alloys of silver and other elements having melting points in the range from 610°C to roughly 760°C.

Another commonly employed method of metal joining is to use a brass comprising 50 per cent each of copper and zinc as the solder. This process is known as brazing and the brass wire used for the purpose is known as brazing spelter. This is not a commonly used process in the amateur's workshop, since brass becomes liquid only above 875°C and a great deal of heat is required.

Brass is also not ideal for use as a solder, and alloys of silver, and other elements, with copper and zinc, create alloys which become liquid at lower temperatures than brass and which

have better properties generally for this type of work.

Due to the long use of the term brazing, this persists in modern parlance and the term silver brazing is used, as well as hard soldering and silver soldering, all to describe the same operation.

Solders are most commonly supplied as rods or wires. These are used with separately applied fluxes which may be in the form of liquids or pastes, although one very common form of soft solder is manufactured as a hollow wire which is filled with flux.

Use of solders in modelling

Soldering is a very useful technique for modelling, since it permits assemblies which represent forgings or castings on the prototype to be built up from smaller components, allowing complex shapes to be produced. Since the solder used for soft soldering is an alloy of weak metals, the resulting joints do not have great strength, but if an assembly is decorative,

Figure 9.1 Soldered fittings inside a water tank filler lid.

rather than functional, soft soldering may be adequate.

Since a solder can be made to flow throughout a joint, it is a useful way to complete an assembly which needs to be sealed, and if the item is small, the solder might be used both for the purposes of assembling the parts and forming the seal. This might be the case for small water, petrol or oil tanks for models.

The ability to make assemblies is exemplified by the water tank filler lid shown in Figure 9.1. The lid has a locating hoop on the upper surface which locates the closing hinge strap, and this has been pushed through two drilled holes and then soldered into the lid, on the inside. The hinge strap is held into the lid by a headed brass pin which is secured into the lid by a brass washer soldered to its inside end. Although the lid is functional, no great strength is needed from these joints and they continue to serve the function after some years of use.

The tank filler has been soldered into the tank top for this model, and the tank was sealed by solder, although its component parts were screwed together.

On the same model, the rear lamp irons shown in Figure 9.2 have an integral rib on the left-hand side. On the model, this has been rep-

resented by a short length of copper wire soft soldered into a shallow sawcut on the upstanding part of the iron.

The examples of soldered items described above are not highly stressed joints, and they have been made by soft soldering. If a strong assembly is needed, it should be silver soldered, as noted above, since silver solders are much stronger than soft solders.

The buffer stock of Figure 9.2 is one item which might be made by silver soldering together its component parts. To make the stock, a square of mild steel can be silver soldered to a short length of round steel bar, leaving just the minimum amount of material to be removed from the face of the square flange, and the outside diameter of the round bar. Making the complete stock from a length of square steel requires much material to be machined away, material which, after all, one has had to purchase, and the silver-soldered assembly is a more viable way to create these items.

There are other tasks for which silver soldering is the natural choice, principally the manufacture of model boilers and their fittings. The silver solders used have high melting points, making silver-soldered assemblies suitable for

Figure 9.2 Details on the rear bunker of a 5-inch gauge pannier tank. The lamp irons have a bar soft-soldered to their sides and the buffer stock is the type of item which might be made by silver soldering a square plate to a short stub of round steel bar.

use with high-pressure (and therefore high-temperature) boilers and steam circuits.

The soldering process

Soldering consists of melting a material (a solder) into a clamped-together joint. The solder forms an alloy with the metals in the joint binding them together. The intention is not usually to fill gaps in the joint, and the items to be joined must be in close contact.

When making a soldered joint, the solder melts at a lower temperature than the metal items to be joined. To introduce the solder to the joint, it is heated until the solder melts when touched on it. The solder flows through the joint, forming a bond with the hot materials and uniting them when the heat is removed, allowing the joint to cool and the solder to solidify.

To achieve an adequate bond between the solder and the metallic items in the joint, the metals must be clean. The components in the joint must be de-greased and all tarnish removed before the soldering operation commences, and freedom from formation of tarnishing compounds must be maintained during soldering. Unless the metal is exceptionally dirty, simple mechanical cleaning by use of emery cloth is all that is required, but any deposits of oil or grease must be removed first by use of a suitable solvent. Cleaning must be performed immediately prior to soldering and once cleaned, the metal should not be touched.

Cleanliness during the soldering operation is maintained by use of a flux. This is designed to be liquid at the soldering temperature so that it flows over the surfaces within the joint and prevents the formation of oxides and tarnishing compounds. Pre-cleaning, and the use of a suitable flux, are vital to the production of sound soldered joints.

Solders can be obtained which melt at temperatures lower than that of boiling water (100°C), or at around 200°C. There is also a group of solders which melt between 600 and 800°C, and it is possible to use a type of brass as a solder. This melts at 875°C. The range of solders is very wide, and the requirements for heating the joint can be just an electrically operated soldering iron, or may require the use of a propane or oxy-acetylene torch.

Basic procedure

Essential steps

The process of soldering is quite straightforward if carried out correctly, but to be successful, the process must be carried out as follows:

1. The material in the joint must be cleaned, and must remain clean throughout the heating and soldering operations. This means adequate pre-cleaning and the use of an appropriate flux.
2. The joint must be firmly held together during soldering and the gaps must be within the range recommended for the solder used. In practice, this generally means close fits.
3. The metal in the joint must be hot enough (right through the joint) to melt the solder which is in contact with it.

This last point is probably the most important of all. If the job is not sufficiently hot, solder cannot flow through the joint.

The normal method of carrying out the soldering operation when using solder in stick form, is to heat the items to be joined, testing their temperature periodically by applying the solder stick to them, just outside the area being heated, until the solder melts and flows when touched on the surface. In this respect, it is helpful to heat the joint from one side, but to apply the solder stick to the other, on the basis that this ensures that the joint is hot right through.

The close mechanical fit of the parts, and the cleanliness of the metal must be maintained during the whole operation. If the metal is dirty, solder cannot flow over the surface since the dirt acts as a barrier, preventing contact between the solder and the base metal which, in turn, prevents the formation of a sound joint. If the gap between the items to be joined is too large, the solder will not form a bond since it is not generally designed to bridge large gaps.

Heating the work

Soldered joints made by model engineers generally require quite large amounts of heat to bring the work to the temperature at which the solder will melt when touched on the surface. A small silver-soldered joint requires the work to be brought to red heat, and a soft-soldered joint might be, for example, to seal the water tank for a large-scale tender or locomotive.

A blowlamp (or blowtorch) is likely to be the heat source. The most convenient general type is the propane torch since a range of burners is available, capable of burning gas at different rates and of providing large or small amounts of heat. A description of a typical torch/burner system is given in Chapter 4.

Small silver-soldered joints might be made on a small tin-box hearth, of the type illustrated in Figure 4.24, but the assembly of a large structure such as a locomotive or traction engine boiler will require a metal bench, or brazing hearth of the type shown in Figure 4.23, in view of the large amount of heat which is required for the later stages of assembly.

There is also a need for plenty of room around the hearth in view of the large amounts of radiated heat from the structure and the throw of the flame produced by the larger burn-

ers. It goes without saying that there should be no combustible material adjacent to the hearth.

Holding while heating

Holding the pieces to be joined firmly together during the heating and cooling process is vital. Larger items are usually screwed or riveted together, which provides adequate security, but the components of smaller assemblies need to be located firmly together, otherwise movement can occur, especially as the water-mixed fluxes used for silver soldering dry out. The water is driven off as steam, and the flux melts as the temperature rises, and these events can easily displace small items which are not adequately located.

For smaller assemblies which are to be soldered, the basic requirement is for some simple, cheaply made clamps. If a blowlamp or torch is being used to heat the job, the clamps are subjected to exposure to the naked flame, and they may discolour. They are also likely to become contaminated with flux, and are rapidly spoiled. A few clamps should be made up in the workshop and reserved especially for use when soldering.

For soft soldering, wooden spring clothes pegs can be used. They are ideal for use as temporary clamps, especially once their ends have been cut away, but they are not suitable when using a naked flame and must be reserved for small jobs which can be heated by a soldering iron.

There are also some small clamps which go under the name of curl clips. They are ideal for holding smaller items, or for use where space is limited. One blade comprises two parallel arms but the other has a cross bar near the outer end. The clips are absolutely ideal as clamps and, although metallic, do not have too large a mass to prevent their use for soldering when an iron is used. Being made in aluminium, or an aluminium-based alloy, their use for silver soldering is restricted since a propane burner will quite easily melt these materials, but they are occasionally available in stainless steel and this type can be extremely useful.

Soft soldering

Common types of soft solder

Soft solders are alloys of tin and lead, usually in the proportions of 60 per cent tin and 40 per cent lead. The actual proportions do differ between different grades of soft solder, conferring different characteristics on the alloys in the region of the melting point.

Most solders (and this applies to silver solders also) do not have a melting point in the sense of a single temperature above which they are liquid and below which they are solid, although it is possible to prepare alloys which do have a single, identifiable melting point. Such alloys are called eutectic alloys and a soft solder comprising 62 per cent tin and 38 per cent lead is eutectic and melts at 183°C.

Alloys containing other proportions of tin and lead do not have a single melting point but are said to have a melting range i.e. two temperatures are specified. The lower of these is called the solidus and is effectively the freezing point since below this temperature the alloy is solid. The higher temperature, the top of the melting range, is called the liquidus since above this temperature the alloy is liquid. Between solidus and liquidus the alloy is in a 'pasty' state – neither fully liquid nor solid.

For applications in which good spreading of the solder is required, that is, where the solder is required to flow readily through the joint, a short melting range is required. This is produced by utilising proportions of the alloying elements close to the eutectic point, that is,

close to 62 per cent tin. For example, 64 per cent tin, 36 per cent lead has a melting range from 183°C (solidus) to 185°C (liquidus) whereas 60 per cent tin and 40 per cent lead produces an alloy having a melting range from 183°C to 188°C.

Solders with a long melting range are very useful when gap filling or surface filling are required and were used for car body filling before the days of glass-reinforced resin fillers and also for making 'wiped' joints in lead pipes for plumbing. Reducing the tin content to 50 per cent produces an alloy having a melting range from 183 to 212°C and an alloy containing only 30 per cent tin, with 70 per cent lead, has a melting range of 183 to 255°C.

As might be expected from the above descriptions of typical solder characteristics, a wide range of alloys is available having differing properties. The two major classifications are tinmans' solder, having a short melting range, and plumbers' solder, intended for wiped joints and for use as a body filler, and therefore having a wide melting range. The 30 per cent tin alloy referred to above is of this latter type (183 to 255°C melting range).

The most familiar form of tinmans' solder is cored solder used for electrical work. This comprises a hollow wire into which a non-corrosive, rosin-based flux is loaded so that solder and flux are introduced into the work simultaneously. Soft solders are also available as solid sticks or bars. These are used with separately applied fluxes.

High-temperature soft solders

There is frequently a need to carry out two or more soldering operations on the same assembly and there is value in having solders available with quite different solidus temperatures.

The addition of a small amount of silver to a tin–lead alloy significantly raises the solidus temperature, creating a high-temperature soft solder which is not particularly expensive. The alloy most frequently stocked by model engineers' suppliers is *Comsol* (from Johnson Matthey Metals Ltd.) which has 1.5 per cent silver, the remainder being tin and lead. This alloy is eutectic, passing from the solid to the liquid state at 296°C. It is intended for the soldering of armature windings on electric motors but its availability makes possible two-stage soldering using tinmans' solder as the second stage.

Comsol was also occasionally recommended for the final caulking of stays in copper boilers which have otherwise been assembled by use of silver solder, but this practice is not now very common.

Low melting-point alloys

By incorporating a quantity of cadmium or bismuth into a tin–lead alloy, the melting range (or indeed melting point) can be lowered by about 40°C. An alloy containing 50 per cent tin, 32 per cent lead and 18 per cent cadmium is eutectic and has a melting point of 145°C. An alloy of 49 per cent tin, 41 per cent lead and 10 per cent bismuth has a melting range of 142 to 166°C.

Alloys of these types are not generally available from local suppliers and modellers usually understand the phrase 'low melting-point solder' to mean an alloy used for soldering white metal castings which themselves have a melting point of around 200°C. Clearly, the use of normal tin–lead solder on these alloys means there is a very real danger of melting the castings. Consequently, fusible alloys (their commercial description) are utilised for joining white metal components and are much used in small-scale railway modelling.

Alloys are produced having different melting ranges, utilising tin and lead in differing proportions together with bismuth and/or cadmium. Many alloys bear the names of their inventors. One of the best known is *Wood's Metal* which contains 50 per cent bismuth, 25 per cent lead and 12.5 per cent each of cadmium and tin. The most interesting feature of *Wood's Metal* is its melting range which is 70 to 72°C. It thus melts in not-quite-boiling water.

Several of the fusible alloys melt at temperatures lower than 100°C and are used for filling pipes and tubes before bending due to the relative ease with which they may be melted into place and afterwards removed. A commercial version of a low melting-point alloy, sold specifically for pipe bending, is *Cerrobend*.

These alloys are well adapted to the soldering of white metal and they are produced in stick form and sold as low melting-point solders under various commercial names. Their use as solder demands a fairly strong flux since the molten alloy tends not to wet the base metal very well. It is usual to employ an acid flux (see below) when using bismuth-based alloys. Solder producers generally market a suitable liquid flux which is sold under a commercial name.

Alternative solders

Over the last decade or so, small-scale railway modelling has undergone something of a revolution and there is much emphasis nowadays on really fine scale models. This desire for a scale appearance in the smaller scales has resulted in the extensive development of kits for locomotives, wagons, buildings, signals and accessories of all sorts, the parts for which are etched in thin sheets of brass or nickel silver.

The wide availability of etched kits, which are best soldered together, has resulted in a demand for a wider range of solders which can be used for kit assembly. There is naturally a need for solders with different characteristics and melting points (or melting ranges) and the increased demand for specialised products has resulted.

To avoid the need for a metallurgical degree in order to determine the best solders to use, suppliers have developed simple means of identification and generally publish cheap booklets which describe the solders and the related fluxes in detail.

One supplier whose products are widely available markets the following soft solders. *Number 70* described as 'very low melting point', *Number 188* described as having 'a short melting range', *Number 224* described as having 'a wide melting range' and *Number 243* described as 'containing silver'.

You need to consult the supplier's booklet to determine exactly what the difference is between the four types, but it seems likely that the names actually identify the melting points, or allude to the liquidus temperature of the solder.

If you have need of these specialist solders, you are more likely to find them in a model railway stockist, rather than a model engineering supplier.

Forms of solder

For most soldering work, a solder having good spreading characteristics and a short melting range is required. For soft soldering, this is satisfied by tinmans' solder. The most useful form is a 60–40 tin–lead alloy in wire form with a number of cored longitudinal holes filled with rosin-based flux. This is known as cored solder. In use, flux and solder are introduced into the joint almost simultaneously, the flux arriving first since it flows at a slightly lower temperature. The flux is corrosion free and the residues are readily removed using white spirit.

Cored solder wire is available in different diameters, denoted by swg numbers, 18, 22 and 24 swg being readily available. It is best to use the thinnest wire which is available, preferably 24 swg (.022in., 0.56mm). In any event, it is the cheapest per unit length and since it is easy to introduce too much solder to the joint, the small diameter is to be preferred. Cored solder should be bought in as large a quantity as can be afforded since it is expensive in small quantities.

If a large job demands a much larger diameter of solder, a stick of uncored tinmans' will probably serve, since this is available in rough-cast form about ³⁄₁₆in. (4.5mm) square, but it must, of course, be used with a separately applied flux. For most modelling, a 16 swg cored solder will generally be adequate.

Solders are also available in the form of paint i.e. a suspension of solder particles in a flux base. These are excellent for small jobs, providing, as they do, a means of introducing the solder before the joint is assembled. For larger joints, their disadvantage lies in the fact that much of the fluid is flux, and there may be insufficient solder present to effect a really sound joint. Paints are considered in more detail below, under the heading 'Getting solder to the joint'.

Tinning

Tinning is a method of preparation which is used extensively for the leads of electronic components and printed circuit boards.

Tinning simply means bonding a layer of solder to two pieces of metal before bringing them together to make the joint. The layer of solder is microscopically thin so it does not prevent the correct butting together of the two parts. Solder is, however, placed in the joint before starting. The advantage of tinning is that the solder does not readily tarnish and the process of making a subsequent joint between two tinned components becomes a quick and simple operation.

The actual tinning is carried out simply by heating clean metal, possibly already pre-fluxed, melting solder and flux on to the surface and wiping off the excess solder whilst it is still molten. Rosin-cored solder is very convenient for this, but if a more active flux is required, or if it is desired to use a non-cored tinmans' solder stick, the flux must be applied before heating commences. The wiping off of the excess solder can be done with almost any damp, clean cloth, but a piece of damp synthetic sponge, as used in soldering iron stands, is essentially lint-free and leaves no fibres behind to impede the subsequent soldering.

The advantage of treating the parts separately in this way is that the solder is applied, and the bond with the base metal is formed, while all is in sight. Cleaning of any flux residues from the tinned parts is also easy before the joint is assembled and a more vigorous flux can be used for the tinning operation. Since the tinned surface is not very susceptible to tarnish formation, the tinning also assists maintenance of a clean surface at the time of soldering, which is not the case for brass, for example, unless the cleaning is carried out immediately prior to soldering.

The tinning operation produces a very thin layer of solder bonded to the base metal which for all practical purposes can be assumed to be non-existent (dimensionally, that is). When two items have been tinned in this way, it is only necessary to coat them with flux, press them together and heat up the pair until the solder melts, adding just a little additional solder to the joint.

Fluxes for soft soldering

Cleanliness during all soldering operations is maintained by use of a flux. This dissolves any residual tarnish from the metal (which must be well cleaned first) and also prevents formation of oxides on the surface of the molten solder. The selection of a flux depends principally on the metals to be joined. Fluxes are classified into different chemical groups and may be organic or inorganic, acids, salts, rosin-based and so on. Classifying each as a good or bad flux requires an estimate of its tarnish-removal capability, its corrosiveness and its stability and effectiveness over the range of temperatures at which it is expected to be effective.

Acids tend to have good temperature stability, be good or very good at tarnish removal but have a high or moderate corrosiveness. Removal of residues after soldering is very important.

If no corrosion can be allowed, as for electrical work and much modelling, a rosin-based flux is normally used. Such fluxes have relatively poor temperature stability and have only fair tarnish-removal properties. Nevertheless they are used extensively for electrical and electronic work where their non-corrosiveness is of supreme importance and where tinned surfaces are normally used. These are much less susceptible to the formation of oxides than the base metals and do not require the use of such an active flux. Tinning is an important way in which the strength of a soft-soldered joint can be improved and since it brings also a freedom from the formation of surface contaminants, it is a valuable technique to employ.

Non-corrosiveness of the flux is also an important consideration for modelling, so rosin-cored solders and rosin-based fluxes are frequently used. Due to their relatively poor tarnish-removal properties, the base metals must be well cleaned, or tinned, before soldering begins.

The main constituent of rosin-based flux is water-white rosin. This is distilled from pine sap and consists of 80 to 90 per cent abietic acid (also known as sylvic acid) which is a mild organic acid. It has only poor tarnish-removal properties and will not penetrate a heavy tarnish layer, nevertheless it works efficiently to protect tarnish-free surfaces during soldering and can be considered as corrosion-less. In a practical rosin-based flux, additional activators are sometimes included to improve the tarnish-removal properties and this gives rise to the term 'activated rosin flux'. These are the most widely used fluxes for electrical work and are used for forming the flux cores in cored solder.

If rosin is overheated, it turns a dark brown colour and loses most of its tarnish removal properties. If a problem is experienced in making a joint, the result is that the heating may continue for longer than usual as an attempt is made to make the solder 'take', causing the flux to overheat thereby rendering it useless from the point of view of tarnish removal. If the joint is allowed to cool and a second attempt made to solder it without first cleaning all of the overheated flux from the base metal, the overheated (and useless) flux cannot perform its function of tarnish removal and a failure again results. As can be imagined, the effect can be cumulative. If a problem of this sort is experienced, it is best to separate the items, clean down to new metal completely and start again. Some rosin-like fluxes are now available which have a higher charring temperature and this allows some latitude if there is a likelihood of overheating.

Residues from rosin-based fluxes should be removed by using white spirit or methylated spirit since they are not water-soluble as are the residues of some other fluxes.

For soldering metals which readily form oxides, such as copper, brass and steel, a flux which is more active than rosin is desirable. If the work can be thoroughly cleaned afterwards, the flux known generally as 'killed spirits' can be used. The active ingredient in this is zinc chloride (formed by dissolving zinc in hydrochloric acid) and although it is highly corrosive it can be neutralised by washing thoroughly with water. One commercial form of killed spirits is marketed as *Baker's Fluid*.

Provided that the joint is accessible and can be properly washed, a flux based on zinc chloride is safe to use. If making closed assemblies, such as water tanks, it is best to use a really active flux (for copper or brass) only for an initial tinning operation on the plates, afterwards washing thoroughly before soldering the assembly together using a less-corrosive flux.

Another active flux is phosphoric acid which is used mainly for stainless steel. Various commercial fluxes for this material are based on this acid, but since it is not very efficient at wetting the surface it is sometimes combined with other chemicals and supplied in the form of a paste. Commercial fluxes have developed greatly over the last few years and are now much more widely available. They can be relied upon to give good results when used over the recommended temperature range and with the metals for which they are designed. Some are corrosive, but many are water-soluble and corrosive residues are simply removed by washing. On no account should the washing or neutralising operation be omitted.

Silver soldering and brazing

Hard solders

Hard soldering can be considered as two separate operations – brazing and silver soldering. Nowadays, brazing is not much practised but it is nevertheless a useful method of metal joining and it was at one time widely used for joining steel components. Brazing simply consists of using a 50–50 brass wire (brazing spelter) as a solder, melting it into a clamped-together joint, in the presence of a flux. Borax was traditionally used as the flux used for brazing.

Brazing spelter has almost entirely given way

to alternative hard solders, the most common of which are the silver-bearing alloys, generally known as silver solders. These are closely controlled alloys having precise characteristics which are available having different melting ranges, and spreading characteristics. The changes are brought about by changing the silver content of the solder, together with the proportions of other alloying elements, to change the melting range and the temperature below which the alloy is fully solid (the solidus).

By choosing solders with different solidus temperatures it is possible to carry out two or three silver soldering operations on one item without any real danger that heating for the later stages of assembly will melt previously applied solders. This is very important when assembling a model boiler made in copper, for example, since by its nature, the assembly must be performed in several stages.

Table 9.1

| Name | Melting Range (°C) | | BS 1845 |
	Solidus	Liquidus	Designation
Easy-flo No. 2	608	617	AG2
Argo-flo	608	655	AG3
Silver-flo 16*	790	830	–
Silver-flo 24*	740	780	–

* Silver-flo 16 and Silver-flo 24 were previously known as B6 and C4 alloys, respectively

Silver solders are most readily identified and defined by using the British Standard designations (from BS 1845) but in practice these are not often used. This is due to the wide availability of solders from one manufacturer, Johnson-Matthey Metals Ltd. The practice has developed of using this manufacturer's designations rather than the BS references. The most commonly available silver solders from this manufacturer are shown in Table 9.1.

Table 9.1 shows four silver solders having different compositions and exhibiting different melting ranges. *Easy-flo No. 2* and *Argo-flo* are alloys of silver, copper, cadmium and zinc, the former having 42 per cent of silver and the latter 38 per cent. *Argo-flo* has the wider melting range and is used when larger fillets are required or when wider gaps need to be filled, since it is suitable for gaps up to .010in. (0.25mm). *Easy-flo No. 2* is suitable for gaps up to .006in. (0.15mm).

The same manufacturer also supplies *Easy-flo No. 1*, corresponding to the BS 1845 designation AG1. This alloy contains 50 per cent silver and is consequently more expensive than *No. 2*, but it is slightly more fluid when molten and produces finer fillets than the cheaper alloy. *No. 1* has a melting range from 620 to 630°C. The *Easy-flo* and *Argo-flo* alloys are recommended for all general work.

For the first stage of any two-stage silver soldering which is required, *Silver-flo 16* or *Silver-flo 24* are usually specified. For these alloys, the silver content is indicated by the appended reference number. These alloys are, however, cadmium-free.

Cadmium-free silver solders

General-purpose silver solders such as *Easy-flo No. 1* or *No. 2* and *Argo-flo* are alloys of silver, copper, cadmium and zinc. These materials developed out of the use of a simple brazing spelter which is a 50:50 brass comprising an alloy of copper and zinc. The addition of some silver to such an alloy reduces the liquidus temperature and creates an alloy that is significantly better than a plain brass for use as a solder. Further improvements in the alloy are brought about by the addition of some cadmium which further reduces the melting point and also brings other benefits. These include an improvement in the ability of the molten solder to wet the joint surfaces and a reduction in the surface tension, giving the alloy a better capillary action so that it is more readily drawn into the narrow clearances in the joint. Cadmium is also cheaper than silver, thereby reducing the cost of the solder.

The problem with cadmium, and to a lesser extent zinc, is that they are both elements which are characterised by low melting and boiling points. Cadmium, for example, melts at 321°C and boils at 767°C, while zinc melts at 419°C and boils at 907°C. Both elements are naturally molten during the soldering process and although the boiling point of neither element should ordinarily be reached during the process, some cadmium and zinc vapour is normally given off.

The vapours which are produced combine readily with oxygen to form cadmium and zinc oxides, both of which are toxic. By far the most dangerous of these is cadmium oxide and exposure to quite low levels of this compound can be fatal. All silver soldering must, therefore, be carried out in an area which has good ventilation. Heating should be rapid and not prolonged more than is absolutely necessary to effect the joint.

To avoid the major hazard posed by the presence of cadmium, the higher melting point alloys such as *Silver-flo 16* and *Silver-flo 24* are formulated without the incorporation of cadmium and are therefore described as cadmium-free. Even when using these alloys, adequate ventilation should be provided and the work should be brought up to soldering temperature as rapidly as possible and the joint completed quickly.

The flux may be affected, and lose its active properties, if it is raised to too high a temperature or even if it is held within its working temperature range for too long. Heating must be rapid (the temperature of the work must rise quickly through the melting range of the solder) particularly when using alloys having a long melting range, otherwise some separation of the elements forming the alloy may occur. This effect is known as liquation.

Forms of silver solder

The most commonly available types of silver solder are those listed in Table 9.1. Different forms of these alloys are available, normally comprising rods, wire and strips in various standard sizes. Industrially, thin foils are available in some grades but these are not generally to be found in modellers' suppliers. Rods and wires are generally the same thing, except that the description 'wire' is applied to the smaller diameters below 1/16 in. (1.5mm). Wire is generally supplied in coils containing a few metres of one size. Rods, on the other hand, have larger diameters and are sufficiently rigid to be supplied in straight 24in. (600mm) lengths, hence the description 'rod'.

Strip in 600mm lengths is also available, being of rectangular cross-section, from about 1.5mm × 0.6mm to 5.0mm × 1.0mm. These are quite useful as an alternative to large-diameter rods where reasonable quantities of solder need to be introduced into the joint.

Due to the relatively high cost of the alloys, the smaller cross-sections are usually preferred, but it is a mistake to believe that only a small-diameter wire is required since this is not very stiff and it is difficult to control when being applied to the joint. Due to the high temperatures needed for silver soldering, and the large sizes of some assemblies (boilers and the like) it is usually necessary to operate at some distance from the workpiece. A long, stiff rod is advantageous and allows precisely controlled dabs of solder to be applied economically. Naturally, if much small work is undertaken, a fine wire can also be kept in stock.

If there is a likelihood that two or three different types of silver solder will be kept in stock, they should be kept separately, and identified in some way, so that it is not possible to mistake one for another. The difference in the appearance of the different solders is hardly noticeable, and the bundles should be labelled in some way to prevent confusion. Figure 9.3

190

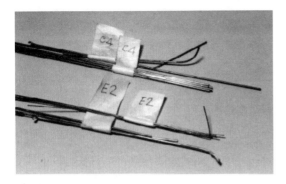

Figure 9.3 Always label silver solders otherwise they are impossible to distinguish.

shows labels made by winding masking tape around bundles of solder, folding it back upon itself, and identifying the type using a ballpoint pen. Notice that partly used sticks have been put back into the bundles.

Silver solder can also be obtained in the form of paints and can be used to place the solder in the joint before assembly, as described below.

Fluxes for silver soldering and brazing

As for soft soldering, flux selection for hard soldering is determined by the characteristics of both the solder to be used and the metal to be joined. The melting range of the solder must be matched by the active temperature range for the flux and the activity of the flux must suit the metal to be joined, and its freedom, or otherwise, from oxidation at the soldering temperature.

With the exception of the old-fashioned brazing operation using 50:50 brass, for which borax was the standard flux, fluxes for hard soldering are usually purchased against particular trade names, the products of Johnson Matthey Metals Ltd. once again being commonly available. If two-stage hard soldering is required, as for model boiler work, two fluxes

will be required, one which suits the *Easy-flo* melting range (around 600°C) and a second which is active at 800°C, corresponding with the melting ranges of *Silver-flo 16* and *Silver-flo 24*. There are some fluxes which are suitable for use over the full temperature range which is required.

Easy-flo flux, having an active range of 550 to 800°C, complements the *Easy-flo* silver solders and is available in a general-purpose type or one specially formulated for use on stainless steel.

Although *Easy-flo* flux remains active up to 800°C, it is usually recommended that *Tenacity Flux No. 4A* is used with the *Silver-flo* alloys since it remains active up to 850°C. For stainless steel, *Tenacity Flux No. 5* is recommended, which remains active between 600 and 1000°C.

Easy-flo and *Tenacity* fluxes are available in paste or powder forms, the latter being the more commonly available type from modellers' suppliers. Before use, powder fluxes must be mixed with clean water to form a paste having the consistency of thick (not whipped) cream. The flux must be allowed to stand for a few minutes after mixing and should be given a thorough final stir before use.

The flux is painted onto the cleaned surfaces of both plates prior to their assembly since it is too thick to be drawn into the small clearance gaps which must be employed, even when using the *Silver-flo* solders with their longer melting range. Any pre-soldering riveting or bolting together must be undertaken after the metal has been cleaned and coated with flux. A good fit of the parts, together with the minimum number of rivets or screws, is beneficial. A further coating of flux should be applied to the assembled joint crevices prior to heating up for the soldering operation.

Figure 9.4 shows a model boiler being prepared for the backhead to be silver soldered into position. The backhead and the inside of

191

Figure 9.4 The backhead of a locomotive boiler secured into the wrapper and generously fluxed. At this stage, the firehole ring has not been fluxed, and the nipples for the stay ends remain to be fitted.

Figure 9.5 The central hole in the hearth permits the boiler to be rested on the front of the firebox for silver soldering the backhead into position.

the outer wrapper have been cleaned and coated with flux, and the assembly screwed together using gunmetal screws. Flux has been generously applied all around the main joint and it remains to fit the two nipples for the upper stays and add flux to these and to the outside of the firehole ring tube.

For this operation, the boiler barrel is positioned in the central hole in the hearth, and the throatplate is supporting the assembly on top of the hearth. This is shown in Figure 9.5. The central hole in the floor of the hearth is also useful when the front of the boiler is being soldered, since the boiler can be supported below the hearth, allowing the smokebox end to project into the working area as shown in Figure 9.6.

Figure 9.6 The smokebox tubeplate can be silver soldered by supporting the boiler on its firebox below the hearth and allowing the smokebox end to project through the central hole.

In both of the above examples, the boiler is packed around with firebricks to provide insulation and concentrate the heat of the torch.

When the heating commences, the lighter elements in the flux (the water or the liquid medium in which the flux particles are suspended) soon evaporate, leaving a coating of flux particles on the surface. As the temperature is raised, the flux becomes molten and flows over the surface, cleaning it and preventing oxidation by excluding the air. Once the plates are hot enough, silver solder applied to the surface melts and runs over the clean metal, displacing the molten flux. When using a solder with a short melting range, such as *Easy-flo No. 2*, it will be seen to 'flash' round the hot joint, over quite long distances, almost instantaneously.

Flux residues

When the job cools, flux residues remain as a solid coating which must be removed before the joint can be examined. Different fluxes leave residues presenting different degrees of difficulty in their removal, some being soluble in water but others requiring a weak acid or alkali solution if chemical removal is desired.

The *Easy-flo* fluxes form residues which are slowly soluble in hot water so that a simple immersion in the domestic sink, together with some mechanical assistance (scraping) will usually suffice to remove the residues. *Tenacity Flux No. 4A* forms a residue which is virtually insoluble in water and a 10 per cent sulphuric acid solution (in water) is recommended as the best means of removal. This ensures complete removal without any mechanical assistance and the acid bath also cleans the base metal, dissolving any scale (oxide) and producing bright, clean metal. The solder is also cleaned, giving a bright finish to the joint and permitting a proper examination to be made.

Figure 9.7 The appearance of a correctly made silver-soldered joint. Solder has run through the joint, from the other side, around each tube to present a significant fillet of solder.

Figure 9.7 shows the appearance of the locomotive boiler of Figures 9.4 to 9.6 at an earlier stage in its completion. The inner and outer assemblies have just been united, but the illustration shows the results of the earlier soldering during which the tubes were soldered into the firebox tubeplate. For this operation, the solder was applied to the opposite side of the tubeplate, and the solder seen around each tube has flowed through during the soldering operation, to provide a bright ring on the inside, confirming good penetration of the solder, and indicating the soundness of the joint.

Pickling

The process of cleaning using a 10 per cent solution of sulphuric acid in water is generally

193

known as pickling, the dilute acid being referred to as the pickle. A 10 per cent acid solution is easily prepared from commercial sulphuric acid in the ratio of nine volumes of clean water to one volume of commercial acid. **ON NO ACCOUNT MUST WATER BE ADDED TO ACID**, since the first drops to touch the acid are likely to cause a chemical reaction not unlike a small explosion, thus spreading acid in all directions. **ALWAYS ADD ACID TO WATER.** This way, the procedure is quite safe, the only dangers really being due to the corrosive nature of the acid itself.

The primary requirement is for a large, safe container in which the acid solution can be stored. Pickling in a ten per cent solution of sulphuric acid is normally reserved for use on brass or copper assemblies and is usually associated with boilermaking. A reasonably large container is required, the minimum size being such as will allow a model boiler at least to stand on its back or front and be half immersed in the pickle.

Figure 9.8 shows a large chemical vat which I use. This has a sealed lid which is held in place by a steel hoop which is tightened onto the top of the vat by a lever toggle. The vat is about one-third full of dilute acid and meets the requirement to immerse about half of the boiler.

Figure 9.8   A safe container is essential if acid pickling is to be undertaken.

In many instances, a builder's plastic bucket will serve for the actual pickling and a screw-top plastic barrel will be suitable for storage. Since the acid is corrosive, even in its dilute state, there is some advantage in having a storage container which is also large enough to use for the pickling operation since this avoids the minor splashes caused by the transfer between vessels.

It is worth noting that, if clothing is splashed by the acid, a hole will result. This may be contained to some extent by washing immediately with water, but a hole will surely appear in due course.

It goes without saying that any splashes on the skin must be washed off with plenty of water, but particular care must be exercised when handling the concentrated acid since this is extremely corrosive and should only be handled when wearing rubberised gloves and protective goggles.

When the concentrated acid is poured from its container, which is likely to be a narrow-necked, plastic 'Winchester', care must be taken not to allow the acid to gulp in the neck of the bottle, as air is drawn in, since this causes an unsteady stream of acid to be poured, in turn causing droplets to be expelled in random directions. Pouring must be slow and steady, allowing air to enter the neck of the bottle continuously.

There are also possible dangers in the pickling operation itself. Some of the earlier writings on the subject of pickling have given rise to the belief that it is necessary to place the job in the pickle while it is still very hot. Since chemical reactions are usually speeded up as the temperature increases, placing a very hot workpiece in the acid produces a spectacular cleaning of the metal and removal of the flux residues. Unfortunately, it also produces a spectacular fizzing and splashing, accompanied by the emission of fumes and this markedly increases the likelihood of damage to clothing.

It is also difficult to avoid inhaling the fumes, and there is an overall danger to health.

It is perfectly safe to place a warm workpiece in the acid solution, but the general advice to allow a large silver-soldered boiler to 'cool to black' and then pickle, often repeated over the years, should be regarded with circumspection since it is definitely not sufficient just to allow the redness to go out of a large copper assembly and then place it straight into the pickle bath. Let it cool at least to being fairly warm. The cleaning of the metal will not be so rapid, but it will be safe.

After pickling, the work should be cleaned by a thorough washing in plenty of water and the joint crevices scoured with kitchen cleaning powder (e.g. *Vim* or whatever is available) and a stiff nailbrush, afterwards washing thoroughly once more. This allows the joints to be inspected without fear that pockets of acid remain in, or on, the job, which might damage clothing, tools and so on, in the workshop.

Thorough cleaning is also necessary to ensure complete removal of flux residues, since these are usually mildly corrosive.

A silver-soldered joint

Preparation

The making of the joint is not essentially different from soft soldering, except that the flux is supplied as a powder which needs to be mixed with water before use, and the soldering operation takes place at a much higher temperature than is required for soft soldering.

The large size of some assemblies which need to be silver soldered means that there is a need to provide large amounts of heat, but if a range of burners is available, a suitable one can be chosen for the job in hand.

After making the parts of the assembly and deciding the method of locating them together, immovably, during soldering, the first operation is to ensure that the mating faces are really clean. Any burrs left by the cutting and shaping processes should be removed and any deposits of oil or dirt removed by wiping with a clean cloth or tissue.

The next process is to clean the surfaces thoroughly using a medium grade emery cloth, afterwards avoiding contact with the cleaned surfaces with the fingers in order not to deposit any grease on them.

Flux must be spread on the joint surfaces before assembly, so the next task is to mix up some flux. If the flux is available in the form of a ready-mixed paste, this can be used directly, but if a powder flux is being used, which is the most convenient form for the occasional user, the powder must be mixed with clean water until the right consistency is achieved.

Flux is expensive, but it must be present throughout the joint before heating commences, otherwise the work will oxidise and a sound joint will be impossible to achieve. The mixed flux needs to be rich in flux, and not a thin, watery mixture.

Mixing should commence by putting what you perceive as enough powder for the job in hand into a clean container. Water needs to be added to the powder, but less water, by volume, is required than there is powder, so water is added only in small quantities.

For a small job, only a small amount of flux powder, say, of the order of half to one teaspoonful, is needed. Add two drops of water (only) to this amount of powder and mix using a medium-sized artists' paint brush. Add one or more drops of water if the mixture is not fluid enough, but on no account add more than a drop at a time, otherwise the mixture will certainly become too wet.

If this happens, more flux powder can be added, but it is amazing how much is needed to correct even a small over-addition of water, and it is much better to err most definitely on the side of too little water and add it drop by drop.

When the work is clean, and the flux mixed to the right consistency, which is always described as 'the thickness of cream', the flux-and-water mixture can be painted onto the joint faces. The mixture must wet the work. If the pool of liquid contracts into a small pool when the brush is removed, the metal is dirty and the mixture is not wetting the surface. If wetting does occur properly, the mixture will stay where it is put on the metal.

When both surfaces have been coated with flux mixture, the joint can be assembled. The fact that wet flux is now present means that assembly by riveting is not really a viable option. A screw or clamp is to be preferred, or a rivet (or rivets) may be used to provide location while relying on the mass of the parts to hold them together. However, positive location is vital, so that accidental disturbance cannot occur during heating and cooling.

Once the joint components are located adequately, the work can be transferred to the hearth and firebricks placed around it to contain the heat yet allow access for the torch. In this respect, it is worth noting that conventional burners, which have air holes in the burner tube itself, will not burn correctly if pointed into a restricted corner. So, enough space needs to be allowed around the object to be heated.

The torch must now be made ready for heating. This means fitting the torch handle assembly to the hose and connecting the required neck tube and burner. With the outlet valve on the gas bottle turned off securely, the hose and hose failure valve, or regulator, is connected up (left-hand thread for propane cylinders) and with the valve on the handle assembly turned off securely, the valve on the cylinder is switched on and the hose failure valve (if in use) set to allow gas to pass to the handle and burner.

It is valuable to add a little more flux around the joint to be certain that there is an adequate supply present in the region(s) in which the silver solder will be applied.

A reasonable length of solder should be taken from stock, bearing in mind that the work and the hearth will become hot and an ungloved hand will not be able to approach the work very closely.

Making the joint

After all of this preparation, the burner can be lit! The flame needs to be adjusted by setting the valve in the handle assembly so that the burner is operating correctly, after which the flame can be applied to the work.

The initial aim is to raise the temperature of the joint generally so that the water is evaporated from the flux mixture. This causes the flux to regain its white appearance, but it coagulates and should envelope the joint if there is sufficient present. If the amount looks rather sparse, it may be wise to stop at this stage, let everything cool, clean up thoroughly and start again.

With sufficient flux present, heating continues until red heat is approached. The flux begins to melt, and takes on a clear, glass-like appearance, surrounding the joint.

At this stage it is possible to see that the flux is doing its job in preventing oxidation, since the joint area is (or should be) bright and clean-looking, while the area outside the immediate region of the joint, where there is no flux, will be black, or obviously discoloured.

When the work has reached red heat, the flame is moved away slightly and the stick of silver solder touched onto the joint surface, typically the intersection of the two overlapping parts. If the work is hot enough, the solder melts on contact with the joint. If a solder with a short melting range is in use, such as *Easyflo No. 2*, it will run along the joint instantaneously. The flame is moved along the joint, assuming that it is long, and the process repeated. When the whole joint, or all of the parts of the joint, have been soldered, the burner is turned off, and the job allowed to cool.

If, when the solder is applied to the joint, it melts but does not run, the work is not quite hot enough and the flame must be brought back once more onto the joint until the solder bead which has been left on the work melts, and runs through the joint.

The solder cannot be placed directly in the flame since it melts immediately and it must always be applied to the work with the flame removed. It then acts as a thermometer, confirming, or not, that the work is hot enough to allow the joint to be made.

A demonstration observed

If all of the above seems very precise and fussy, it is because the adoption of this procedure will ensure the best chance of a sound joint. As an example of how quick and easy it can be, let me relate the story of a demonstration once seen in a nightschool classroom.

A student had a small, forged spring which needed a temporary repair, and we were promised that the course tutor would "see to it, all in good time".

On the evening in question, the tutor sauntered down to our end of the room and stationed himself at the end of the large conference table around which we all sat for the evening's work.

"If anybody's interested" he said, "I'm just going to do a spot of silver soldering" and he placed a piece of firebrick and a DIY blowtorch, with gas canister attached directly to the handle, on the tabletop in front of him.

He asked the young lady for the pieces of her broken spring and then produced from a pocket a tiny piece of emery cloth or carborundum paper, about 1 in. (25mm) square. He cleaned up the broad ends of the two pieces of spring with the abrasive paper, then laid one piece carefully on the firebrick with the clean surface uppermost.

He carried a little spittle on one finger and placed it on the clean spring end and then equally carefully placed the other part of the spring on the firebrick, overlapping the first by a small amount. Further spittle was added to the overlap.

Out of his shirt pocket, he took a small twist of paper which he opened very carefully to reveal a very small quantity of white powder, some of which he sprinkled carefully on to the spittle-coated overlap.

Having twisted the scrap of paper up to seal the packet once more, he placed it back in his shirt pocket and withdrew a short length of silver solder.

Fortunately, there was a smoker present, so the blowtorch was lit and adjusted, and the flame applied to the overlap of the spring. In no time, the small spring was red hot, and one deft touch with the silver solder stick caused the solder to melt and run through the joint.

The job was done – no fuss, no bother, and all the time talking about general aspects of silver soldering, but letting the assembled students draw their own conclusions about the procedure, and the ease with which it might be done. Quite a difference between this demonstration and my first attempts at silver soldering a 5-inch gauge locomotive boiler.

Getting solder to the joint

General

In modelling, the type of joint frequently required means that relatively large areas of metal are butted together and the resulting joint is soldered by heating the two parts and applying solder and flux at the edges of the joint. In applying the solder externally, a significant fillet is sometimes built up on the outside of the joint. This fillet is often not desired and it must be removed afterwards by use of a scraper or some other mechanical means.

One way to reduce the size of this fillet (it may not be eliminated completely) is to introduce the solder into the centre of the joint directly.

The most obvious way to introduce the solder is to place it in position before the items in the joint are brought together. If the joint is to be soft soldered, tinning of the parts prior to assembly is the most convenient method, the technique for which is described above.

Another technique is to apply solder and flux in the form of paint, prior to assembly of the joint.

Direct access to the joint

Alternatively, access holes can sometimes be provided to allow the solder to be introduced into the centre of the overlapping joint, directly.

All that is required is a small access hole drilled in one of the components of the joint, roughly central on the overlap if it is small. For a larger joint, several holes should be drilled. Once the two items have been brought up to temperature, solder can be introduced through the hole(s), right into the middle of the joint, from where it can spread outwards. This tends to eliminate any unsightly external fillet, although if the solder spreads as it should, some will reach the edge of the joint, and a small fillet will appear. Provided that too much solder is not introduced, the fillet is usually small and a much neater job results.

Solder paints

A common commercial method of placing solder in a joint before assembly is found in the plumbing trade for which joint fittings are made which already incorporate a ring of solder. To make a joint, a length of pipe is simply cleaned at one end, fluxed and inserted into the joint piece. Heating the pipe-and-joint assembly melts the solder, which flows through the cleaned and fluxed joint, thus sealing the union.

This method is not exactly ready-made for modelling, but solder paints can be obtained which comprise solder particles suspended in a liquid flux. The well-mixed paint is spread onto the cleaned surfaces of the joint materials, the two parts of the joint clamped together and the whole joint heated. The solder particles melt and flow throughout the joint, solder normally appearing as a small fillet at the external boundary. Heat is then removed and the clamps taken off when the joint has cooled and the solder solidified.

The method has the great merit that solder is introduced directly into the centre of the joint, where it can do most good, and only small fillets show at the edges of the joint. Paints are available containing a suspension of either silver solder or soft solder particles and the method can be used to make both soft- and hard- soldered joints.

Since the paint is only partly solder, only a small quantity is introduced into the joint, hence the better appearance of the finished joint. However, the paucity of solder does mean that the method is not ideally suited to the sealing of joints and should, therefore, be used with circumspection on water, petrol or oil tanks, or pressurised joints.

The solder particles in a paint rapidly settle to the bottom of the jar and a thorough stirring is essential before use. Although solder paints include gelling agents to help prevent rapid settlement of the particles to the bottom of the jar, it is best to stir regularly if a series of joints is to be made.

The lighter elements of the suspension medium evaporate progressively from solder paint and stock needs to be reconstituted after a time. Provided that the paint is not completely dried out, tap water can be used, but if a crust has formed on soft solder paint, it should be removed and discarded since the solder particles will be oxidised and useless for their purpose. Some suppliers of solders and fluxes recommend periodic addition of an appropriate flux from their range to prevent excessive drying out.

Adhesives

Types of adhesive

Three groups of adhesives are widely used in engineering for the joining of similar or dissimilar materials. These are Epoxy Resins, Anaerobic Compounds and Cyanoacrylate Adhesives. These different groups of compounds have different basic characteristics: the as-supplied and final (set) forms and appearance, the bond strength, setting or curing time, suitability for joining different groups of materials and the environmental capability (temperature and relative humidity) which the adhesive is designed to withstand when set. Since each type of adhesive uses different compounds, and basically produces different results, each can be considered separately.

However, one general point is worth making – plastic materials do present some problems in relation to gluing operations. There are several basic types, of which the Polyolefins, Polycarbonates, Styrene-based and Nylon/Acetal types are the most common. There are also other materials, such as glass- or carbon-fibre reinforced materials which must be glued with

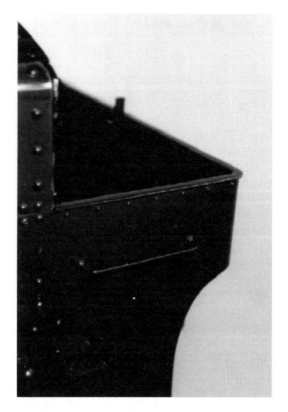

Figure 9.9 This bunker side illustrates what might be done to arrange attachments. The upper part of the tee section beading around the top edge is glued into position, and the handrail is a steel hoop to which two brass collars have been silver soldered. This assembly is soft soldered to the bunker side.

particular adhesives if satisfactory bonds are to be created. It is necessary to know reasonably precisely what type of material you are dealing with before attempting to glue plastic components.

An example of the use of adhesives is shown on the locomotive bunker in Figure 9.9. The top edge of the side platework is trimmed on the prototype with a tee-section beading. On the model, this was represented by riveting a brass strip around the top edge and then gluing the radiused top of the tee into position using a 2-part liquid glue.

As an aside, the handrail was made by silver soldering two collars on to the rail and then soft soldering these to the bunker side

Epoxy resins

Epoxy adhesives are supplied as two components, one part being the epoxy resin and the second being the hardener. These components are supplied as viscous (thick, or slow-flowing) pastes. To use the epoxy, a 1:1 mix of resin and hardener is prepared by stirring together the two parts and the mixed adhesive is then coated onto cleaned joint surfaces. The items to be joined are then clamped together and the joint set aside to allow time for the adhesive to harden, or cure.

Hardening times may be as long as 24 hours at room temperature for standard epoxies, but rapid or quick-set variants are available in some ranges. These have a hardening time of roughly 10 minutes at room temperature which is advantageous if the joint cannot be clamped together conveniently. The chemical reaction which causes hardening to occur commences immediately the two components are mixed together, thus establishing a working or handling time for the mix. For standard epoxies, this is 1 to 2 hours at room temperature, whereas for the rapid-setting types it may be as short as 5 minutes.

The setting time of standard epoxies can be significantly reduced by curing at an elevated temperature, the curing times being typically 24 hours at 20°C, 3 hours at 60°C but only 20 minutes at 100°C. The heat generated by a reading lamp can be beneficial in this respect and two hours under a 60-watt reading lamp placed immediately above the work is usually sufficient to achieve an initial cure so that the job may be handled. Curing at room temperature requires 3 to 5 hours before the joint may be handled.

If an epoxy resin is heated to assist curing, this reduces its viscosity, allowing it to flow more readily, and more easily fill the joint, thus conferring added strength to the cured joint.

Surface preparation of the joint materials prior to using epoxy adhesives is normally by a simple 'abrade and clean' process. That is, by use of an abrasive paper to produce a roughened surface, followed by a dusting off of the material loosened from the surfaces. Chemical cleaning is not usually required, but the materials to be joined must be clean and tarnish-free and this presents some difficulties in relation to brass and aluminium alloys which form tarnish coatings very rapidly in air. Difficulties may therefore be experienced with these metals.

Epoxies are generally well suited to forming metal-to-metal joints or for joining metal to other materials such as wood, plastics and even glass. The difficulty with glass is the prerequisite abrasion, without which the bond strength is relatively poor. Plastics also present other problems since different forms of epoxy resin do not form bonds of similar strength with all plastics and the manufacturers' data sheets should be consulted for detailed information.

Once fully cured, epoxies have the capability to sustain the bond strength exceedingly well at temperatures as low as –50°C and up to 100°C. The quick set or rapid versions of these adhesives do not usually show such good high-temperature performance and for some of these also, the strength achieved, particularly the impact strength, is significantly worse than the regular types. If good strength is required, it is best to use a standard epoxy resin adhesive and to design the joint so that clamping during the long curing time will be straightforward.

Some epoxies are available in high strength variants, as also are some 'loaded' epoxies which incorporate a metallic filler to provide increased hardness in the cured state. Alternatively, epoxies may be loaded to provide electrical conductivity when set. Metal-loaded epoxies are suitable for the repair or filling of castings, but if they are intended for filling rather than for use as adhesives, the two components frequently take the consistency of putty. Mixing of these is by kneading the putty-like components together, again by taking equal quantities of the resin and the hardener.

Anaerobic compounds

Anaerobic compounds are adhesives which solidify in the absence of air, 'anaerobic' simply meaning 'having no oxygen from the air'. Provided that they are excluded from contact with air, these compounds will solidify and the normal method of use is simply to introduce them into a close-fitting joint. This allows setting to commence in the centre of the joint and there is no further requirement for a setting agent or hardener.

These types of adhesive comprise a wide range of materials which have different strengths and characteristics when set. This allows their use for the formation of permanent bonds or, at the other extreme, to provide a sticky, gap-filling compound used for locking ordinary screwed fastenings to prevent their slackening under vibration, yet allow normal disassembly. This method of thread locking is described in Chapter 8.

Since a range of compounds has been developed, different materials are provided for specific duties, the ranges normally being described as high strength or medium strength retaining compounds, and the low-strength, or screw-locking types. The high-strength materials are used for permanent bonds. These descriptions over-simplify the nomenclature however, since even at the low-strength end of the range (the thread-locking compounds) various strengths are available for different uses.

Low-strength anaerobic compounds are intended for thread locking. Two basic types

are available, the first being intended for application to the threads before assembly, while the second type has sufficiently good creep to be applied to the threads after assembly. The *Loctite* range of anaerobics is perhaps the most readily available and this provides a total of four thread-locking compounds. These are *Loctite 221* (Screwlock), *Loctite 241* (Nutlock), *Loctite 270* (Studlock) and *Loctite 290* (Penetrating Adhesive). *Loctite 270* has the highest strength, being intended for the locking of studs which will not normally require removal, *241* (Nutlock) has about half the strength of Studlock and *221* (Screwlock) is half the strength again and is intended for low-strength locking of screws which may require adjustment or where easy disassembly is required. It is also suitable for use with fasteners made in soft metals.

For permanent or semi-permanent bonding of cylindrical parts, anaerobic compounds provide a more convenient alternative to the press or interference fits which were previously used. As described in Chapter 8, these forms of fit require accurate machining of the components and careful pressing together in order to achieve correct results. The advantage of using an anaerobic adhesive is in the wider machining tolerance which the gap-filling capacity of the adhesive allows. The machining of the parts is less critical, and is therefore usually quicker.

Retaining compounds, as the high-strength anaerobics are called, are formulated for two specific duties – the permanent fitting or assembly of cylindrical items which will not require disassembly, and the fitting of such items as ball and roller races which are subject to wear and may therefore eventually require removal. In the *Loctite* range, *601* (Retainer) is the high-strength (permanent) type and *641* (Bearing Fit) is the lower-strength alternative. *Loctite 641* has a strength slightly less than Studlock while *601* (Retainer) is about twice as strong.

The gap-filling qualities of these two compounds allow the items to be slip fitted so there is no question of pressing parts together. This is especially beneficial when fitting ball or roller races since there is no possibility of squeezing in the races and thereby damaging the bearing. *Loctite 641* (Bearing Fit) can be used with radial gaps up to .006in. (0.15mm) whereas *601* will accommodate a radial gap of .004in. (0.1mm).

Although described as being of use for permanent fits, bonds made using *Loctite 601* (Retainer) can be disassembled should the need arise. The working temperature range of the cured adhesive extends to beyond 150°C but at this temperature the bond strength is between 30 and 50 per cent of that at room temperature. Warming up the joint will normally make it possible to break the joint bond by pressing the two items apart. Heating the cured material does not significantly affect the bond strength which is retained when the joint has cooled to ambient temperature once again and the pressing operation must therefore be undertaken while the joint is still hot.

Surface preparation of the joint prior to application of anaerobic compounds is dependent on the grade to be used. The thread-locking grades should be used on clean threads, but will cure effectively even if the parts are coated with a rust-preventive coating, provided that this is not excessive. In industrial applications this means that threaded fasteners are used in the as-received condition, without any special pre-cleaning.

For retaining compounds, a degree of cleaning is desirable, the aim being to produce clean, bare metal. Cleaning should be performed using the solvent recommended by the manufacturer. Most anaerobics are soluble (in the uncured state) in chlorinated solvents such as trichloroethylene and these materials are usually recommended for use as surface cleaners prior to application of the adhesive. Some increase in ultimate strength is provided by roughing the surfaces and a treatment with medium grade abrasive cloth, prior to use of the solvent cleaner, helps to provide a key within the joint.

Although the anaerobics harden naturally when deprived of oxygen, the speed at which curing takes place is affected by the material in the joint, curing occurring less rapidly on aluminium, cadmium, zinc and stainless steel than on iron or mild steel. Primers are consequently sometimes available to accelerate the curing process. Even without a primer, curing still takes place on the less active metals, but the times to achieve handling and full strength (normally about 15 minutes and 3 to 5 hours, respectively) are increased. For the amateur worker, only stainless steel is likely to create problems in this respect.

The anaerobic adhesives are designed for use in metal-to-metal joints and although they will bond to some plastics, there are many for which anaerobics are not suitable. If this type of application is required, the manufacturer's literature should be consulted.

The anaerobic compounds include a range of materials intended for use as replacements for paper, cork or fibre gaskets which are used for sealing pressurised joint faces. These can be extremely useful for modellers since their use avoids the need to cut and punch small gaskets in conventional joint materials. Sealants are usually supplied in a tube or syringe with applicator nozzle built-in. A thin bead of sealant can consequently be readily applied around the joint face and the parts assembled. Once the joint is closed up, the sealant cures from the centre in the normal way. Surface preparation prior to sealing is usually by use of a suitable solvent.

Cyanoacrylate adhesives

The most common form of cyanoacrylate adhesive is the familiar *Loctite Super Glue*. Like most of the cyanoacrylates, this hardens in a few seconds on almost any material and no heat or long-term clamping is required. The hardening time can be as short as 10 seconds, hence the description instant glue. This feature brings with it the danger that accidental bonding of the skin may occur, so these adhesives must be used with great care.

It is worth noting that, although hardening occurs in just a few seconds, full strength is normally only achieved after 12 hours.

Although suitable for use with many materials, some types of plastic are not readily joined by these adhesives and special variants are available to suit these more difficult applications. Porous surfaces, such as wood and ceramics also present possible problems and an activator is frequently recommended in these instances. This acts as an accelerator for the hardening process. The activator is normally applied to one surface to be joined and the adhesive to the other but it is sometimes necessary to coat both surfaces with activator, depending on the materials to be joined.

Some specialised applications do require specially formulated versions of the adhesive, particularly those which require the use of an activator. These are catered for by special packs, usually sold under descriptive names which make their intended use obvious. Among available types are included those suitable for glass-to-metal joints, especially for use on toughened glass, as used for car windscreens, and some for use on rubber and plastics. These are provided with full instructions and a general specification of the materials for which they are suitable.

Cleaning of the joint materials using a chlorinated solvent is sometimes a requirement prior to bonding but a clean and dust-free surface is frequently all that is required. Once fully hardened, cyanoacrylates are generally satisfactory at temperatures up to 80°C and if service use is likely to exceed this, a high-temperature variant should be chosen. These allow satisfactory use up to 120°C. Since the maximum in-service temperature is relatively low, the bonds may be broken down by heating the joint above the maximum and peeling it apart. After heating, the glue residues are generally in the form of a tacky strip which may be removed by scraping.

10

Hole production

Introduction

The production of holes, normally referred to simply as 'drilling' is an absolutely basic requirement of any engineering manufacturing activity. For smaller holes, the normal tools used are twist drills which are available in a wide range of closely spaced sizes. The range of sizes likely to be of interest will be from about .0135in. (0.35mm) to ½in. (12.7mm) or so in diameter although the occasional use of larger drills may be required, depending on the facilities which are available and the type of work being carried out.

Certainly, for holes above 1in. (25mm) in diameter and perhaps from ½in. upwards, alternative means of hole production are normally employed. These will take the form of tools having single-point cutters, such as the trepanning tool, some of which require a 'pilot' hole to be drilled prior to their use to ensure positional accuracy. An adjustable boring head may be available for the production of larger holes, this being a single-point cutter mounted in an adjustable holder and having a fine adjustment of the diameter being cut and which consequently is very versatile in use. These types of tool definitely take us out of the drilling types of operation.

The production of holes is not restricted to just plain drilling or boring but encompasses also the related operations of countersinking, counterboring and reaming, each of which is described separately below.

Holes may, of course, be produced on the lathe as a drilling or boring operation, or perhaps on a vertical mill or mill/drill, but the most usual method is to use a drilling machine.

As noted above, the amateur worker may have little interest in drills beyond ½in. or so in diameter since the slow speeds required for larger drills are not normally available on the smaller drilling machines. The power on a smaller machine is also normally less than is required for the larger drills and the normal maximum chuck size is likely to be ½in. or around 12.5mm.

The drilling machine

Twist drills are nowadays such a universally available tool that there can be few people who have not used them for producing (drilling) holes. The usual method of use is to fit a drill into the chuck of a DIY electric drill, and drill away. If you have the luxury of a 2-speed, or variable-speed drill, you will be aware that larger drills need to run at a slower speed and, as an experienced 'driller', you will appreciate the general need for squareness when drilling holes. However, DIY and related drilling is frequently quite forgiving in terms of squareness, and the materials to be drilled are often relatively soft, apart from concrete and brick, for which special techniques are used.

If you have tried to drill a large hole in a hard material, you will know that considerable force is needed to push the drill into the work and if much drilling of hard metals is required, an arrangement which gives some leverage (mechanical advantage) is advantageous. The need for squareness is also much more stringent in engineering and there needs to be a re-think of the drilling method.

One of the simplest ways to effect the change is to use a bench drill adaptor for the DIY drill. Such adaptors mount the drill into a frame which slides on a vertical column mounted into a casting which forms the base, or pedestal, of the tool. The frame is usually bi-ased upwards by a spring which counterbalances the mass of the drill.

A long lever is pivoted to the column, or an attachment to it, and is linked to the frame so that pulling on the handle presses frame and drill downwards and into the work, which is fixed to, or resting on, the base casting. The closeness of the lever pivot to its attachment at the drill frame means that the pull on the handle is magnified at the drill and considerable pressure is applied to the drill point to force it through the work. The adaptor also satisfies the need for squareness by supporting the drill to the column and maintaining alignment with the work throughout the whole range of movement.

Figure 10.1 A bench-mounted drilling machine fitted with a ½in. (12.5mm) capacity chuck. Modern drills are fitted with a transparent guard around the chuck.

A more regular type of bench drill is shown in Figure 10.1. This is typical of the medium-sized machines which can be purchased from a wide variety of sources. The drill is built around a column of ground mild steel, 2in. (50mm) in diameter fitted to a substantial cast iron base which can be screwed or bolted to the bench. At the head of the column, a large casting provides a rear mounting for the drive motor and at the front, provides a bearing for a sleeve which contains a spindle.

The spindle is driven by the motor through a vee-section belt, both spindles being provided with a 4-step pulley, as shown in Figure 10.2. This machine thus has four speeds. The spindle is driven from its pulley cluster by a key which locates in a keyway running down the spindle,

Figure 10.2 The drill spindle is driven through a 4-speed drive arranged between a pair of pulleys.

which can therefore be allowed to slide up and down. The bottom of the spindle carries a drill chuck which can be both rotated and advanced downwards under control of a feed handle.

To maintain the rigidity of the spindle location, it advances in one with a sleeve which provides support even when the spindle is extended downwards. To allow the sleeve to be advanced, it is cut with teeth, like a rack, which mesh with a small gear on a shaft which has a control lever attached to it so that it may be rotated. This is seen in Figure 10.3. Pulling on the lever rotates the shaft and pushes down the sleeve and drill spindle. In Figure 10.3, the sleeve and spindle are fully down, this machine being provided with just over 2in. (50mm) of 'down feed'.

A return spring is fitted to counterbalance the mass of the chuck, spindle and sleeve so that the 'at rest' position of the chuck is 'up'. On the machine illustrated, the return spring is contained within the circular housing enclosing the left-hand side of the down-feed shaft, and is seen in Figure 10.1.

There is a frequent need to drill holes partway through a workpiece and the down-feed mechanism usually incorporates an adjustable stop which can be set to prevent spindle and sleeve advancing farther than is required. On most machines, this takes the form shown

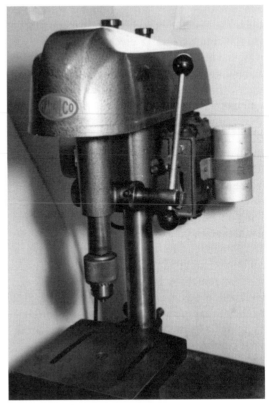

Figure 10.3 The down feed handle rotates through three-quarters of a turn to bring the chuck to the fully down position.

in Figure 10.4, but the machine of Figure 10.1 has the stop incorporated into the right-hand side of the down-feed gearshaft and is in this instance an adjustable collar surrounding the shaft.

Since drills vary in length according to their diameters, and items to be drilled also vary in size, it is essential to have a movable table below the drill head. Drilling machines have an adjustable table, roughly equivalent in size to the base, which can be moved up and down the column and locked in any position. On a medium-sized machine such as the one illustrated, the table is moved up and down by hand.

The main physical characteristics of the drill is the height beneath the chuck, the others being the down feed and the distance between the column and the spindle axis. Further considerations are the size of the chuck which is fitted and the spindle speeds. Both of these features should match one another since the lowest speed really determines the maximum size of hole which can be drilled or bored.

Bench drills may be said to be made in small, medium and large sizes. The one illustrated here is of medium capacity, allowing work up to about 8in. (200mm) to be positioned below a ½in. (12.5mm) drill, which is the capacity of the chuck, and allowing 4½in. (115mm) between the column and the drill centre. The drill actually has 12in. (300mm) clearance between the chuck and the base.

This drill has what might be described as a 'fixed' chuck. That is, it is not designed to be removed from the spindle, but is attached permanently to it. Larger machines naturally have larger spindles and they may be large enough to be bored with one of the standard Morse tapers at the lower end. In these cases, a drill chuck mounted to a tapered arbor can be used in the drill, or taper-shank drills used directly.

The Morse tapers are one of a range of shallow tapers which are described as 'self-holding' i.e. a tapered item such as a drill or drill chuck can be jammed into a matching tapered socket, where it holds itself in. Provided that the drill is not too large, the friction grip is adequate, but larger drills need a more positive drive and the end of the taper is machined with a rectangular tang which engages in a recess inside the spindle. The tapered ends of a drill chuck arbor and a large, taper-shank twist drill are shown in Figure 10.5.

If the spindle is provided with a Morse taper fitting, it is also normally slotted just above the position occupied by the driving tang so that a tapered drift can be inserted for driving out the drill or drill chuck. Figure 10.6 shows a ¾in.

Figure 10.4 This conventional depth stop is fitted to a mill/drill and is fitted with a scale engraved in mm.

The table is drilled with a central hole, and can be set with the hole immediately below the axis of the chuck and spindle to allow a drill to pass completely through the work without damaging the surface of the table. The table can take any position around the column and can be positioned to one side if this provides better support for the work, but there is then the possibility that the drill will mark the table if it pierces the work and something which can be 'sacrificed' must be interposed to prevent this. Since the table can be rotated about the column, it can be swung out of the way in order to place the work directly on the base.

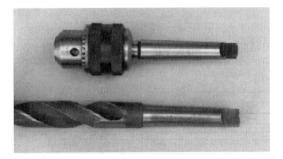

Figure 10.5 Some drilling machines accept drills and chucks which have Morse taper arbors. These can also be used on a lathe.

Figure 10.6 ¾in. (19mm) drill directly mounted into the spindle of a large pillar drill.

(19mm) drill with taper shank, mounted directly into the spindle of a large drilling machine. The slot which provides access for driving out the drill is visible at the top of the spindle.

This large machine is a floor-standing model, a type generally called a pillar drill, built around a column 4½in. (115mm) in diameter, with the chuck roughly 4 feet (1220mm) above the base. The machine head incorporates a 2-speed gearbox and 4-step pulleys to drive the spindle and has some very low speeds, making it suitable for drilling with large drills. It can also be used for boring large holes, using a boring head.

On a pillar drill, the movable table is usually sufficiently large that it requires some form of gearing to lift and lower it on the column and a rack is normally fitted to allow this, a pinion shaft with a large handle being fitted into the rear of the table for winding it up and down.

If you envisage the need to do heavy work, and you have room, a pillar drill is certainly valuable.

At the other extreme, if you envisage a great deal of work with small drills, a high-speed bench drill can be useful. A machine having a top speed of 2000 to 3000 rpm with a small chuck accepting drills up to ⁵⁄₃₂in. (4mm) in diameter is often described as a sensitive drilling machine since the down-feed handle does not provide a huge mechanical advantage but is designed to provide a good 'feel' when using small drills.

These two machines, at the extremes of what is readily available, might suit individual modellers, but for most, a bench drill having a ½-inch capacity chuck (12.5mm) will generally satisfy their needs. It is, however, useful to have a machine with a Morse taper to the spindle which matches the lathe or other machines in the workshop, since this allows taper-mounted drill chucks to be interchanged between them.

Drilling machine accessories

When a drill cuts, it tends to rotate the work, a tendency which must be resisted. The force which the drill exerts can be very high, and positive means need to be taken to resist it. For this purpose, the work is held in a small vice, suitable for use on the drill table, in which the part may be held while drilling. A vice suitable for this task is known as a machine vice.

A typical vice intended for use on a drill is shown in Figure 10.7. The vice is machined flat on the underside of its base and the upper surface of the base is machined parallel to it, maintaining parallelism with the drill table. The base casting incorporates an upstanding, fixed jaw, and a movable jaw slides in the central slot in the base, being retained by a plate which is recessed into the underside. The moving jaw is positioned and tightened by a knurled sleeve with an internal thread which engages a screw retained in the moving jaw.

When the work is being drilled, the tendency for it to turn is resisted by holding the vice handle firmly to prevent rotation. The jaws of this vice are fitted with hardened face pieces which are machined with vee-shaped grooves that allow circular bars to be held securely, but it should be noted that the sharp corners of such grooves do mark circular work which is

Figure 10.7 A machine vice for the drilling machine.

large in relation to the groove width and this must be remembered if damage to already prepared surfaces is to be avoided.

The vice shown has jaws 3¼in. (83mm) wide and 1¼in. (32mm) deep, and provides a maximum opening of 2in. (50mm). It suits my ½-inch (12.5mm) capacity bench drill which has a table 8in. (200mm) square. If the workpiece is very large, and the hole being drilled is small, for example, if ¹⁄₁₆in. (1.5mm) diameter rivet holes are being drilled in a locomotive running board 30in. (750mm) long, it is sufficient to resist the tendency for the work to turn, when the drill is cutting, by hand pressure, since the 'hand hold' acts as a sufficient radius for the tendency for the work to turn (the torque) to be resisted easily.

For large holes, say, from ½in. (12.5mm) diameter upwards, hand pressure may not be sufficient to resist the torque exerted by the drill, and there is then the need to bolt down the work, or the vice with the work in it, to the drill table. The base of a machine vice is provided with slotted lugs so that it can be bolted down when required, the drill table being slotted to allow this.

For a small drilling machine for which the table has through slots, plain nuts and bolts are sufficient for clamping items to the table, but the tables of larger drilling machines may have a special type of slot in which the table is not pierced through, preventing the use of ordinary fastenings for holding down the work or the vice. In these cases, the slots are machined to an inverted tee shape and special nuts, or bolts with specially shaped heads, are used for holding down. These are known as T nuts or T bolts, and they have to be machined to a section which matches the slot. Tee slots are also commonly used on lathes, and the requirements for these special attachments are described in Chapter 16.

A further essential accessory for the drill is an accurate pin chuck. This is needed to hold

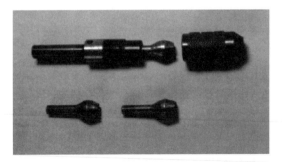

Figure 10.8   A pin chuck having a turned spigot is used for holding small drills in larger chucks.

small drills since a ½-inch or ⅝-inch capacity chuck (12.5 or 16mm) cannot be expected to hold the smallest sizes, and there are often occasions when these are used.

Pin vices or pin chucks for use on the drilling machine have a smooth-turned finish to the shank so that they may be held truly in the drill chuck. A typical pin chuck is shown in Figure 10.8. The body comprises a turned shank with a screwed-and-bored lower end, onto which a knurled sleeve can be screwed. The body has a hollow end which accepts one of three split collets, each of which has a head which is coned both ways so that screwing the sleeve on to the body closes up the collet, causing it to grip the drill which is inserted down the centre.

The three collets, together, cover the range from virtually zero to about ³⁄₃₂in. (2.4mm) diameter. The minimum size of drill that an ordinary ½-inch or ⅝in. drill chuck will grip, is normally ¹⁄₃₂in. or 0.8mm, so a pin chuck extends the gripping capability of the main chuck down to the very small sizes.

Figure 10.9 shows the pin chuck in use for spotting through the holes in an attaching angle into a frame plate on a model locomotive. In this case, the small size of the chuck has allowed the work to be approached, even though there is a large motion plate riveted to the angle.

Twist drills

The twist drill is the most widely used general type of drill. It is available in two basic forms – those with a plain (parallel) shank and those having a Morse taper shank. The parallel- or plain-shank type is the most widely used (and useful) type although there are occasions when the availability of taper-shank drills for direct mounting in the tailstock or mandrel tapers of the lathe is extremely useful.

A selection of parallel-shank twist drills is shown in Figure 10.10. Two types are shown: some general purpose, or jobbers' drills and the shorter stub drills, in the same range of sizes.

Figure 10.9   A pin chuck also allows easier access to the work in those instances in which a large chuck would foul the work.

Figure 10.10 Jobbers' drills are the normal ones used, but the shorter stub drills are sometimes used for drilling directly in the work, without centre punching the location of the hole.

Jobbers' drills are the ones most frequently used, and many workshops manage with only this type.

A twist drill has two spiral flutes machined up the body and at the tip, the drill is ground to a cone shape so that two cutting edges are formed. The flutes are deep in order to create large cutting edges, and allow the cut material to be ejected easily away from the cutting lips. This means that the drill is not very strong and is easily bent if subjected to any side force, so that the drill is inclined to 'wander' away from the truly axial path and this sometimes leads to positional inaccuracy. If the hole is deep, drill wander can cause the hole to be drilled out of square with the work's surface.

A deep, out-of-square hole naturally bends the drill, and since it is also quite hard and relatively brittle, this means that breakages can occur if the side loads become large. The shorter stub drills are designed to minimise the bending, since the shorter an item is, the less it bends for a given force. Stub drills are beneficial when positional accuracy is required.

Drills are made in an extremely wide range of sizes, from virtually nothing, to very large indeed, in closely spaced sizes. The preferred range of British Standard drill sizes is now based on a metric series although some fractional imperial drills are contained within the standard. If you are setting up a workshop from the beginning, metric drills will have to be purchased except for the few imperial sizes which are allowed by the standard and therefore remain in the catalogues.

Within the previously used British Standard drill sizes, three reference systems were in normal use. These were the fractional sizes, rising in increments of ¹⁄₆₄in. from ¹⁄₃₂in., and the number and letter drill sizes, ranging from number 80 (.0135in. diameter) to number 1 (0.228in. diameter) and from letter A (0.234in.) to letter Z (0.413in.). The size increments within the number and letter ranges were quite small, the difference between numbers 59 and 60 (.040in.) being only .001in. and between letters Y and Z .009in. Tables 10.1 to 10.3 list the previous letter and number drill sizes.

To assist in selecting metric drills as equivalents of the previous standards, a recommended set of sizes is also given in the table. Since imperial fractional sizes are still among those recommended, the list of equivalents includes some of these where they provide a closer alternative to the previous sizes.

In the range of metric drills, the increments are also small and sizes between 3mm and 10mm diameter are available in increments of 0.1mm. In addition, within this range, the intermediate sizes 0.25mm, 0.5mm and 0.75mm are available. Between 1mm and 3mm diameter, the increment is .05mm. Above 10mm preferred and second preference sizes are recommended, these latter being somewhat more expensive than the preferred range which rises in the sequence 10.00, 10.20, 10.50, 10.80, 11.00 and so on. The second preference range provides the 0.10mm increments in addition to 10.25mm, 11.25mm etc.

Table 10.1 Drill Gauge (Number Series) Drills, Sizes 1 to 60.

Size (Drill Gauge)	Diam. (in.)	Recommended Metric Alternative	Size (Drill Gauge)	Diam. (in.)	Recommended Metric Alternative	Size (Drill Gauge)	Diam. (in.)	Recommended Metric Alternative
1	.2280	5.80mm	21	.1590	4.00mm	41	.0960	2.40mm
2	.2210	5.60mm	22	.1570	4.00mm	42	.0935	3/32in.
3	.2130	5.40mm	23	.1540	3.90mm	43	.0890	2.25mm
4	.2090	5.30mm	24	.1520	3.90mm	44	.0860	2.20mm
5	.2055	5.20mm	25	.1495	3.80mm	45	.0820	2.10mm
6	.2040	5.20mm	26	.1470	3.70mm	46	.0810	2.10mm
7	.2010	5.10mm	27	.1440	3.70mm	47	.0785	2.00mm
8	.1990	5.10mm	28	.1405	9/64in.	48	.0760	1.90mm
9	.1960	5.00mm	29	.1360	3.50mm	49	.0730	1.90mm
10	.1935	4.90mm	30	.1285	3.30mm	50	.0700	1.80mm
11	.1910	4.90mm	31	.1200	3.00mm	51	.0670	1.70mm
12	.1890	4.80mm	32	.1160	2.90mm	52	.0635	1.60mm
13	.1850	4.70mm	33	.1130	2.90mm	53	.0595	1.50mm
14	.1820	4.60mm	34	.1110	2.80mm	54	.0550	1.40mm
15	.1800	4.60mm	35	.1100	2.80mm	55	.0520	1.30mm
16	.1770	4.50mm	36	.1065	2.70mm	56	.0465	3/64in.
17	.1730	4.40mm	37	.1040	2.60mm	57	.0430	1.10mm
18	.1695	4.30mm	38	.1015	2.60mm	58	.0420	1.10mm
19	.1660	4.20mm	39	.0995	2.50mm	59	.0410	1.00mm
20	.1610	4.10mm	40	.0980	2.50mm	60	.0400	1.00mm

Note: Imperial sizes are those specified for the drill gauge.

Table 10.2 Drill Gauge (Letter Series) Drills.

Size (Drill Gauge)	Diam. (in.)	Recommended Metric Alternative	Size (Drill Gauge)	Diam. (in.)	Recommended Metric Alternative	Size (Drill Gauge)	Diam. (in.)	Recommended Metric Alternative
A	.2340	15/64in.	J	.2770	7.00mm	S	.3480	8.80mm
B	.2380	6.00mm	K	.2810	9/32in.	T	.3580	9.10mm
C	.2420	6.10mm	L	.2900	7.40mm	U	.3680	9.30mm
D	.2460	6.20mm	M	.2950	7.50mm	V	.3770	3/8in.
E	.2500	1/4in.	N	.3020	7.70mm	W	.3860	9.80mm
F	.2570	6.50mm	O	.3160	8.00mm	X	.3970	10.10mm
G	.2160	6.60mm	P	.3230	8.20mm	Y	.4040	10.30mm
H	.2660	17/64in.	Q	.3320	8.40mm	Z	.4130	10.50mm
I	.2720	6.90mm	R	.3390	8.60mm			

Note: Imperial sizes are those specified for the drill gauge.

Table 10.3   Drill Gauge (Number Series) Drills, Sizes 61 to 80.

Size (Drill Gauge)	Diam. (in.)	Recommended Metric Alternative	Size (Drill Gauge)	Diam. (in.)	Recommended Metric Alternative	Size (Drill Gauge)	Diam. (in.)	Recommended Metric Alternative
61	.0390	1.00mm	68	.0310	1/32in.	75	.0210	0.52mm
62	.0380	0.98mm	69	.0292	0.75mm	76	.0200	0.50mm
63	.0370	0.95mm	70	.0280	0.70mm	77	.0180	0.45mm
64	.0360	0.92mm	71	.0260	0.65mm	78	.0160	0.40mm
65	.0350	0.90mm	72	.0250	0.65mm	79	.0145	0.38mm
66	.0330	0.85mm	73	.0240	0.60mm	80	.0135	0.35mm
67	.0320	0.82mm	74	.0225	0.58mm			

Note: Imperial sizes are those specified for the drill gauge

Sharpening twist drills

Introduction

Twist drills are sharpened by grinding. For larger drills, this means using the bench grinder, but smaller drills below about ⅛in. (3mm) in diameter can be ground by hand on a slip stone or oil stone. Accurate drill grinding is an art which many people find difficult to acquire and over the years drill grinding aids have been developed both for home manufacture and for commercial sale. Do not despair before you start – as noted elsewhere, the whole of model engineering requires skills which can only be gained by practice and drill grinding should just be regarded as one of the skills that you must acquire.

Drill point angles

The drill point takes the form shown in Figure 10.11. It is ground to a flattish cone shape in such a way that two cutting lips are formed, one at the base of each flute. The precise shape of the drill point is important in promoting good cutting and in helping the drill to cut at its correct size. Drill grinding is, therefore, one skill which must be practised until consistent results can be achieved.

The angle of the cutting edges should be 59

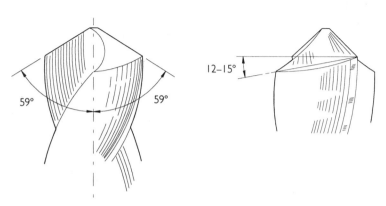

Figure 10.11   A detailed view of the point of a twist drill.

degrees each side of the centre line. This angle is naturally a compromise, chosen to maximise the general usefulness of the drill. A sharp angle at the point helps to centre the drill in the starting hole better than a larger one, but making the included angle at the point smaller lengthens the cutting edges and requires more power to drive the drill into the material being cut. A shallow-angle point drills faster than a more acute point, even though better centring is obtained with the latter.

Both lips of the drill must be ground to the same angle and must be of the same length, otherwise the point of the drill (in reality, a line where both flanks of the drill meet in the web) will not be in line with the centre of the drill body. This causes the point to describe a circle when the drill is rotated, throwing the workpiece around and making the drill difficult to start in a pilot hole, or on the centre punch mark.

If one flank of the drill is longer than the other (the point is not at the centre) the drill produces an oversize hole. The longer of the two cutting lips also absorbs more power and this unbalances the load on the drill. Since the body is deeply fluted, only a relatively small amount of material holds the drill together – this is the central core, or web. To provide as much strength as possible, the flutes become narrower towards the shank, but at the tip, the deeper flutes leave only a small web and if an uneven load is placed on the drill lips, a breakage can result. Any difference in lengths of the cutting lips on the two sides of the drill, whether due to unequal angles or to a misplaced point, must be avoided.

The cutting edge

If the lips of the drill are to cut, it is vital that there is clearance behind them. This means that the coning of the point must be done so that the lips touch (and cut) the work but the drill point surface behind the lip does not. This then allows the lips to slice their way into the work and the material is drilled. This clearance angle is generally made between 12 and 15 degrees, as shown in the side view of the drill point, Figure 10.11.

In creating the clearance angle, material is ground away immediately behind the cutting lip thereby removing material which provides strength in this important position behind the cutting edge. The clearance angle cannot therefore be made excessive since removal of too much material seriously weakens the drill and also reduces the heat conduction capacity of the cutting region, causing heating of the drill point. Large clearance angles cannot be used on account of both of these effects.

If clearance is too small, or non-existent, the point of the drill merely rubs on the work and overheating is the result. Carbon steel drills are less common now than was once the case, so overheating is less of a problem (it is possible to draw the temper of a carbon steel drill by overheating it) but too low a clearance angle stops the drill cutting and turns the process of hole-making into one of heat production. Clearance angles must be maintained in the range between 12 and 15 degrees.

Grinding smaller drills

Commercially, the flanks of the drill point are ground as the flanks of a cone i.e. the flanks are slightly convex when viewed from the side. This convexity is quite noticeable on large drills, around ½in. (12.5mm) in diameter, but if examined closely, the surface will actually be found to be not far removed from being flat. On smaller drills, the convexity is barely discernible and it will be quite satisfactory to sharpen small drills with the flanks of the point ground flat. Point angle and clearance at the cutting lip must, of course, be maintained, but the use of flat flanks simplifies the achievement of the desired shape.

For small drills even the fine stone of the grinder is too fierce and something gentler is required otherwise much of the drill will be removed when all that is required is a fine skim off the point. A slip stone, or general-purpose carborundum stone may be used and the drill sharpened by hand, it then only being necessary to hold the drill at the correct (compound) angle and to rub it gently backwards and forwards on the stone. In order to maintain the required angle, a watchmaker's glass should be used and the drill point examined after every few strokes since, even at the low speeds of hand grinding, only light touches are usually required unless the drill has been broken or damaged through contact with hardened vice jaws or some such object.

For small drills, a gauge to check the angle of the point (as described below for the larger drills) is not very valuable unless it is made to fairly fine limits. The eye is remarkably good at detecting faults in shape however, especially if symmetry is really the only criterion, and it will not be found difficult to see when the drill point is not quite right and what needs to be done to achieve the desired shape.

Grinder safety

For sharpening drills, a double-ended bench grinder having coarse and fine wheels, of the type described in Chapter 14, is normally used. For drill grinding, only small amounts of material need to be removed and the fine wheel is entirely satisfactory.

A grinding wheel is designed to be used on its periphery and it is strongest when loaded on the edge. It is relatively weak when subjected to a load on the side, and may shatter, with disas-trous consequences, if subjected to a large side load. Heavy work, such as grinding and sharpening lathe tools must therefore only be undertaken on the edge of the wheel.

Although described as being made in high-speed steel, drills are less tough than lathe tools and only very light touches of the drill on the grinding wheel are required. In any event, there is a need to remove only small amounts of material from the drill tip and very short and light contact is all that is needed.

The arrangement of the wheel guards on a bench grinder leaves the sides of the wheels exposed. Since it is normal to approach the grinder with the edge of the wheels presented to the operator, the most natural approach for grinding the end of the drill at an angle is by presenting the end of the drill to the side of the wheel. In this position, the hands are relaxed and comfortable, and are in the position most suited to control the drill accurately. Consequently, it was normal to teach apprentices to sharpen drills in this way and it still remains the natural method for those who were so taught, including the modeller who was invited to demonstrate drill grinding in my workshop, so that photographs could be taken.

Note particularly, however, that this is definitely not the preferred method of using the wheel, and act accordingly. Safety goggles should be worn whenever any grinding is performed and the wheel guards and safety guards must be in position at all times. In Figures 10.12 and 10.13, the safety guards have been removed to allow the photographs to show clearly the movement of the drill that is required.

It is very important that the tool rest is adjusted so that there is only a very small gap between it and the wheel, to ensure that nothing can be pulled down into the space between the wheel and the guard. If this does happen, the wheel is very likely to be broken by the subsequent jam-up.

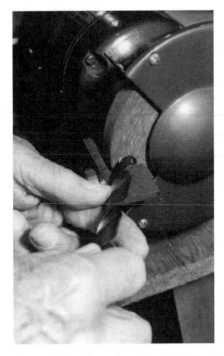

Figure 10.12 The starting position for grinding a large drill on the side of the grinder wheel.

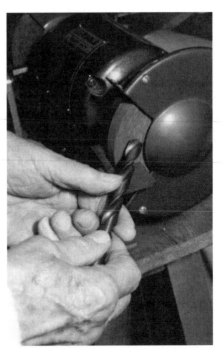

Figure 10.13 The position of the hands and the drill after passing one lip over the surface of the grinding wheel.

Grinding larger drills

As noted elsewhere, in relation to the filing of curves and compound shapes, it pays to have a clear idea of what is required before commencing. This can best be obtained by making up a template. Since drill grinding will be a continuing activity, a basic template of drill point angle should be made up in a suitable piece of steel sheet – 20 swg or 1mm is about right. A template should not be necessary for estimating the clearance angle since the actual value of this is not all that critical and may be estimated by eye.

To grind the drill and create the cutting lips, a compound movement of the drill against the grinding wheel is required.

The side of the wheel is much the most convenient to use, and the movement of the drill which is required is shown in Figures 10.12 and

10.13. Figure 10.12 shows the position as the cut starts, and Figure 10.13 the finishing position, indicating the change in approach angle which the drill must make in order to create the clearance.

Initially, the drill is held at 31 degrees to the line of the grinding wheel, with the cutting lip horizontal. To grind the flank of the point, the drill must be rotated and, at the same time, the shank lowered, in order to create the clearance angle behind the cutting lip.

The drill is held loosely between the thumb and middle and index fingers of the left hand with the thumb located in one flute, and the hand supported on the rest on the grinder. The drill is controlled by the right hand, being pushed forward and rotated, at the same time, lowering the right hand to create the clearance angle.

For drill grinding, very little material normally needs to be removed and the fine wheel of the grinder is consequently satisfactory.

Grinding and sharpening aids

Small drills can quite adequately be sharpened by hand on a small slip stone, grinding off the lips of the drill to a flat profile rather than the commercially produced cone shape. To grind the cone-shaped facets on a large drill on a grinding wheel, a compound action of the hands is required, as described above. This action can be mimicked by a relatively simple attachment for an ordinary grinder, and these attachments are available commercially or as parts and drawings for home manufacture.

The principle of operation of a drill grinding jig (the usual name) is illustrated in Figure 10.14. The drill is placed in a carrier or holder which has a pivot rod which can rotate in a bearing in the top of a stand. The carrier can be swung to pass the end of the drill across the side face of the grinding wheel. The carrier is angled so that the drill lies at 30 degrees below the horizontal and the face of the lip presented to

the wheel is ground at the required angle. The feed of the drill towards the wheel is controlled by a feedscrew attached to the carrier, against which the end of the drill rests.

To generate the 12 to 15 degrees relief angle behind the cutting edge, the bearing for the carrier is not vertical but is inclined towards the wheel. The bearing has a short, spiral slot machined in it, and a cross pin in the pivot shaft engages with this slot so that the carrier is lifted out of the stand as it rotates. This means that the carrier and the drill advance towards the wheel as they rotate, grinding progressively more material off the point and automatically backing off the cutting edge.

To ensure that the cutting edge is correctly ground, there needs to be a stop at the end of the carrier, against which the flute of the drill bears, so that the cutting edge is vertical at the start of the cut, otherwise the backing-off cone is not correctly aligned on the end of the drill.

The attachment ensures that the two principal angles on the drill point are automatically maintained during grinding and the operator need only adjust the cut on the two lips to ensure that they are of equal length.

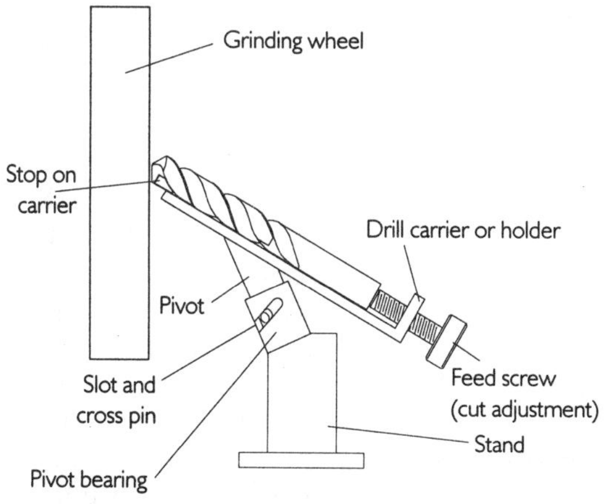

Figure 10.14 Arrangement of a drill grinding attachment for the bench grinder.

Drilling speeds

Recommended cutting speeds for drilling are substantially the same as those used when turning and drill speeds may therefore be inferred from the information given in Chapter 14 (Tables 14.2 and 14.3). When drilling, the cutting speed is that calculated for the drill diameter and the speeds to use may be read directly from Table 14.3. However, although there is seldom a need to turn items much smaller than ⅛in. diameter, there are frequent occasions on which very small holes are required and it is useful to reproduce Table 14.3 here (as Table 10.4), confining it to the range of sizes below

½in. (12.5mm) and extending it downwards to show the speeds required for drills down to the smallest normally required, the No. 80, or 0.35mm.

Notice particularly from Table 10.4 the very high rotational speeds which are required in order to keep small drills cutting at the correct speed. In practice these very high speeds are seldom available, the limit for a normal bench drill usually being of the order of 3500 rpm. It can, however, be useful to adopt higher speeds on some occasions and if very much drilling is likely to be undertaken in the smaller sizes, the construction or acquisition of what is usually called a sensitive drilling machine might be contemplated. This type of drill is specifically designed for use with small drills and it provides a very sensitive feed arrangement and utilises only a small chuck in conjunction with high rotational speed.

Table 10.4 Drilling Speeds, Diameter and rpm.

Drill Diameter		Cutting Speed ft/min				
(in.)	(mm)	300	100	85	70	60
No. 80	.35	60260	20100	17100	14100	12060
No. 70	.70	39300	13100	11150	9200	7870
¹⁄₁₆	1.50	18340	6120	5200	4280	3670
⅛	3mm	9170	3060	2600	2140	1835
¼	6mm	4580	1530	1300	1070	920
⅜	9mm	3060	1020	865	715	610
½	12mm	2290	765	650	530	460

The following cutting speeds should be used:
brass, nickel silver, aluminium alloys, 300 ft/min
BMS, copper, gunmetal, phosphor bronze, 85 to 100 ft/min
cast iron, 70 ft/min; stainless steel, silver steel, 60 ft/min.

Starting the drill

Centre marking

An ordinary jobber's drill cannot be expected to make its own centre, and for two reasons: first, it has only a thin web holding it together,

and it is therefore relatively flexible, and second, the rather flat point of standard drills (118 degrees included angle) makes the drill poor in starting the drilling operation.

Short stub drills, with their much shorter fluted length than jobbers' drills, will start a hole immediately below their axis, and if the need is just for a hole, or if some other means is available to determine its position, there is no need for the marking-out and centre-punching operations.

For these reasons, and to ensure positional accuracy, the centre position of a drilled hole is marked by use of a centre punch. This produces a small, shallow, vee-shaped dimple which not only allows the operator to see where the hole should be, but also permits a small twist drill to find the position when drilling commences. This initial drilling must, however, be with a small-diameter drill. A larger drill, with its thicker web and larger point, is difficult to start directly in a centre punch dimple.

A small drill more readily finds the punched centre and will even bend significantly to commence drilling in the centre if the operator has not located the dimple immediately below the drill's axis. Obviously, any significant bending is likely to break the drill if it is excessive or is allowed to persist during any reasonable depth of penetration into the work and so must be corrected immediately by adjusting the position of the job in relation to the drill's axis.

If a drill does find the centre, but remains bent, it is not operating in its designed condition and cannot be expected to produce an accurately positioned hole which is square to the face of the workpiece. If the tip of a small drill is brought into contact with the work and cannot find a centre, it bends and describes a wide arc on the work and will break or bend if pressure is applied.

For drilling small holes, it is usually not necessary to clamp the workpiece to the drill table, although it is desirable that the job should be mounted in a machine vice if its size and shape allow. With loose mounting of the work and the vice on the drill table, the operator can adjust the position of the work to prevent the drill bending thereby avoiding the risk of drill breakage and ensuring that the hole is correctly positioned. The initial drilling operation is normally by use of a small drill which can be expected to find the small centre punch dimple to create a good centre (or initial hole) should a larger drill eventually need to be put through.

Sometimes, really accurate location of drilled holes is required. In these instances, it may be that two or more holes are required in some accurately defined relationship, in which case consideration should be given to co-ordinate marking out or a centre drilling operation which might perhaps use the lathe's feedscrews and dials as the means of obtaining the required accuracy. If position is required accurate to what may be discerned by eye, it is usual to use a small drill simply to enlarge the centre initially, without attempting to drill to any depth. The position of this enlarged centre may then be viewed with a watchmaker's glass (loupe) or magnifier relative to the marking-out lines to judge its positional accuracy. Figure 10.15 shows a small brass plate in which the centres of holes near the corners have been enlarged using a small drill.

Figure 10.15 Centre punch dimples enlarged using a small drill as a preliminary to drilling at the required size.

Correcting a misplaced centre

If an enlarged centre does not correspond with the desired (marked) position, it must be 'drawn over' in the required direction. Various dodges can be employed for this operation, and the method to use depends upon the magnitude of the error and the relative size of the finished hole in relation to the existing enlarged centre. Workpiece thickness also affects the choice of method.

Assuming that the error is small, the simplest way is to set up the workpiece at an angle below the drill so that further drilling causes the centre to move in the desired direction. Once sufficiently deep, the central position can be expected to have moved by a discernible amount and its position may then be examined as before to determine whether the amount of movement has brought the centre into line. It is then necessary to put in a larger, stiffer drill with the work supported directly to the drill table to establish a centre or hole which is square to the base of the work.

If the initial centre enlargement has revealed a gross error in position, it is probably wise to re-establish the correct position by drilling through from the other side (if a through hole is required) since this produces a good hole. To assist the location of the hole, it is generally best to drill through any associated holes, if this is possible, to provide a reference for the hole's position.

If drilling from the reverse is not possible, or the hole is blind, an approach must be made from the marked-out side. It is usually best in these cases to take out the drilled depression by chipping away the adjacent material to produce a flat surface on which to start again. A milling operation (refer to Chapter 16) may be used to machine a small flat, if this is necessary.

If a misplaced hole position has been established in thin material, but the drill used for centring is small in relation to the final size, it is

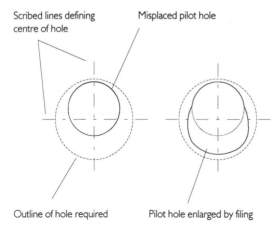

Scribed lines defining centre of hole

Misplaced pilot hole

Outline of hole required

Pilot hole enlarged by filing

Figure 10.16 If a hole is drilled in the wrong position in thin material, it can sometimes be corrected by filing the hole to bring its centre to the correct location.

best to drill through and pull the hole over by filing, assuming that a small file is available to enter the hole. This is illustrated in Figure 10.16. The hole is filed out in the required direction so that the effective centre is moved by the necessary amount. Enlargement is required in both directions in order to allow a larger drill to centre itself on the required axis to drill a pilot hole in the required position.

As can be inferred from all of the above, it is generally beneficial to drill the initial pilot hole using a small drill, small, that is, in relation to the final diameter. Provided that the initial centre punch dimple has been correctly positioned in relation to the marking-out lines, ordinary observation of a small drill as it finds the centre allows adjustment of the workpiece position, as drilling starts, to achieve a correctly positioned hole.

The foundation of good positional accuracy for holes lies in accurate marking out and centre punching of the position. Careful attention needs to be paid to the centre punching, to ensure that a clean dimple is created, exactly at the right position. If the dimple has been punched several times in order to attempt to

locate it correctly, it will almost certainly not be ideally shaped.

Using a centre drill

In some situations, marking out and centre punching of hole positions is not desirable on the basis that it cannot be performed sufficiently accurately. However, if the work can be accurately positioned below the axis of the drill, a short, stiff drill may be used to drill its own centre, without prior marking out being necessary. The suitability of stub drills for this is described above, but another type of short drill is also suitable, and may be preferred. This is the centre drill. Its primary use is for creating centres in the ends of work which is to be mounted between centres on the lathe. Its use for this purpose, and for drilling a centre prior to the use of a twist drill in the lathe, is described in Chapters 13 and 15.

The centre drill, being short and stiff, is ideally suited to the creation of centres on the drilling machine, given some means of establishing the workpiece in the correct position. One such situation is shown in Figure 10.17. A

Figure 10.17 A simple indexing head bolted to the table of the drilling machine so that the work can be indexed round to spot the fixing hole positions.

circular workpiece, still mounted in the lathe chuck, is fitted to a simple indexing spindle, the mounting base of which is bolted to the drill table.

With a centre drill fitted into the drill chuck, the work is indexed to the required number of positions and the centre drill used to drill the centres directly, the work and indexer having been set so that the holes are drilled at the correct diameter. The centred holes are afterwards opened up to the required diameter using an ordinary twist drill, without difficulty.

If the lathe can be adapted so that the workpiece can be mounted on the cross-slide, a similar arrangement is possible for the centring of holes having linear rather than angular displacements, the cross-slide feedscrew being used to provide accurate positioning. Should a vertical slide be available, calibrated movement in two dimensions is possible. Similar methods are, of course, used on vertical milling machines and mill/drills.

Drilling

Clearing the drill of swarf

Once the drill has commenced drilling, steady, but not excessive, pressure must be applied to keep the drill cutting. When drilling mild steel or aluminium alloys, and also some bronzes, the cut material comes away in long curls which rotate with the drill and represent a hazard. Pressure on the drill needs to be released from time to time to cease cutting temporarily and cause the curls to come away in short lengths, otherwise they may become caught up on the operator's hands or around the drill and chuck.

When drilling brass, cast iron or gunmetal, the cut material is generally created in the form of chips. In a deep hole (deep in relation to the diameter of the drill) the chips progress only slowly up the flutes of the drill, under pressure from the newly formed chips at the cutting edges, and the flutes can become blocked due to chips not being ejected sufficiently quickly. The drill must be withdrawn completely from the hole quite frequently to allow material to fall away and clear the flutes. Without this, small drills are likely to break due to build up of material in the flutes.

Breaking through

If a through hole is being drilled, the point at which the drill breaks through must be approached with some caution. At the instant of breaking through, the hole is not quite complete. The drill point has penetrated the work, but on the lower surface the hole has not yet reached the full diameter of the drill.

The flutes of the drill may be regarded as a very quick screw thread. Given a chance, the drill is inclined to screw itself into the work and will do this if, for example, one flute becomes jammed against some obstruction in, or on the edge of, the hole.

As the drill breaks through, some material is usually pushed out of the hole ahead of it, assuming that there is space below the workpiece. As the point breaks through progressively, this 'pushed ahead' material forms a ring around the end of the hole, as shown in Figure 10.18. If the drill is finally forced through this ring of material so that it cuts through in less than one half turn, the ring of material is not removed completely, leaving the material in the condition shown in Figure 10.19.

The drill point has been forced through the hole and the cutting edges cannot remove material since they are outside the hole. The edges of the flutes, which are not designed for cutting, are now caught on the unremoved

Figure 10.18 A ring of material is pushed out of the hole ahead of the drill. Its size is dependent on the material, the sharpness of the drill, and the pressure which is applied. Such a large burr as this should not ordinarily be produced.

Figure 10.19 As the drill breaks through, it can get caught up in the ring of material which it is pushing ahead of it. If the material is tough, and the drill is forced through under great pressure, the drill can become jammed in the work, with nasty consequences.

material and the drill screws itself into the hole very rapidly.

When drilling under power in the drilling machine, three possibilities exist: either the job rotates with the drill, if not held down sufficiently well, or the job remains stationary and the drill screws itself rapidly into the hole, or the drill stalls in the hole and the chuck keeps turning, spoiling the shank of the drill as the hard jaws rotate against it.

Backing up the work

All of the above situations are hazardous and are to be avoided. Preventing breaking through is perhaps the most sensible precaution to take since it avoids the problem entirely. This can be achieved by backing up the workpiece i.e. by clamping the work firmly to a piece of scrap material, preferably of the same type as that being drilled. The backing piece needs to be reasonably thick, otherwise the drill may break through it, with the same consequences.

The use of any sort of backing is generally beneficial. Offcuts of soft materials are frequently used for this purpose, aluminium alloys being good in this respect, but any reasonably parallel block can be used. Composite materials such as *Tufnol* are useful but a dense chipboard will be found satisfactory and is readily available, as scrap, in small pieces. Figure 10.20 shows a block of chipboard in use for backing-up a small brass plate which is being drilled near its corners.

Use of backing also removes the need to ensure that the drill will pass correctly through the clearance hole in the drill table. This allows the table to be positioned to support the work to best advantage, which is beneficial, for

Figure 10.20 To assist the drill to break through the work cleanly, it is helpful to position some backing below the work. An offcut of flooring grade chipboard has been pressed into service here.

example, when drilling on the edge or end of a large workpiece.

Size and finish

Having established a suitably sized centre, the actual operation of drilling a hole by use of a twist drill is straightforward, but there are several points which need to be borne in mind. First of all, a drill must not be expected to drill a hole exactly at its nominal size. Even if the drill is perfectly ground, it will cut holes of different sizes in different materials. For example, a sharp ½in. (12.5mm) drill might cut a hole .004in. to .006in. (0.1mm to 0.15mm) larger than its nominal diameter in steel but within .001in. or .002in. (0.025mm or 0.05mm) of its own size in copper. A newly sharpened drill also tends to cut bigger than a drill which has been dulled at the corners.

The second point to bear in mind is that a drill does not necessarily provide a good surface finish inside the hole, particularly if the drill is not newly sharpened. The rough finish results from the fact that the drill is designed to cut on its end and material is torn out of the sides of the hole rather than cut in a clean fashion, the effect usually being more obvious if the drill is not really sharp, or the material is particularly tough.

If a correctly sized hole having a good finish is required, an alternative procedure known as reaming is utilised, the cutter (a reamer) being designed to cut on its outside diameter in order to produce the required surface finish and size in the hole. Its action is quite different from a drill. Reamers and reaming are described below.

Progressive opening out

The technique of starting the drilling operation with the work held in the machine vice, but not otherwise held down, is described above. This allows easy positioning of the work to bring the centre 'pop' immediately below the axis of the drill by observing that the drill does not bend when finding the centre. For small holes, say, up to ¼in. or ⁵⁄₁₆in. (9mm) in diameter, a hand-held machine vice will adequately resist the torque produced by drilling. For larger drills, the torque exerted by the drilling process becomes significant and it is necessary to consider clamping the machine vice, or the work itself, to the drill table. Should clamping be considered appropriate from the outset, a tapered spindle held in the drill chuck provides the means to locate the hole centre accurately below the drill spindle axis. This is similar to the method of centring the four-jaw chuck with reference to the tailstock, described in Chapter 13.

Once the work is clamped into position, successively larger drills may be put through progressively to bring the hole to its final size, the clamping preventing any displacement of the hole during drilling. Enlarging the hole progressively in this way also helps to avoid an oversize hole since the fact that all except the first drill are cutting only on their outer edges tends to avoid errors caused by faulty grinding of the drill point. It also reduces the power required to drill the hole and may improve the finish inside the hole. Usually, progressively doubling the diameter produces good results, so a ⅛in., ¼in., ½in. (or 3mm, 6mm, 12mm) sequence is satisfactory.

Large drills are not very well suited to finding a small centre, due to their thicker web (larger point). Also, a large drill requires a slower speed, which is again not conducive to good centring. There are benefits all round in the progressive approach to the final size. Having the work clamped below the drill axis also helps to avoid difficulties when attempting to open out a hole with a large drill since it guarantees that the drill, if correctly ground, cuts equally on both edges. If the work is not clamped down, but positioned by eye, one lip of a large drill sometimes catches in the pre-drilled pilot hole and throws the workpiece (and machine vice) around, chipping uneven lumps out of the work as it does so.

A similar effect can also be obtained when attempting to ream a hole to final size with the reamer mounted in the drill chuck, and it is advisable to clamp the work securely during both drilling and reaming otherwise the desired good finish in the bore may not be achieved.

If a drilled hole is needed close to a particular diameter, then that size of drill is naturally used to create it. A drill generally produces a hole nearest to its nominal size if it removes only a small amount of material from an existing hole, and the enlarging sequence should, therefore, allow for the final drill to remove only .020in. (0.5mm) of material, and the previous drill chosen with this in mind. This method of sizing a hole is recommended when an accurate size is required, but a fine finish to the hole is not necessary, for example to create a close-fitting hole when riveting.

Removing burrs

Most machining operations do not usually take place absolutely cleanly, since some material is not cut away, but is deflected out of the way. When drilling, some material is pushed outwards on the entry side and some is also pushed out, ahead of the drill, on the exit side. This means that burrs form on both sides of the work. The size of the burr is a function of several factors, but the most important is the bluntness of the drill, burrs becoming larger as the drill becomes blunter. So, the first essential is to keep drills sharp. Nevertheless, burrs do form and must be removed, otherwise the fit and the appearance both suffer.

The traditional way to remove burrs is to use a slightly larger drill to put a small chamfer on the edge of the hole. It is usually sufficient to do this by hand, particularly if the drill is small, since it can just be twirled in the fingers, and the burr is easily removed. For models which are miniatures of some real object, you may feel that the method is inappropriate, since even a small chamfer at $\frac{1}{12}$ scale, represents something huge on the prototype. There may also be situations where, even on real items such as tools, the chamfer formed may appear unsightly. In these cases, it is often better to remove the burr by rubbing the surface of the work with a smooth file. If the burr is small, this usually removes it adequately, but it may turn the burr over into the hole. In these cases, the drill should be put into the hole again, by hand, and twirled in the fingers, when it will remove the vestiges of the burr, and leave a nice, sharp corner.

For large holes, it is best to use a scraper, or one of the specialised de-burring tools.

Triangular holes

A problem which is sometimes encountered, particularly when opening out holes in thin sheet materials, is that of the 'not-round' hole. The habit of the cutting edges of a large drill catching on the rim of a small pilot hole is described above. This principally occurs if the drill axis and the hole are not aligned when contact first occurs.

Since the drill is running eccentrically to the pilot hole, it tends to throw the workpiece around in a circular motion. However, the drill may still be cutting the work somehow, and the combination of the two motions causes the cutting edge (or that part of the edge which is cutting) to describe a triangular motion relative to the work. The result is a hole with a triangular form, as shown by the example of Figure 10.21. Since the problem is caused by the work moving about the drill axis, the solution is to

Figure 10.21 The start of two triangular holes in a steel plate. These have been caused by the larger drill not being correctly aligned with the pilot hole.

bolt the work to the drill table (under the drill axis) and thus prevent movement.

The problem is likely to occur also, whenever the drill approaches the work without proper guidance from its point or its outside diameter. It is likely to be worse, therefore, for larger holes drilled in thin sheet materials, or if a pilot hole is being opened up, as described above. A palliative may be to back up the work, since this gives the drill point something to cut, and as it forms, the dimple guides the drill point and provides a steadying action.

The problem is heightened when drilling softer materials such as aluminium alloys, since the drill is then able to cut even though it is not in the ideal situation. Thin alloy sheets are the worst offenders in these respects.

Drilling and punching thin sheets

General

Small holes in sheet materials are usually among the most straightforward drilling opera-

tions. The holes are frequently required only for fitting attaching parts (rivets or screws) and position is often not critical since what matters is that hole positions should match those in the mating item, through, or into which the fastenings pass. There are naturally situations in which hole positions are important, in clock plates for example, but whenever the hole is small in relation to the material thickness, there is usually no difficulty in actually producing (drilling) the hole.

As drill size increases, the depth of the drill point, with its standard, 118-degree included angle, naturally also increases. If the material is thin, this means that the drill point breaks through the work before the drill is cutting at its full diameter. This leads to the situation in which the cutting edges of the drill are liable to catch in the edges of the hole, in the same manner that this can occur when opening out an existing hole with a larger drill, as described above.

Under these circumstances, there is little guidance for the drill in the material, since the point has broken through, and the drill is not yet cutting at its full diameter. In any event, if the material is thin, there is little material to act as a guide, and it becomes very difficult to achieve positional accuracy.

There is also another problem. If the material is thin, it has little strength, and when the drill starts to cut, material tends to 'climb' up the drill rather than the drill cutting into the material and the result is often disaster.

One way to mitigate these effects is to clamp the work securely between two pieces of scrap, preferably of the same material as the workpiece. All three items must be drilled at the same time, and their combined thickness must suit the diameter of the drill being used. The top packing (at least) must be sufficiently thick and strong to withstand any tendency for the thin workpiece to climb up the drill.

Modifying the drill point

Another way to approach the problem is to re-grind the end of the drill so that the point is flatter. There is a limit to this, however, since a flat-ended drill naturally has no point to provide guidance, and as the included angle of the point increases the force needed to push the drill through the work also increases. A compromise can be adopted by using an included angle of 145 degrees or so, 72 degrees each side of the drill's axis as frequently recommended in some workshop practice books.

Using this drill point angle means that a $^5/_{32}$in. drill has a point 'depth' of .025in. and a $^1/_{16}$in. drill, a depth of .010in. so, if thin sheets do need to be drilled, the hole must be quite small for the drill to be cutting at its full diameter before breaking through. In metric terms, the above figures equate with a 4mm diameter drill, and 0.64mm thick material, or a 1.6mm diameter drill, and 0.25mm material, all of these figures relating to an included angle at the drill point of 145 degrees.

Deforming the workpiece

A further way to cope with thin materials is to make the material thicker (effectively) by deforming it into a more suitable shape. This can be done by placing a block of hard wood below the drill point, either bolted to the drill table, or held in a machine vice which is itself bolted down. The drill can then be used to drill a shallow depression in the block, just deep enough to cut to the drill's full diameter.

The situation when drilling the thin sheet material is shown in Figure 10.22. The work is pre-drilled with a small pilot hole and this is placed immediately below the drill point. The rotating drill is brought down firmly onto the work, guided by the pilot hole and the pressure forces the thin metal into the depression before

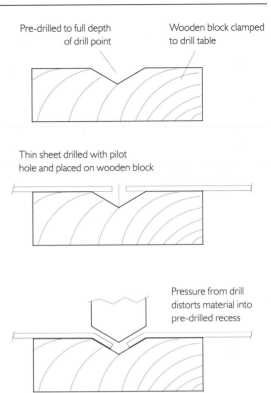

Pre-drilled to full depth of drill point

Wooden block clamped to drill table

Thin sheet drilled with pilot hole and placed on wooden block

Pressure from drill distorts material into pre-drilled recess

Figure 10.22 One method of drilling a thin sheet.

the drill starts to cut. The deformed material, which has the same diameter as the drill, acts as a guide during the drilling operation, and the drill always cuts at its full diameter.

Punching holes

If very thin sheets need to be pierced, it may be best to abandon the idea of drilling and instead use a punch. These are available commercially but they can be made up fairly easily. A case-hardened mild steel block can be used in association with a punch made from silver steel which has been hardened and then tempered to dark purple. If only a few holes are needed, it is not essential to harden the mild steel block which acts as the die.

Drilling brass and gunmetal

Reference to Chapter 14 will show that the cutting point of a lathe tool is created by grinding clearance angles on the tool. These are known as top rake, side rake and front clearance. For turning brass or gunmetal, top rake is normally ground at zero degrees, or may even be negative, the tool being ground downwards towards the point.

For a conventionally sharpened drill, the equivalent of top rake is determined by the helix angle (spiral) of the drill, as shown in Figure 10.23. Drills are available having different helix angles, normally described as having a 'standard', 'quick' or 'slow' helix, but unless the flute is straight, there must always be a positive rake at the cutting edge.

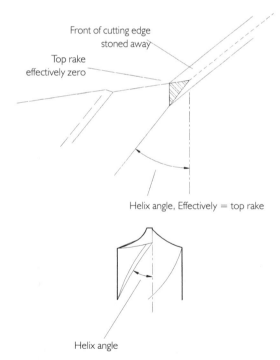

Figure 10.23 A close-up view of the cutting edge on a drill, showing how a small flat can be put on with a slip stone before using the drill in gunmetal or cast brass.

The tendency for the drill to screw itself into the work when breaking through is described above. A cutting tool with positive rake shows a great propensity for digging in when cutting brass or gunmetal. That is, there is a tendency for the cutting edge to get under the surface and force itself into the work. On the lathe, this is not usually a problem since the tool feeds are ordinarily through screws which can resist the pulling action which the work has on the tool.

When using the drilling machine, the situation is quite different. The down feed is via a rack, which can just as easily be pulled from below as pressed down by the handle. In addition, considerable force is normally required when drilling and there is consequently considerable pressure on the feed handle. Should the drill dig in, and be pulled into the cut, it will 'walk away' rapidly, screwing itself into the workpiece and most probably out of the other side, totally out of control. During this process, the drill is ripping material out of the hole, screwing itself in at a rate determined by the helix angle, virtually cutting itself a thread in the manner of a tap. The finish in the hole is therefore not good, size is not likely to be correct and if you wanted a blind hole, it may now be too late. Positional accuracy may also suffer and the work is possibly spoiled.

There are, of course, dangers posed by the dig in. The drill is caught in the work, which is trying to rotate with the drill and considerable torque is necessary to resist this. If the holding force is not sufficient, the job rotates and the fingers, or whatever is in the way, may be damaged. The work itself, or even the drilling machine may also be damaged.

Since prevention is better than cure, the remedy is to modify the cutting edges of the drill so that they have zero rake. At one time, straight flute drills were produced especially for use in brass, but these are no longer available. Fortunately, however, the drill's cutting edge may easily be modified to produce the desired rake.

The enlarged view of the drill's cutting edge shown in Figure 10.23 shows that the edge has a top rake equal to the helix angle. If the edge is modified, by removing the shaded portion, the top rake is reduced to zero degrees. To achieve this change, only a small amount of material needs to be removed, creating a flat just a few thou wide. This can be achieved quite simply by rubbing a fine slip stone down the cutting edges a few times, parallel with the axis of the drill. Once modified in this way, the drill will cut normally in brass and gunmetal and will exhibit none of the grabbing and digging in which will otherwise occur.

After modification of the edge, the drill needs re-sharpening before use in other materials.

All drills for use in brass do not need to be modified. When drilling ordinary through holes in thin sheets, drills up to ³⁄₁₆in. (5mm) or so in diameter are quite satisfactory if conventionally ground, especially when drilling half-hard sheet (the normal temper). However, brass or gunmetal castings are a different proposition, due to their different composition, and great care must be exercised when drilling even small holes in these materials if the drill point has not been modified.

Cross drilling

The cross drilling of circular rods is a frequent requirement. It can be carried out by eye but this doesn't always produce good results and some form of drilling jig is frequently employed. However, if the need is only an occasional one, it may not be worthwhile making a jig and the 'by eye' approach may serve.

Assuming that the item must be machined flat on the end, it must at some time be in position in the lathe. The tool point is set at centre height for facing the end and once the turning operation is complete it may be used to scribe a line across the diameter on the end. Without rotating the chuck, the tool can also be used to scribe a line along the rod exactly corresponding with the marked diameter thereby providing a reference for the hole position.

Sometimes, a turning tool does not produce a very fine line when used for scribing the reference lines (it depends on the tool and material, which way the tool moves against the stationary work and how much pressure is applied). It may be preferable to position a rectangular bar in the lathe toolpost, adjusting its height with reference to the centre of rotation which is visible on the turned end. Once set, the bar may be used to scribe the diameter and longitudinal line.

Marking out for the hole position along the length allows a centre punch dimple to be put in and the work can then be mounted in the machine vice ready for drilling. The work needs to be positioned so that the diametral line across the end stands vertical with respect to the drill table. It might be possible to check this with a square standing on the table, but if the workpiece is short, or the base of the machine vice has integral lugs, like the one illustrated in Figure 10.7, this may not be possible.

A machine vice without such lugs is a useful alternative for small work, since it allows a

Figure 10.24 A circular workpiece with a centre line scribed on its end, can be aligned for cross drilling using a square.

Figure 10.25 A small drill might be used for aligning the work for cross drilling.

closer approach to the work, as illustrated in Figure 10.24.

If a square cannot be used to align the work, it is best to align the rod's diameter with the drill axis by placing a small drill in the chuck and adjusting the rod's position in the vice so that the point moves parallel to the scribed line as the chuck is lowered. This is shown in Figure 10.25. You may find it is easier to do this using the point of a pocket scriber, since the finer point makes proper alignment easier to detect.

An alternative method of holding small work is by use of a pair of vee blocks. These are rectangular blocks which are machined with one or more vee-shaped grooves in which a circular bar can be rested. Some types have rectangular grooves machined along both sides and are supplied with a stirrup clamp which can be used to hold the work, as shown in Figure 10.26. A long workpiece can be clamped to one block and be supported by the other, providing a rigid arrangement and allowing the drill and chuck to approach the work.

A vee block is again not helpful if the work is short, and another method must be sought. A simple alternative to a full-blown cross-drilling jig can be created if a bush is made up having the same outside diameter as the rod to be drilled and itself having a true axial hole down its centre. Workpiece and bush can then be set up in the machine vice, as shown in Figure 10.27, and the bush used to guide a small pilot drill.

Should a jig be thought desirable, it can best be arranged as a vee block incorporating a workpiece clamping arrangement but also having a bridge piece which can accept a guide bush centrally over the vee. Something specially made is therefore required, provided with inter-

Figure 10.26 A pair of vee blocks is the preferred method for holding round work.

Figure 10.27 Cross drilling can be simplified by using a drilled bush as the guide, but bush and workpiece need to be exactly the same diameter and the machine vice must have parallel jaws so that it grips both the bush and the work equally.

229

changeable bushes to act as guides for differently sized drills. Numerous designs have been published over the years, and making one of these can be an interesting and useful exercise.

Reamers and reaming

A twist drill, even if correctly ground, is not capable of drilling a hole to an exact dimension and will, in general, cut larger than its nominal size. One way of testing whether the drill is cutting near its nominal size is to drill a hole in some piece of scrap material and then test the fit of the shank in the drilled hole. This should reveal any gross oversize in the drilled hole and allow any fault to be rectified. This test, although widely practised, should be regarded only as an informal check, since the shank of a drill is always finished to a smaller diameter than the drill itself.

The method of drilling a hole slightly smaller than that required, and following up with a drill of the required size, to produce an accurate-sized drilled hole, is described above. However, since a drill is designed to cut on its end, the finish imparted to the bore is, in general, rather rough. A drilled hole is not suitable for use as a bearing or whenever an accurately sized smooth bore is required, and reaming is employed whenever a good finish and/or a specified size are required.

Figure 10.28 shows a group of reamers up to ⅝in. (16mm) in diameter. The cutting edges of a reamer are formed on the outside diameter by cutting a number of flutes along the length. These usually progress in a slow spiral from tip to shank, but some reamers have straight flutes. There are from four to ten flutes cut on the reamer, depending on its size.

The reamers illustrated in Figure 10.28 are designed to produce parallel-sided holes and are passed completely through a pre-drilled hole to clean up its internal surface. Since parallel-sided holes are produced, these cutters are known as parallel reamers.

Although described as a parallel reamer, a small amount of taper is generally incorporated in order that it may enter a drilled hole slightly smaller than the nominal diameter. Since the reamer cannot readily be sharpened while still remaining at the same size, it is advantageous if it removes only small amounts of material, and the initial drilled hole is, therefore, only slightly smaller than the final hole size.

There are essentially two types of parallel reamer – hand reamers and machine reamers. Either type can have a parallel shank, with squared end, so that a tap wrench can be used to turn and drive the reamer, or may have a tapered shank, making it suitable for holding in a machine tool.

The essential difference between hand and machine reamers is that hand reamers are significantly tapered in their length. A hand reamer is used by drilling a hole slightly smaller than the finished size, but large enough for the small end of the reamer to enter. The reamer is then fitted with a tap wrench, entered into the drilled hole and turned by hand so that the drilled hole is enlarged to the size cut by the reamer.

Machine reamers have very little taper, but are sharpened to cut on the end, in addition to

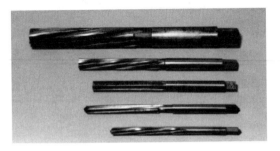

Figure 10.28 A group of reamers, up to ⅝in. (16mm) in diameter.

the sides, and can therefore be forced into a hole which is smaller than the reamer's end diameter. Accurate reaming in this manner requires that the reamer is accurately in line with the pre-drilled hole. This means that the work must be clamped below the drilling machine spindle (or in another machine tool) so that drill and reamer follow the same path successively, and there is no tendency to force the reamer to one side and cause uneven cutting in the drilled hole.

There is quite a wide variation in the amount of taper which is ground on a reamer. Among my stock of reamers, extending from ⅝in. (16mm) down to 1⁄16in. (1.6mm), some 20 reamers in all, five have tapers between .015in. and .022in. (0.4mm to 0.6mm), five have no taper and the remainder have tapers between .003in. (0.1mm) and .007 in. (0.2mm). The tapers on the hand reamers are remarkably consistent throughout the range of sizes, for example the ⅝in. being tapered .017in., the ⅜in. by .022in. and the 5⁄32in. by .015in.

All of these reamers have been bought just as reamers, from various sources over many years, and are perhaps not typical of current production. By way of contrast, a new 12mm (15⁄32in.) reamer from a well-known British manufacturer, identified on its protective box as a hand reamer, has a tip diameter of 11.5mm, making the taper the equivalent of .020in. The maximum diameter of this reamer is 12.01mm and the plain, squared shank is ground to 11.99mm.

When using a reamer, the initial drilling prior to reaming should be arranged so that the reamer removes only a small amount of material. If the reamer is tapered, the normal practice is to measure its tip and select a drill which just allows the reamer to enter the hole. The reamer is then put into the drilled hole and turned under pressure to enlarge the hole progressively to the reamer's size.

The cutting edge of the reamer flute is ground with a parallel portion, which may be from .005in. to .020in. (.13mm to .50mm) wide, immediately behind the cutting edge. There is considerable rubbing action during the cutting process and, as a consequence, reamers are inclined to chatter. The reamer must be prevented from merely rubbing and must, therefore, be pressed steadily into the work.

There are real benefits in clamping the work rigidly to the drill table and performing both the initial drilling and final reaming without disturbing the clamps. The reamer must then be in line with the drilled hole and square to the work, and steady pressure can be exerted from the feed handle. One possible problem is that the flutes of the reamer, which can be very shallow on small hand reamers, can become clogged with swarf. This causes the reamer to jam in the hole, always with unfortunate consequences. This is one reason for drilling the initial hole close to the final size, so that there is little material to remove.

Ideally, the reamer should pass once into, and once out of, the hole, so it is unfortunate if it has to be withdrawn partway through, just to clear the flutes.

When reaming under power in the drill (or lathe) speed should be reduced to roughly a quarter of that used for drilling at the same size.

Reamers should never be turned backwards as this tends to turn the cutting edge over and rapidly blunts the tool.

Since new reamers are ground with a slightly oversize maximum diameter, they generally cut slightly oversize when new. If reaming is being carried out to create the bore of a bearing, this is usually of no consequence, but if holes are being reamed to accept items for press or interference fits, this must be borne in mind when making the corresponding parts.

Counterboring, countersinking and spot facing

Introduction

Associated with drilling there are several frequently performed operations. These are the counterboring and countersinking of the mouth of a drilled hole and the local facing of the workpiece around a drilled hole, referred to as spot facing. This operation may precede the drilling of the hole but can also often be performed afterwards.

These operations may be performed using commercially available cutters but are frequently undertaken using home-made tools since commercial products tend to be rather expensive and the process of manufacture is easily carried out using high-carbon steel (silver steel).

Figure 10.29 A rose countersink compared with an 8BA screw.

Countersinking

Countersinking is perhaps the most common of the three operations. This is often carried out using a rose countersink, as shown in Figure 10.29. This is a specially made cutter designed for creating the 90-degree cone in the mouth of an existing hole to accept a countersunk-head screw. For the modeller, the rose bit has some in-built disadvantages, the principal of which is the large number of cutting edges which is normally provided.

The depth of the flutes reduces towards the point and in this region there is little space in the flutes to accommodate the cut material. When used in small holes, the flutes quickly become blocked and the bit ceases cutting since the edges are masked by the jammed-in material. Frequent clearing of the flutes is therefore necessary. The problem is emphasised by the presence of an 8BA screw in Figure 10.29. This thread is .086in. (2.2mm) in diameter, and is certainly not small by modelling standards.

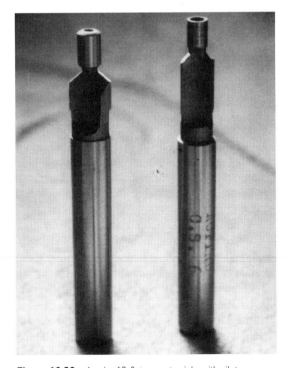

Figure 10.30 A pair of 2-flute countersinks with pilots.

A more useful form of cutter, particularly for small holes, has fewer flutes and a two-flute cutter, not unlike a centre drill, but having a 90-degree cone, is a much more useful form of the countersink. Figure 10.30 shows a pair of commercial cutters of this sort, both of which are provided with a plain pilot portion, to steady the cutter in the hole, made to suit a particular screw size. Each suits just the one size and is described as a '2BA countersink' or whatever. The use of only two (or sometimes four) cutting edges normally prevents build-up of cut material on the edges, except when cutting aluminium alloys, which are almost always a problem in this respect.

The use of a pilot on the countersink tends to reduce chatter since it acts to steady the cutter's position at the mouth of the hole and gives a smoother finish to the cut. The ordinary rose bit has a great tendency to chatter if firm pressure is not maintained, and the rough finish can be difficult to eradicate once it has occurred, although a change of speed can be beneficial in this respect since it changes the rate at which the cutting edges contact the chatter marks.

A useful form of countersink cutter can be made as shown in Figure 10.31. This is made by turning the end of a piece of silver steel to a 90-degree cone and then filing or machining away a short length of the coned end to half the diameter of the rod, less .002in. or .003in. (.05mm to .07mm). Hardening and tempering to light straw produces a tool which is chatter-free when cutting, but does need firm pressure. If, instead of having a plain shank, the cutter is given a knurled handle, it can be turned by hand and used for de-burring holes.

Counterboring

Counterboring is normally carried out for the same reason that countersinking is employed; to recess the heads of screws or bolts below the surface of the workpiece. The requirement here is to produce a flat-bottomed enlargement of the mouth of a clearing-size hole, for a particular size of screw, to create a recessed seating for the head.

A flat-ended cutter having a parallel pilot is essential for this operation. Once again, cutters of this type are available commercially and normally take the forms shown in Figure 10.32. A flat-ended cutter with the correct size pilot is

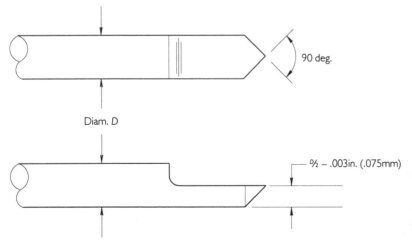

90 deg.

Diam. *D*

²⁄₂ – .003in. (.075mm)

Figure 10.31 A countersink made from silver steel

233

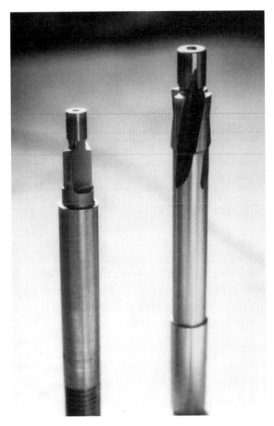

Figure 10.32 Two commercially made counterbores.

Figure 10.33 A counterbore made in silver steel.

ground to provide two or four cutting edges and can be used as a second-stage cutter to create the counterbore at the mouth of an existing hole.

Such cutters can readily be made in the home workshop, using silver steel, which is then hardened and tempered to pale straw, as described in Chapter 3. Figure 10.33 shows a close-up of a home-made counterbore. The main diameters are turned in the lathe, following which the cutter is filed down on both sides to match the width of the pilot. To create the cutting edges, the resultant flats on each side of the pilot are backed off, again by filing, to provide clearance behind the cutting edge. This needs to be carried out carefully so that, on both sides, a minute portion of the faced shoulder is left as a witness. To provide the equivalent of a lathe tool's top rake, a shallow groove is filed across the cutting face, using a small round or half-round Swiss file. This also needs to be done carefully so that the cutting edge is sharp and the witness on the face remains. The groove cannot be too deep otherwise the cutting edge is weakened and the tool's possible life is reduced.

Some backing-off of the outside diameter, just to provide clearance, may also be beneficial, since if the outside of the cutter remains at the full diameter it rubs on the sides of the counterbored hole and rapidly heats up, causing further jamming as it expands. Once again, the filing must be done with some care to avoid spoiling the cutting edge and leave a witness at the full cutter diameter.

Once filing is completed, the cutting edge must be polished (but not rounded or blunted) and the tool can then be hardened and tempered as described in Chapter 3.

Spot facing

Spot facing is carried out whenever the workpiece surface is not truly square to the axis of a drilled hole and it is required to fit a bolt or screw so that it abuts the surface adjacent to the hole. This can happen if the drilled hole is not at right angles to the surface or if the surface is in the 'as cast' condition and has not yet been machined. This is likely to be the case where a casting is provided with bosses at the positions of attaching-bolt holes and needs machining only across the tops of these bosses. This might be done by a milling operation but it is frequently simpler to use a counterbore which will provide a local seating for the nut and/or washer, the seating being cut following the drilling operation so that the tool may incorporate a pilot to centre the faced area around the hole. No great depth of penetration is required, it being necessary to clean up the surface to provide the seating. A cutter exactly like a counterbore is required, but slightly larger, in order to accommodate the washer and allow clearance for the nut or bolt head.

Boring larger holes

Once the required hole size becomes larger than the capacity of the drilling machine's chuck, or beyond the capacity of the spindle taper to hold a taper-shank drill, the drilling operation becomes that of boring. Various types of adjustable cutter are used for boring (or cutting) larger holes, ranging from variants of the old-fashioned, carpenters' brace-mounted tank cutter, through fly cutters to the more precise boring heads. A boring head is simply a body which provides a mounting for a cutting tool holder which can be offset from the axis of rotation under precise control. A home-made boring head is shown in Figure 10.34. The tool is angled in the end of a cylindrical holder which is mounted in a block which can move along a slideway formed in the body. The block is tapped to engage an adjusting screw which is free to turn in the body endplate, but is retained

by an inserted key. Rotation of the screw allows the block and tool holder to be positioned in the slide, thus adjusting the radius at which the cut is taken. For this boring head, the screw is $\frac{3}{8}$in. BSF (20 tpi) providing an increase in diameter of 0.1in. per turn.

The block can be locked in the slide through the action of an Allen grub screw which presses a brass pad onto the block. This is visible in Figure 10.34.

An alternative form of boring head is shown in Figure 10.35. This has a circular body which provides rather less variation in diameter than the one shown in Figure 10.34, but it has a 40 tpi adjusting screw and may be set more precisely than its 'big brother'. In this instance, the cutter takes the form of a forged-and-ground boring tool, not unlike the type used on a lathe. This provides the means to bore a deep hole, or

Figure 10.34 A large adjustable boring head fitted to a No. 2 Morse taper arbor.

Figure 10.35 A small adjustable boring head on a No. 1 Morse taper arbor.

a hole deeply recessed inside the workpiece. There is obviously a limit to how long the cutter might be, since additional length decreases the rigidity, but some form of extension is frequently necessary. However, as with all machining operations, overhang should always be reduced as much as possible.

The boring heads illustrated are mounted on Morse taper arbors, and are suitable for mounting directly into the spindle socket of the drilling machine, if the spindle can accommodate it, but this needs to have a sufficiently low speed to enable larger holes to be cut. Cutting a 4-in. (100mm) diameter hole requires a spindle speed of 80 rpm in mild steel (85 ft/min (26m/min) cutting speed).

A boring head mounted on a Morse taper arbor may also be mounted in the mandrel taper in the lathe and can be used for boring items which are secured to the cross-slide. However, the drilling machine spindle taper ordinarily incorporates a recess to engage the driving tang of a standard taper but this is not usually provided in the lathe mandrel. An alternative type of Morse taper arbor is more useful if use in the lathe is envisaged. This has a tapped hole at the small end so that a threaded drawbar may engage the arbor to pull and hold it into the mandrel taper. Without such assistance, the taper cannot be expected to hold except for very light cuts, and even these may be too much at larger diameters. An arbor of this type is fitted to the boring head of Figure 10.34 and its threaded end is shown in Figure 10.36.

Various other forms of tool with an adjustment of the cutting diameter are available, less sophisticated (and therefore less expensive) than a boring head. They can easily be made up in the home workshop. Figure 10.37 shows one which I made up, and Figure 10.38 provides a dimensioned sketch.

The cutter (a ⅛in. diameter, high-speed tool bit) is carried in a 1⅜in. diameter carrier, ⁹⁄₁₆in. thick, the cutter bore located ⅜in. off the axis

Figure 10.36 Tapped hole for a drawbar in the end of a No. 2 Morse taper arbor.

Figure 10.37 An adjustable hole and washer cutter.

of the carrier. A 4BA grub screw holds the cutter in place and the carrier is split and recessed so that two 4BA Allen screws may be fitted to allow the carrier to be clamped to an arbor.

The arbor itself has a ⅜in. diameter shaft, 1⁵⁄₁₆in. long, and a two-step location for the carrier which is offset from the shaft by ⅛in. A ⅛in. diameter hole for a pilot is drilled on the axis of the shaft. The ⅝in. diameter flange is made ¼in. long leaving the ½in. diameter to be ⁵⁄₁₆in. long.

The arbor should be made first and the carrier bored to be a nice fit for both diameters, afterwards being split and tapped for the clamps and reamed for the cutter bit.

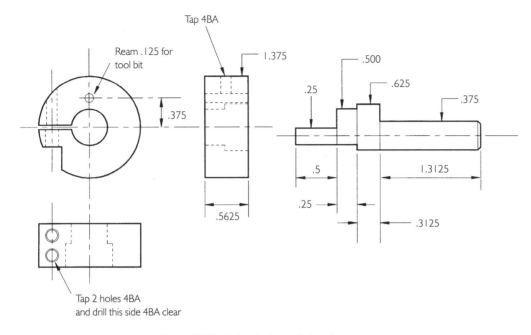

Figure 10.38 A drawing for the hole and washer cutter.

The dimensions suggested here provide a cutter which will cut holes from ⅝in. to 1⅛in. diameter, but if you need different sizes, then change the dimensions to suit.

The original cutter was made to cut holes of ¾in. and 1in. diameter in ⅛in. thick steel. To cut a 1in. diameter hole, a speed of about 325 rpm is required in mild steel, and the drilling machine must be capable of rotating the cutter at a sufficiently slow speed, as shown in Table 6.3. Do bear this in mind before building a cutter for 4-inch holes!

Although the hole in the shaft for the pilot is only ⅛in. diameter, it is better to use something larger in practice, so a ¼in. pilot with one end turned down can be secured with *Loctite*. There is quite a side thrust on the pilot due to the out-of-balance mass which will wear if not hardened. A case-hardened mild steel pilot, or one made from hardened silver steel, is preferred.

The cutter is shown in a configuration in which it is suitable for cutting holes, but if the cutting tool is ground appropriately, it may equally well be used for making washers.

Screw threads

Introduction

Everyone must be familiar with the use of nuts and bolts for fastening things together, and with the basic concept that nut and bolt must be made to fit one another. Colloquially, nut and bolt are said to be 'the same size'. However, this is not literally true since the nut must be a free fit on the bolt and clearances have to be allowed for when making both items. Before cutting one's own threads it is important to understand the basic concepts behind the manufacture of mating nuts and bolts (for want of a better phrase).

When referring to threads, it is usual to refer simply to the nominal outside diameter of the bolt or screw. In reality, four distinct diameters are involved – the maximum, or major diameter, and the minimum, or minor diameter – for both the bolt and the nut. These are shown in Figure 11.1. The minor diameter is also referred to as the core diameter.

The most common forms of screw thread are those having a simple vee form of the type illustrated in Figure 11.1. Since it is essential to ensure that the major diameter of the nut is

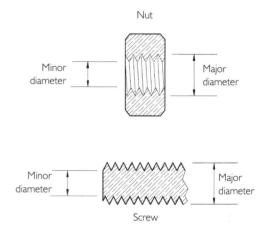

Figure 11.1 Screw thread terminology.

larger than that of the bolt, and the minor diameter of the bolt is smaller than that of the nut, the thread crests and roots are not usually formed as sharp vees but are flattened or radiused to create the necessary clearances.

Figure 11.2 shows a nut and bolt mated together. The major diameter of the bolt is reduced by flattening the crests of the thread and the minor diameter of the nut increased by the same means. Nut and bolt therefore mate

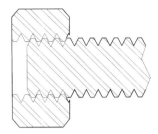

Figure 11.2   The fit between a nut and a bolt.

correctly and there is no interference between the major and minor diameters of the two parts. Flattening or radiusing of the crests is the normal means by which the clearances are created and a vee-form thread does not normally show the full shape of the vee at the crest or root. A small reduction in the theoretical thread height is usually specified.

The major and minor diameters for nut and bolt are defined in a relevant standard for the particular thread series. Together, they naturally define the height or depth of thread which is a useful measure when threads are to be cut using a single-point tool on the lathe. The height (depth) of the thread is also used to define other characteristics of the shape.

It goes without saying that nut and bolt must have the same pitch of thread (the distance between adjacent crests or roots) and the nominal bolt or screw diameter, together with the pitch and shape of the vee form, define the basic characteristics of the thread. The full height (full vee) of the thread is defined by the vee angle and the pitch. Rounding or flattening (truncating) at the crests and roots is usually specified as a fraction or proportion of the full vee-form height.

Every thread system (Whitworth, Metric, British Association etc.) has a form which is defined by an appropriate specification. In the UK, such specifications are issued by the British Standards Institute. Relevant extracts from the standards are widely published in engineers' reference books which should be referred to for fuller information about thread forms. Full details about particular threads are only likely to be needed when screwcutting in the lathe, but an appreciation of the general characteristics of thread form is also helpful when using the more convenient taps and dies for thread production. Details of the more common thread series are given below, followed by a description of the use of taps and dies.

For modelling activities, the smaller and finer pitch threads such as British Association and the fine pitch Whitworth form are likely to be of most interest, but since machine tools and accessories utilise some larger threads, the more common of these are also described below.

Thread series

Whitworth and BSF threads

In the early days of the Industrial Revolution, no standards existed for screw threads. Such 'ironmongery' as was required was locally made, each machine builder using his own standards. In 1841, Joseph Whitworth determined to establish a standard for screw threads and decided to measure samples of nuts and bolts from various sources. From the results obtained, average values for diameter, pitch, and nut and bolt head sizes were calculated.

This, the first attempt at standardisation, resulted in specifications being generated for all of the major dimensions for a complete thread system. This provided a range of nominal diameters, and corresponding pitches, definitions of bolt head and nut dimensions, the shape of the vee form and the tolerances on major and minor diameters to ensure interchangeability of threaded items from different makers.

The pitches chosen by Whitworth, although based on average pitches in use in his day, eventually came to be considered as rather coarse

239

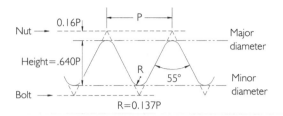

Figure 11.3 Whitworth and BSF thread form.

and with the improvement in techniques generally, a finer series of pitches seemed desirable. The original Whitworth scheme had been formally adopted by the British Standards Institute and became known as British Standard Whitworth (BSW) and the fine pitch series was issued as British Standard Fine, or BSF.

Both BSW and BSF thread series use a vee form having a 55-degree included angle, as shown in Figure 11.3. The form has radiused crests and roots, the radius being related to the pitch of the thread (P). Since the vee-form angle is constant, the depth of thread is related to the pitch. For the Whitworth thread, having a 55-degree vee form, and the standard rounding of the crests, the depth of thread is numerically equal to $0.640 \times P$. In most thread series (although there are exceptions) it is normal for the pitch to vary with the diameter so that larger diameter screws use coarser threads. In the scheme finally developed from Whitworth's proposals, nominal diameters from ¹⁄₁₆in. to 6in. were standardised, having pitches from .0167in. or 60 threads per inch (tpi) to 0.4in. (2½ tpi).

As specified, the thread has one standard form, with one major and one minor diameter specified. For a nut, the major diameter is increased by a factor equal to 0.16P and for a screw, the minor diameter is decreased by the same amount. A nut therefore has radiused crests at its minor diameter while a screw is radiused at its major diameter.

As an alternative to the radii shown in Figure 11.3, the standard allows a flat-topped, or truncated, thread form. The amount of trunca-

tion is related to the pitch but is effectively a flattening of the crests corresponding with removal of the alternative radius to the point where it just meets the straight vee form.

As for all standards, tolerances are applied to ensure that there is always interchangeability between nuts and bolts, to allow the production tooling a degree of wear. The tolerances which are provided broadly allow the major diameter of a nut to be greater than, but not less than, the standard size and a similar tolerance is applied to the minor diameter of a screw.

As for some other thread forms, the published standards for BSW and BSF allow for both nuts and bolts to be manufactured to 'close', 'medium' and 'free' limits. These different fits are represented in the standard tables by progressively wider tolerances applied to the major and minor diameters. Most ordinary nuts and bolts are manufactured using the medium or free fit limits since these allow wider tolerances and hence greater tool wear before out-of-tolerance items are produced. These different fits are unlikely to be of interest except to makers of 'production' quantities of screwed items.

Table 11.1 Common Imperial Threads

Unified Whitworth Diameter	55-degree vee		60-degree vee	
	Unified BS Fine tpi	Coarse tpi	Fine tpi	tpi
¹⁄₁₆	60			
³⁄₃₂	48			
⅛	40			
⁵⁄₃₂	32			
³⁄₁₆	24	32		
⁷⁄₃₂	24	28		
¼	20	26	20	28
⁵⁄₁₆	18	22	18	24
⅜	16	20	16	24
⁷⁄₁₆	14	18	14	20
½	12	16	13	20
⁹⁄₁₆	12	16	12	18
⅝	11	14	11	18
¾	10	12	10	16
⅞	9	11	9	14
1	8	10	8	12

Table 11.1 shows the standard pitches adopted for both Whitworth and BSF threads. As noted above, both thread series utilise a 55-degree vee form and adopt the standard shapes shown in Figure 11.3. The BSF series does not extend to sizes below ³⁄₁₆in. in diameter since the BSW sizes below this utilise acceptably fine pitches.

Unified and American Standard threads

The Unified thread series was established jointly by the USA, Canada and Great Britain with the objective of creating a standard which would allow interchangeability of threaded items between the three countries. The resultant Unified Series does not quite cover the same range of sizes as the American Standard Series, but all of the unified threads are included in the American Standard.

Two separate series, Unified National Coarse and Unified National Fine (UNC and UNF) are provided for, roughly equating with BSW and BSF, but utilising a 60-degree vee form rather than the 55-degrees of the British Standard. Neither series caters for fractional diameters below ¼in. although there are some smaller diameters designated by number references (see below). Table 11.1 shows the adopted pitches from ¼in. to 1in. diameter. For the Unified Fine Series the maximum specified diameter is 1½in. at 12 tpi, but the Coarse Series extends to 4in. diameter (4 tpi). These larger sizes are not likely to be of interest to the modeller.

Both UNC and UNF thread series include a range of fine-pitched threads extending downwards below ¼in. in diameter, both series allowing a smallest basic diameter just below ¹⁄₁₆in. These sizes are shown in Table 11.2.

In order to allow a distinction between these numbered threads and others such as the BA series and the whole number metric sizes, these threads are almost always referred to by size and pitch. Thus, the references are 2–56 UNC

Table 11.2 UNC And UNF Small-diameter Threads

Size	Major Diameter	Turns per inch	
		UNC	UNF
0	0.060	64	80
2	0.086	56	64
4	0.112	40	48
6	0.138	32	40
8	0.164	32	36
10	0.190	24	32

and 2–64 UNF etc. the coarse or fine pitch series usually being identified also.

The thread forms specified for the Unified Series are very similar to the Whitworth and BSF form shown in Figure 11.3 (except that the vee form is 60 degrees) with truncation of the crests to create the required clearances.

Model Engineer threads

When considering the simple case of nuts and bolts, it is usually the case that larger diameters utilise coarser pitches. The coarser (and deeper) threads, being basically stronger than the finer (and shallower) ones, naturally suiting the larger diameter bolts and screws. For the same reason, coarse pitch is normally used if aluminium alloy components need to be threaded since this provides more strength due to the thicker root of the thread.

In cutting screw threads for models, conditions are frequently quite different from those in the 'real' world. There are particular problems to solve such as cutting threads on thin-walled tubes, and arranging fastenings for miniature items. Fine-pitch threads are extremely useful in both cases (the depth of thread is small) although rather larger diameters with fine pitches are usually needed.

There is also a frequent need to utilise items which are threaded both internally and externally, to be screwed onto (into) both threads simultaneously. This requires threads of different diameters, but the same pitch. There is therefore value in having threads with finer pitches relative to their diameters, and also thread series having the same pitch throughout a range of different diameters, usually called a constant-pitch series.

Fine- or special-pitch threads are allowed by the BSW specification which contains recommendations for standard depths of thread for all pitches from 4 tpi to 20 tpi (rising in steps of 2 tpi) and also for 24, 26, 28, 32, 36 and 40 tpi. From these recommendations there have developed the special, fine-pitch, Model Engineer threads which provide two distinct, constant-pitch series having pitches of 32 and 40 tpi. The standard allows any size of bolt to be threaded with any of the special pitches, but what has developed is a range of taps and dies for cutting threads from ⅛in. in diameter to ½in. in increments of either ¹⁄₃₂in. or ¹⁄₁₆in. as shown in Table 11.3.

For the standard form, the depth of thread may be calculated from 0.64 divided by the turns per inch. Thus the depth of thread for 32 tpi is .020in. and for 40 tpi is .016in. Comparison between Tables 11.1 and 11.3 will show that ⅛in. × 40 tpi and ⁵⁄₃₂in. × 32 tpi are standard BSW threads while ³⁄₁₆in. × 32 tpi is from the BSF range. The 40 tpi range is most interesting since the thread depth is .016in. making the core diameter just about ¹⁄₃₂in. smaller than the nominal size. It is thus possible to tap a ⁷⁄₃₂in. diameter hole with a ¼in. × 40 tpi thread, although this does not mean that ⁷⁄₃₂in. is the correct size hole to drill before cutting a ¼in. × 40 tpi thread. See below.

British Brass thread

Although not covered by a British Standard, there is a well-established thread which has been used for many years for gas fittings, brass tubes and general work. The origin of the thread seems to have been lost, but it is still widely used for gas fittings in sizes above ¼in. and is also used in Germany for the imperial sizes ¼, ⁵⁄₁₆, ³⁄₈, ⁷⁄₁₆, ⁵⁄₈, ¾, ⁷⁄₈ and 1 inch diameter.

For all diameters, the Brass thread uses a Whitworth-form (55-degree) of 26 tpi. It therefore complements the 32- and 40-tpi Model Engineer threads and provides a further fine-pitch series. The minor or core diameters for the range of Brass threads are shown in Table 11.4, the depth of thread for the Whitworth 26 tpi thread being .025in.

Table 11.4 British Brass Thread (55-degree Whitworth Form, 26 tpi)

Size	Minor Diameter
¼	.201
⁵⁄₁₆	.263
³⁄₈	.326
⁷⁄₁₆	.388
½	.451
⁵⁄₈	.576
¾	.701
⁷⁄₈	.826
1	.951

Metric threads

The International Metric Thread System, usually known by the abbreviation SI (from the French Systéme Internationale) was established

Table 11.3 Whitworth-form Model Engineer Threads

Size (in.)	Minor (Core) Diameter	
	40 tpi	32 tpi
1/8	.093	.085
5/32	.124	.116
3/16	.155	.148
7/32	.187	.179
1/4	.218	.210
5/16	.281	.273
3/8	.343	.335
7/16	.401	.398
1/2	.468	.460

following an International Congress held in Zurich in 1898. Metric threads have since been adopted widely, the principal non-participant being the USA where the Unified Thread System is still widely in use.

The current standard is that known as the ISO Metric Thread Series (from the International Organisation for Standardisation) represented in the UK by British Standard 3643. This specifies a range of standard diameter-and-pitch combinations ranging through basic major diameters from 1mm to 300mm. The thread takes a 60-degree vee form with flattening of the crests to provide the basic clearances, although rounding of the crests of bolts is allowable provided that the standard truncated outline is not exceeded.

A coarse- and fine-pitch series of threads is specified in BS3643 in a range of diameters which rise in small increments from 1mm. The allowable diameters are classified into first-, second- and third-choice recommendations, providing very close spacing of screw sizes. The preferred, first-choice sizes are shown in Table 11.5.

The second- and third-choice sizes provide threads with major diameters between the sizes shown in Table 11.5, making an extremely wide range of sizes available, however, not all of these are produced commercially.

To distinguish the Metric series of threads from others which use simple number references, the Metric series is usually referred to as M2, M2.5 etc. This practice avoids confusion with other series such as BA and the smaller, number-referenced, Unified threads, although the BA designator is almost always used.

As noted elsewhere, imperial size materials, while still available in some sizes, are slowly being replaced by metric alternatives. It is likely that some items, such as copper tubes, will remain available for the foreseeable future, but the gradual change to metric sizes will eventually mean that imperial threads will become less

Table 11.5   ISO Metric Pitches (mm)

First Choice Diameter	Pitch Coarse Series	Fine Series
1.0	0.25	–
1.2	0.25	–
1.6	0.35	0.2
2.0	0.40	0.25
2.5	0.45	0.35
3.0	0.60	0.35
4.0	0.70	0.50
5.0	0.80	0.50
6.0	1.0	0.75
8.0	1.25	1.0
10	1.5	1.25
12	1.75	1.5
16	2.0	1.5
20	2.5	2.0
24	3.0	2.0

useful, and therefore less widely used.

There will still be a need for a constant-pitch, fine thread series for the 'metric modeller' and three ranges of diameters having pitches of 0.5mm, 0.75mm and 1.0mm, as shown in Table 11.6, have been recommended. These constant-pitch series are not yet widely adopted.

Pipe threads

To allow for the threading of pipes and couplings, standard threads have been developed allowing both tapered and parallel (normal) threads to be employed. In the UK, the British Standard Pipe Thread (BSP) utilising the 55-degree Whitworth form is still widely used. In the USA, the Society of Automotive Engineers (SAE) standard is typical of a range of such threads which utilise the 60-degree vee form adopted for other American threads.

One problem posed by these threads is the size description which is applied, this being the nominal bore of the tube for which the thread is intended. Thus, ⅛ BSP has a major diameter

Table 11.6 Recommended Constant-pitch Metric Threads For Model Engineers

Nominal Diameter	Core Diameters		
	0.5mm Pitch	0.75mm Pitch	1.0mm Pitch
3.0	2.38	–	–
4.0	3.39	–	–
4.5	3.90	3.58	–
5.0	4.39	–	–
5.5	4.90	–	–
6.0	5.39	5.08	–
7.0	–	6.08	–
8.0	–	7.08	–
10.0	–	9.08	8.77
12.0	–	11.08	10.77
14.0	–	–	12.77
16.0	–	–	14.77
18.0	–	–	16.77
20.0	–	–	18.77

of 0.383in. and similarly the SAE Dryseal Taper Thread specified as ⅛in. is designed for a pipe of outside diameter equal to 0.405in. Only the smaller sizes are likely to be of interest to model engineers since they are occasionally specified for boiler fittings, especially the BSP sizes. These small sizes have the characteristics shown in Table 11.7.

Table 11.7 British Standard (Parallel) Pipe Threads

Size (Nominal Bore of Tube)	Pitch	Major Diameter	Depth of Thread
⅛in.	28 tpi	0.383in.	.0229in.
¼in.	19 tpi	0.518in.	.0337in.
⅜in.	19 tpi	0.656in.	.0337in.
½in.	14 tpi	0.825in.	.0457in.

The sizes listed in Table 11.7 are taken from the definition of the parallel form of the BSP thread which is intended for general engineering use. The more common form of the BSP thread is the tapered one, for which the standard taper is 1 in 16. The specifications contain recommendations for the length of 'useful thread' and the size of the smaller diameter end which is the measuring or gauging point. This is the same as the major diameter given in Table 11.7 for the parallel thread of the same nominal size. If tapered threads do need to be cut, reference should be made to an engineer's reference book which will provide full information. Usually it will merely be a question of cutting a thread to fit a commercial plug or coupling and trial-and-error methods may therefore suffice.

British Association threads

One of the most useful thread forms for the modeller is the British Association (BA) Series. Although British by name, this series is based on a metric standard, the basic size being a screw of 6mm nominal major diameter having a 1mm pitch. This is size number 0, known simply as 0BA. This is the largest size.

The other BA sizes are given simple number references, from 1 upwards, each succeeding smaller size having a nominal diameter which is a fixed proportion of the size above. The pitch of thread becomes progressively finer as the nominal diameter decreases, the increase in pitch also being a fixed percentage of the previous size. Table 11.8 shows the BA series of threads, decreasing in size from 0BA, giving the nominal metric major diameters for the screw of each size. The BA thread series uses a vee-form having a 47.5-degree included angle.

Although based on a metric standard, the BA series of threads actually includes one or two which are extremely useful to the imperial modeller. Table 11.8 includes the imperial equivalents of the major diameters of the standard-sized screws. As can be seen, 5BA is extremely useful since its major diameter is just .001in. greater than ⅛in. It is also possible to achieve a reasonable 7BA thread on a ³⁄₃₂in. (.094in.) rod and 2BA can be formed on a ³⁄₁₆in. (0.188in.) diameter.

As originally conceived, the BA series

Table 11.8 British Association (BA) Thread Series

BA Size	Nominal Major Diameter		Nominal Nut Hexagon Across Flats (AF)*	
	mm	in.	mm	in.
0	6	0.236	10.50	0.413
1	5.3	0.209	9.28	0.365
2	4.7	0.185	8.23	0.324
3	4.1	0.161	7.18	0.282
4	3.6	0.142	6.30	0.248
5	3.2	0.126	5.60	0.220
6	2.8	0.110	4.90	0.193
7	2.5	.098	4.38	0.172
8	2.2	.087	3.85	0.152
9	1.9	.075	3.33	0.131
10	1.7	.067	2.98	0.117
11	1.5	.059	2.63	0.103
12	1.3	.051	2.28	.090
14	1.0	.039	1.75	.069
16	0.79	.031	1.38	.056

* = 1.75 × major diameter
Thickness of ordinary nuts = 0.90 × major diameter for 0, 1 and 2BA; = 0.95 × major diameter for smaller sizes
Head thickness for hexagon head bolts and screws = 0.75 × major diameter

included 26 threads from 0BA to 25BA, the very small sizes being intended for watch-making and similar applications. The most commonly used sizes are the even-numbered threads from 0BA to 16BA, although this latter size, close to ⅟₃₂in. in diameter is a relative rarity. Sizes below 16BA have been replaced by alternative horological threads and are no longer available.

As noted on the previous page, 5 and 7BA threads are useful to imperial modellers due to their close approximation to standard rod sizes, and equipment to cut these threads can be valuable. The odd BA threads do, however, have applications to modelling generally since bolts and nuts having one size smaller head than standard are available from modellers' suppliers.

When modelling, you ideally need nuts and bolts which are miniatures of the prototype items. Unfortunately, small threads such as the BA series were not designed with this in mind

and the hexagons used are consequently out of proportion (for model work) with the screw and bolt diameters. An improved appearance can be achieved by using a nut or bolt head one size smaller than normal, for example, using a 7BA head on a 6BA bolt, or a 5BA head on a 4BA bolt. Nuts are also available with the smaller hexagon. Table 11.8 includes the across flats (AF) dimensions of standard nuts or bolt heads to allow selection of stock should it be necessary to manufacture these in the home workshop. The definition of thickness of ordinary (not thin) nuts is also given together with the method of determining the head thickness of hexagon-head bolts and screws. Some modellers' suppliers do stock hexagon bar in the BA nut sizes.

Identification of threads

Naturally, if a thread needs to be identified, with a view to making a matching screwed item, it will frequently be possible to fit a nut to a bolt and thereby mate one with another. However, even if this is possible, the process does not actually reveal which thread you are dealing with since it only confirms that you have a nut and bolt which can be screwed together.

Figure 11.4 Thread pitch gauges.

To identify a thread absolutely, it is necessary to confirm at least one diameter, to measure the pitch and confirm the angle of the vee form. These latter two features are best identified by use of thread pitch gauges. These are available for all of the standard thread series and three are illustrated in Figure 11.4. These are for the Whitworth, BA and Metric (SI) series. Each gauge comprises a set of steel blades, each one cut with a different pitch, at the appropriate thread form, covering all of the pitches of the thread series, or some convenient range. The gauges shown cover the following ranges:

- Whitworth 4 tpi to 62 tpi
- BA 0BA to 10BA (pitches not specified)
- Metric 3.0mm to 0.25mm pitches

Unless a thread gauge is available, the accurate determination of pitch is the major problem, the best alternative being the use of a tap as a comparison. Determination of the vee form may also present difficulties and a gauge which can be used for comparison is the most convenient means to do this. The thread gauge provides for ready identification of both of these features of a thread.

Figure 11.5 shows a thread gauge in use. At A, the correct 12 tpi Whitworth gauge is shown positioned in the vee of a ½in. diameter Whitworth screw. The thread gauge matches perfectly. The 10 tpi gauge is shown adjacent to the screw in Figure 11.5B. The mismatch is obvious – the gauge will not enter the vee of the thread, thus confirming that this is not the correct pitch.

Naturally, in order to select a thread gauge for comparison, it is necessary to decide whether the thread in question is imperial or metric, but a measurement of the outside diameter (major diameter) of a screw usually reveals the basic type. For the screw illustrated in Figure 11.5, the outside diameter measured 0.493in. or 12.52mm, using a dual-system vernier gauge, making it the minimum specified diameter (0.4937in.) for ½ BSW and

Figure 11.5 12 tpi and 10 tpi BSW thread gauges compared with a ½ BSW screw.

therefore not likely to be metric. It may be a Unified thread, but if so it would be 13 tpi (if UNC) or 20 tpi (if UNF). If it were BSF, it would be 18 tpi and there is little doubt in this case that the size is ½ BSW.

Provided that the size allows, identification of an internal thread follows the same procedure except that the minor diameter is measured to provide the essential information concerning the size. Naturally it is not possible to observe the fit of the thread gauge in the thread, so the trial must be made by touch. However, the fit is so good when the correct pitch gauge is used that there is usually little doubt concerning the pitch.

If the internal thread is too small for convenient measurement of the minor diameter, trial-

and-error methods must be used to find a screw which fits, and by inference to determine the size.

Cutting internal threads

Introduction

Thread production is most conveniently carried out using hand tools known as taps and dies. Taps are used to cut threads in pre-drilled holes, or for the making of nuts. Dies are used for cutting threads on circular rods, or for making screws or bolts. Both taps and dies are available in different forms.

Screw threads may also be cut on the lathe and for the larger or more unusual threads this is frequently the method employed. Adaptation of a lathe for screwcutting is described in Chapter 16.

Taps

Put simply, a tap is a hardened screw, cut to the correct vee form, pitch and diameter and machined with cutting edges so that it may be screwed into a suitable hole to cut a thread of the required type. Three types of tap are normally employed:

- a taper (or first) tap which starts the cutting process
- a second tap
- and a bottoming (or plug) tap which completes the cutting.

Two or three taps are generally necessary simply because excessive turning force (torque) is required to cut a thread to its full depth using a single tool, and a specially made tap is used for starting the thread-cutting process.

A set of three taps (for 2BA) is shown in Figure 11.6. Each comprises a precisely shaped

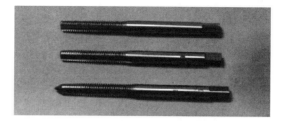

Figure 11.6   A set of taper, second and plug taps.

thread of the required vee form and size, carried on a fluted body, providing three cutting faces. On a taper tap, the thread is ground away progressively towards the tip, at a shallow angle so that the end of the tap is small enough to enter a hole whose size is related to the minor diameter of the thread to be cut. This is the starting tap. The taper tap does, however, incorporate a useful length of full thread and provided that the workpiece allows the tap to pass through the hole completely, a taper tap may cut the thread adequately.

If a blind hole must be threaded, the taper tap can only be used as the first stage, since a full thread extending to the bottom is normally required. The initial thread cut by a taper tap must be cut further to create a full-form thread. A second tap is ground to a much steeper taper than a taper tap, and for a much shorter length, and it is used to open out the thread formed by the first operation.

Bottoming, or plug, taps have no taper as such but are ground with a small 'lead' to assist them to be eased into the cut. A plug tap is used after a second to bottom out and complete a thread in a blind hole.

Hole sizes for tapping

Before a tapping operation can begin, a hole appropriate to the thread to be cut must be drilled in the workpiece. At first sight, it might appear that this should be the minor diameter of the thread, but for several reasons this is not the case. First of all, the cutting action of the tap has to be considered since its action is not simply that of cutting away and removing metal. All metals are ductile to some extent i.e. they can be squeezed or drawn (extruded) into different shapes. Ductility naturally varies between different materials. Copper for example is soft and very ductile but cast iron has low ductility. Steel has intermediate ductility while aluminium and its alloys tend to be like copper.

When a tap forms its thread, some metal is cut and removed, but there is also an action which tends to push the metal out of the way of the tap and squeeze or extrude it into a thread-like shape. If the drilled hole is larger than the minor diameter of the thread, the effect of the tap is to cut at the major diameter but to squeeze metal down into the minor diameter of the tap so that a full-form thread is produced even though the initial hole (the tapping hole) is larger than is theoretically necessary. Since ductility varies, there is an argument to support the view that the tapping drill size should vary according to the material being tapped.

The second question to consider is whether a full-form thread is actually needed. The answer to this is "No", and for several reasons. The concept of close-fit, medium-fit and free-fit nuts and bolts is described above. The different fits are specified by allowing greater tolerances on the major and minor diameters, allowing differently dimensioned items to be produced. The strength of the resulting screwed components is virtually unaffected by these dimensional changes since the strength of nut and bolt together is not changed significantly until about one-half of the engaged thread is removed.

Within limits, the actual depth of thread in a nut can vary without materially affecting the strength. The tapping size hole, and hence the amount of thread which is formed, may therefore be selected for other reasons.

The amount of thread which is present is usually expressed as a percentage of the specified full thread, allowing for rounding or truncation. It is on the basis of the required depth of thread, and the material to be tapped, that the tapping size hole is sometimes determined.

Due to the extrusion effect, which varies for different materials, it is usually recommended that tapping drills should be selected especially for the material to be tapped. The selection is made by choosing a drill which cuts a hole that leaves sufficient material only for a given percentage of the full thread depth. The percentage depth of thread which is aimed for depends upon the ductility of the material being cut (the ease, or otherwise, with which it may be squeezed into a new shape) a typical range of recommendations being:

- Copper and aluminium   70 per cent
- Gunmetal   75 per cent
- Steel and brass   80 per cent
- Cast iron   90 per cent

The force required to turn a tap when cutting the thread depends on the amount of metal being removed and is related to the percentage depth of thread which is to be achieved. For a large and relatively strong tap, it may be possible to aim for, and achieve, a 90 per cent depth of thread but for a small BA thread an attempt to produce as full a form as this almost certainly results in a broken tap. A large percentage depth of thread should not be attempted, especially in the smaller sizes.

Many published tables of tapping drill sizes appear to have been drawn up totally without regard to the percentage depth of thread. Certainly, the percentage is frequently not stated and such tables should be regarded with some circumspection. Lists which do specify percentage depth of thread generally recommend figures between 70 and 85 per cent. These percentages are most definitely on the high side, especially for smaller sizes, and a depth of thread of greater than 75 per cent is

Table 11.9 Recommended Tapping Drills (Mm); Whitworth-form Threads (55 Degrees)

Size (in.)	BSW tpi	BSW Drill	BSF tpi	BSF Drill	British Brass 26 tpi	Model Engineer 32 tpi	Model Engineer 40 tpi
1/16	60	1.25					
3/32	48	1.95					
1/8	40	2.65					As BSW
5/32	32	3.30				As BSW	3.45
3/16	24	3.90	32	4.10		As BSF	4.20
7/32	24	4.70	28	4.80		4.90	5.00
1/4	20	5.30	26	5.50	As BSF	5.70	5.80
5/16	18	6.60	22	6.80	7.10	7.20	7.30
3/8	16	8.00	20	8.40	8.60	8.80	8.90
7/16	14	9.40	18	9.80	10.20	10.40	10.50
1/2	12	10.70	16	11.30	11.75	12.00	12.10
9/16	12	12.40	16	12.90	–		
5/8	11	13.70	14	14.25	15.00		
3/4	10	16.75	12	17.00	18.00		
7/8	9	19.50	11	20.00	21.25		
1	8	22.50	10	23.00	24.50		

Note: 65 per cent depth of thread for 1/4 diameter and below, 75 per cent depth of thread for other diameters

quite unnecessary, and very likely to lead to broken taps.

To keep things simple, it is adequate to utilise only two different values for percentage depth of thread, 75 and 65 per cent, using the lower percentage for smaller threads, say, below 1/4in. (6mm) in diameter and the higher percentage for all other sizes, irrespective of the material being tapped. Tables 11.9 and 11.10 show the recommended tapping sizes for the Whitworth-form threads and also for the BA series based on the above percentage depths of thread. These sizes are calculated without any allowance for extrusion of the material being cut. Only drills from the current British Standard recommendation are listed since these are now the preferred sizes.

When deliberately cutting threads having only 65 per cent depth of thread, it must be borne in mind that the tapping hole must not be oversize – the drills used must be correctly ground so that oversize holes are not produced. Sharpening and testing of drills is described in Chapter 10.

If alternative depths of thread are required, the relevant tapping drill diameter can be calculated from:

$$\text{Drill diameter} = \text{nominal diameter} - (2 \times \text{thread height} \times (E \div 100))$$

where E is the percentage depth of thread.

Thread height is defined by:
- $0.64 \div \text{tpi}$ for Whitworth-form threads
- $0.60 \times \text{Pitch}$ for British Association threads.

Table 11.10 Recommended Tapping Drills; British Association Threads; 65 Per Cent Depth Of Thread

Size	Tapping Drill	Size	Tapping Drill
0	5.20mm	8	1.85mm
1	4.60mm	9	1.60mm
2	4.10mm	10	1.45mm
3	3.50mm	11	1.25mm
4	3.10mm	12	1.10mm
5	2.75mm	14	0.82mm
6	3/32in.	16	0.65mm
7	2.10mm		

For other threads, a standard engineers' reference handbook (such as *The Model Engineer's Handbook* by Tubal Cain, published by Nexus Special Interests) will provide the relevant details.

Tapping operations

With the tapping-size hole drilled, the threading operation may be performed. This is simply carried out by screwing the taps progressively into the hole. To do this, a tap wrench is fitted to the squared end of the tap thus allowing the driving torque to be applied.

As well as needing to be rotated, the first tap to enter the unscrewed hole (normally a taper tap, except for the finer threads) needs to be steered into the hole in order to achieve and maintain squareness. The tap wrench must fit the tap firmly to allow gentle guidance to be given to the tap. Figure 11.7 shows a selection of tap wrenches ranging from a small tee wrench suitable for the BA sizes to a large traditional type that suits taps up to ⅜in. or so in diameter. Personal choice, and the size of the tap, dictate the type to use and a range of sizes such as that shown will be required.

For the smaller BA sizes, even the smaller tee wrench shown is too gross and a pin chuck may be preferred when sizes are very small. Something quite lightweight and flimsy is in order for taps smaller than, say, 8BA since this matches the strength of the tap and provides a better feel during the cutting operation.

When starting the first tap in the plain hole, gentle downward pressure combined with about a quarter turn of the tap will jam it into the hole as it just starts to cut. At this stage, the tap must be checked for squareness to the surface of the workpiece, in two directions at right angles to one another. If your eye is good, no aid may be necessary, but if in doubt, stand a square on the job and check that the tap stands squarely.

If all is not well, turn the tap another quarter turn, putting gentle pressure on the tap wrench (very gentle pressure if the tap is small) in the required direction and check once more for squareness. It is best not to turn the tap further into the work until squareness has been achieved and if the first two quarter turns have not achieved this, it is best to remove the tap and start again, before much metal has actually been removed. Although there is a limit to the number of times that this can be done, it is important that the tap should start squarely as it is impossible to correct misalignment once a significant amount of thread has been cut.

Taper taps do generally tend to lead themselves into the hole quite nicely provided that a reasonably square start is made. Experience will bring confidence in persuading the tap to align itself squarely with the hole axis. If tapping is a new operation to you, start by cutting something reasonably large, say 2BA, ³⁄₁₆ BSF or M5, in Brass about ¼in. (6mm) thick, and steer well clear of blind holes and very small sizes until confidence has been built up.

If the tapped hole is a 'through' hole, a taper tap is adequate to cut a full thread, particularly in thin work. For longer holes, or blind holes, second and plug taps do need to be used to bottom out the thread. There is usually no difficulty in detecting when the tap bottoms in the hole but this position must be approached with some care or the tap may be too highly stressed.

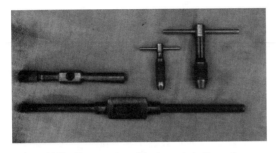

Figure 11.7 Tap wrenches of the two principal types.

Tap breakages do occur and are principally due to attempts to tap holes with too large a percentage depth of thread rather than through hitting the bottom of a blind hole. Indeed, experiments to determine the torque required to cut different thread depths suggest that to attempt to cut threads having 100 per cent depth results in 100 per cent of tap breakages! The use of 65 and 75 per cent depth of thread avoids this problem for the most part but if small threads are required in stainless steel, 50 per cent depth may be a more sensible aim. The mating screws do need to be cut accurately to size however, and it is vital that the tapping drill cuts absolutely the correct size.

Removing a broken tap is usually a problem since they nearly always break off flush with, or just inside, the hole. You may be lucky and find that the broken tap can be screwed out using pliers, but if not it must be broken up or driven out using a small punch. Driving out is only possible if the hole is not blind. Either way, the hole is unlikely to be suitable for tapping at the required size and it must either be opened out to suit the next larger size or opened out significantly so that a plug can be used to fill the hole and allow another attempt to drill and tap the hole correctly.

Blunt taps are one cause of breakages and the workshop's stock should be examined from time to time so that problems can be averted by purchasing replacements before a breakage occurs.

Tapping holes for fine-pitch threads

For threads in the series having 32 tpi or 40 tpi, there is a need only for second and plug taps. If you examine a taper tap for a fine-pitch thread such as ⅜in. × 32 tpi, you will see that the tap is tapered and backed off to such an extent that there is virtually no thread at the end. As such a tap starts to cut, there is little thread being

formed initially and therefore no strength available to hold the tap squarely in the hole. The slightest side pressure tends to 'strip' the small amount of partly formed thread and the result is usually not a tapped hole but a hole which is reamed out to some intermediate size by the front end of the tap.

As a consequence, it is best to start the tapping operation for large-diameter fine threads using a second tap so a taper tap is not required for threads larger than ¼in. diameter in 40 tpi, ⁵⁄₁₆in. in 32 tpi and ⅜in. in 26 tpi.

Avoiding burrs at the mouth of the hole

If a tapping-size hole is drilled and the hole tapped, the operation produces satisfactory results except at the mouth of the hole. The extrusion (metal flowing) effect which tends to cause the tap to push metal out of the way rather than cutting it, can occur more readily at the mouth of the hole since there is open space into which the metal can be readily pushed. This produces an unsightly hump around the hole which destroys the flatness of the surface locally and impairs the fit of whatever will be attached to the tapped workpiece. Figure 11.8A shows the problem.

Various makeshift ways are sometimes adopted to remove the hump, such as filing the surface or de-burring the mouth of the hole using a drill. Both methods destroy part of the thread form however, and it is then necessary to put the tap in again to clean up the thread. All of this can be avoided by drilling the initial hole as shown in Figure 11.8B. After drilling at

Figure 11.8 Counterboring a hole to avoid a burr at the mouth when tapping.

Figure 11.9 Counterbored tapped holes in a boring head body.

the tapping size, the appropriate clearing size drill is entered into the hole, not very deeply, but just enough to equate with two pitches of thread, say. This provides a small internal space in which the tap can create its burr and leaves the surface of the workpiece unblemished by the tapping operation. Figure 11.9 shows a mild steel block tapped in this way.

Cutting external threads

Threading dies

Although there are solid dies, the most common form is the split type, a range of which is shown in Figure 11.10. These are made in a range of standard outside diameters to suit the thread size for which the die is made, the standard nominal diameters being $^{13}/_{16}$in., 1in., $1^5/_{16}$in. and 2in. this range covering dies for threads up to 1in. (25mm) in diameter. The next standard die size above this is $2^1/_4$in. diameter but this is definitely outside the model engineering range of sizes.

A die consists of a hardened steel disc having a central threaded hole of the appropriate size and form. To provide the cutting edges, 3, 4 or 5 holes are machined through the die around the central, threaded hole, creating not only the cutting edges but also the clearance necessary to allow the cut material to be ejected from the centre of the die.

The dies shown in Figure 11.10 are of the split type i.e. they are slit into one of the clearance holes and the slit is machined with a vee-shaped chamfer on its outer edge. A die is hardened and tempered, and retains a little spring and a split die may, therefore, be opened out or closed in a little thus adjusting the size of thread which it cuts. To allow adjustment of the die, a standard holder incorporates three screws which engage the die, one in the slit and the two others, one on each side of the slit.

Dies are provided with two machined recesses into which the two side adjusting screws of the die holder can locate. This is illustrated in Figure 11.11. UK and American practice is different in respect of the positioning of the side recesses, American preference being for positions 90 degrees each side of the slit, but UK practice positioning them at 30 degrees on each side.

The central screw in the die holder engages

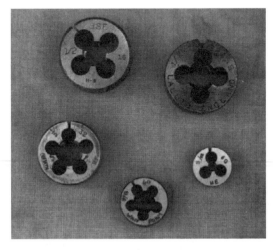

Figure 11.10 Split dies are supplied in a range of standard outside diameters.

Figure 11.11 A split die and die holder, showing the adjusting screws.

the chamfered slit and when screwed in, its pointed end opens up the die causing it to cut a larger thread. Alternatively, with the central screw withdrawn, the side screws may be used to squeeze in the die and hence cut a smaller thread. There is the possibility of producing grossly undersized or oversized threads, and the result of the threading operation must be checked by cutting a thread on a piece of scrap material and testing the result.

If the thread being cut is to mate with a tapped hole, then the best test is to tap a hole in a piece of scrap material and make the screw to fit it. If the die is opened out too much, it cuts a shallow thread, making the minor diameter too large and preventing the screw from entering the test hole. In this case, the die must be closed up (a little!) and run down the thread again, the process being repeated until a proper fit is achieved.

If the thread being cut is to mate with a commercially produced nut, then the nut, or preferably several from the same batch, should be used for the test. For the smaller threads which might be produced, it is very important that the dies are set correctly, and do not cut undersized threads.

Cutting the thread

Starting a die on the end of the workpiece is generally difficult – at least starting it squarely is. It is essential to check the die holder's posi-

tion relative to the rod, particularly by looking across the holder. Achieving squareness is quite easy when comparing the left-hand arm against the right-hand arm and pressure can be brought to bear to correct any lack of squareness here, but not at right angles to the line of the handles. The die must be started, squareness achieved and then the holder turned through a further 90 degrees into the cut and squareness checked again, once more using the line of the handles as a guide.

To assist starting, the die is ground with a significant chamfer to provide a generous lead to the cutting edges. This chamfer is usually ground on the side of the die carrying the identification of its size and thread form and is not always carried out on the back, a practice which can have useful advantages. Starting is also assisted by opening up the die and by providing a good chamfer on the item to be threaded.

Once the cutting has been started squarely, the die holder is simply screwed into the cut. The material which is removed from the workpiece is squeezed into the clearance holes in the die (which create the cutting edges) and is produced in continuous curls if the die continues to be screwed into the cut. To break up the curls, the die should periodically be reversed for a short rotation and the action should be one of cutting for one or one-and-a-half turns and then reversing the die for a quarter of a turn, or perhaps a little more. The breaking of the curl can easily be felt and this type of action considerably eases the job of cutting the thread. For short or small threads, the breaking of the curl is not so important.

A more satisfactory way to cut external threads is to perform the operation on the lathe using a tailstock-mounted die holder. Such a holder maintains axial alignment with the work when starting and cutting the thread and is more likely to produce a correctly aligned thread. A holder which will accept $^{13}/_{16}$in. and

1in. diameter dies will cater for threads up to ⅜in. (10mm) diameter and most definitely gives better results than a manual operation. Use of the tailstock die holder is described in Chapter 15.

The chamfer cut into the leading edge of the die means that it will not cut a full-form thread up to a shoulder on the work. If the face of the shoulder is required to butt against the mating part, one or other of the items must be machined to allow full engagement of the two parts. This may be done in two ways: either by cutting away the oversize, incomplete thread left by the die, or by counterboring the end of the screwed hole to provide clearance. Counterboring to prevent burrs at the mouth of a tapped hole is described above, and illustrated in Figure 11.9. The drilling of a somewhat deeper clearing-size hole at the mouth, perhaps with a small countersink or chamfer, will provide clearance for the unthreaded portion of the screw.

On some dies, the reverse side is not chamfered to any marked degree and it is sometimes possible to form the full thread a little nearer a shoulder by finally putting the die on to the formed thread the wrong way round.

Alternatively the unthreaded portion of the screw should be removed as part of the operation to turn the screw blank diameter and form the shoulder, as shown in Figure 11.12. This operation, known as undercutting, is performed on the lathe. See Chapter 15.

Lubrication during thread cutting

Most cutting processes benefit from the presence of a lubricating and cutting fluid at the

Figure 11.12 A thread undercut at the shoulder.

point at which cutting takes place. Fluid is perhaps most useful when high rates of metal removal are desired since the flow extracts heat from both the tool and the workpiece. If the fluid has an oily base, it also acts as a lubricant and can improve the sliding contact which occurs between the work and the tool thereby significantly improving the finish.

When thread cutting using taps and dies, high rates of metal removal are not normal in the amateur's workshop, even if using the tailstock die holder with the lathe powered, since the threads produced are usually quite short. Nevertheless, there is benefit in providing lubrication for the threading process. The traditional lubricant for threading was tallow but there are now available various commercial preparations specifically designed for slow-speed cutting operations. They are valuable in promoting smooth cutting and preventing the tap or die tearing the material and help to produce the desired smooth surface finish. Model engineers' suppliers generally stock these products which are known by the general name of tapping compounds.

12

The lathe – essential machine tool

What the lathe can do

Provided that it is of adequate size, the lathe is capable of carrying out all of the machining operations which are essential for the amateur worker. Apart from the drilling machine, it is likely to be the only machine for cutting metal which the amateur has at his disposal and it can be developed into a very versatile machine tool.

Basically, the lathe provides for a bar or rod of material to be rotated and a fixed cutting tool to be brought into contact with it. Since the tool is harder than the workpiece, material is removed from the work as long as the tool continues to be fed into the rotating material. This basic operation is shown in progress in Figure 12.1.

The workpiece may require material to be removed from its end face. In this case the tool is turned through 90 degrees in relation to its position when surfacing the outside of a bar, and the metal cutting operation performed on the face, as shown in Figure 12.2. This operation is known as facing.

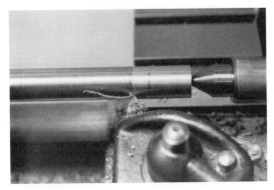

Figure 12.1 A turning operation which is reducing the outside diameter of a steel bar.

Figure 12.2 A facing operation machines the end surface of work which is rotating in the lathe.

The above operations are described as turning; the workpiece is rotating and the cutter is a single-point tool that is fed into the work to remove material. Turning is essentially what a lathe is designed to do, but by suitable adaptation it may be arranged to perform the functions of other machines. Some adaptations of the lathe are described in Chapter 16.

General description

A simple outline drawing showing the essential parts of a lathe is shown in Figure 12.3. The machine comprises a motor-driven spindle, or mandrel, mounted in a robust casting at the left-hand end of the bed of the machine. This end of the machine is known as the head and the assembly that supports the spindle, or mandrel, is known as the headstock.

The foundation of the lathe is its bed. This is usually a heavy iron casting which is machined at the left-hand end to accept and locate the headstock and machined down the remainder of its length, on the top surface and on the edges, in order to locate a carriage, or saddle, which can be moved along the bed.

The machine is strictly known as a centre lathe since its earliest use was for turning long, thin work which was mounted between centres for machining. This method of use is still con-

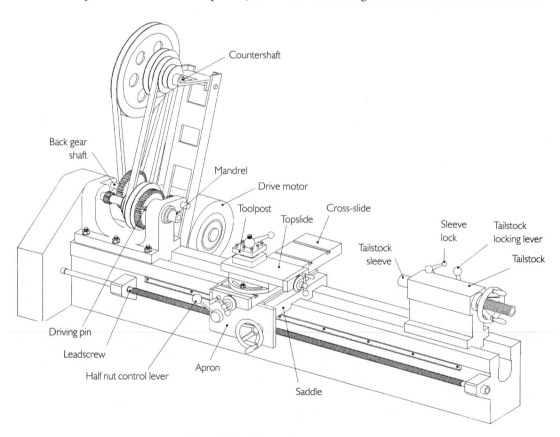

Figure 12.3 The main features of a lathe.

venient for many jobs. The centres are merely spigots with pointed ends which engage in cone-shaped holes in the ends of the workpiece. One centre is mounted in (or on) the end of the mandrel and the other is mounted in (or on) a supplementary casting which is fitted towards the right-hand end of the bed. This end is known as the tail of the machine and the supplementary casting is the tailstock.

The tailstock supports the centre at the same height above the top surface of the bed as the axis of the mandrel. This dimension, known as the centre height, is used to describe the basic capacity of the lathe, although the diameter that can be accommodated by the machine is sometimes specified.

For driving (rotating) work which is mounted between centres, a lathe carrier is fitted to the job at the mandrel end. This is usually a casting which fits over the workpiece and is provided with a set screw so that it may be clamped to it. The end of the carrier acts as a radial extension to the workpiece and is driven by a peg fitted to a plate attached to the end of the mandrel. This is called the catchplate. Figure 12.4 shows work mounted between centres and Figure 12.5 the driving arrangement. The procedure adopted for cutting the female centres in the ends of the workpiece so that it may be supported and driven is described in Chapter 15.

Figure 12.4 Work mounted between centres on a lathe.

Figure 12.5 A carrier used to drive work which is mounted between centres.

For removing material progressively from the outside of a bar mounted between centres, it is necessary to move a tool in and out in relation to the work. The saddle is provided with a machined upper surface so that a supplementary slide may be mounted on it. This is called the cross-slide since it moves across the axis of the lathe bed. The cross-slide is provided with a feedscrew and handle so that controlled and precise movement of a tool mounted on the cross-slide may be achieved, allowing precise control of the cut and hence the diameter of the workpiece.

With a tool mounted on the saddle and cross-slide, controlled movement in two directions is possible, both along and across the end of the workpiece. When work is fitted to the mandrel only, and not supported at the tailstock centre i.e. fixed somehow to the mandrel and not mounted between centres, two cuts may be applied to the work, either along its length (a surfacing cut) or across its end (a facing cut).

The above paragraphs describe the basic and essential features of a plain lathe. However, even a modest practical lathe has additional features which render it more versatile than the simple machine described, and the following paragraphs consider the practical requirements for the headstock, saddle and tailstock individually.

The headstock

Mandrel speeds

In order to be suitable for a wide range of work, the mandrel must be capable of rotating at different speeds. To be really versatile, the speeds need to range from a few tens of rpm to a few thousand rpm, and to accommodate this speed differential some form of gearing is usually incorporated into the speed-change arrangement.

Figure 12.6 The original headstock of a small lathe which was driven by a flat leather belt.

Figure 12.7 The complete drive arrangement on a small lathe, showing the countershaft and the driving belts.

Many modern lathes incorporate an enclosed, all-geared headstock drive but lathes of older design frequently use belt drive. Figure 12.6 shows the driving arrangement of a small lathe designed some years ago for the amateur market. The mandrel is supported by two upward projections in the headstock casting and the space between is occupied by the drive components. These comprise a three-step pulley and two gears. The mandrel drive is here seen in its original form, with flat-belt pulley, before its conversion to vee-belt drive.

Figure 12.7 shows the full drive, with the three-step, vee-belt pulley fitted. A countershaft is fitted above the mandrel and slightly to the rear, and carries a matching three-step pulley. The countershaft carrier is adjustable so

that the belt can be correctly tensioned and has a quick-release lever to assist rapid belt repositioning, although this is not visible in the figure. Drive to the countershaft is provided from the motor, mounted at the rear, with pulley sizes chosen to give a countershaft speed of about 400 rpm.

The completed drive to the mandrel is shown in Figure 12.8. With the pulley cluster and gears occupying virtually all of the space between the headstock bearing housings, there is no room for more than the three-step pulley, and this would appear to limit the range of mandrel speeds to just three. However, the two spur gears fitted to the mandrel provide the means to extend the speed range. The smaller, left-hand gear is mounted in an assembly with

Figure 12.8 The rebuilt mandrel drive arrangement which now uses a vee belt.

- In direct drive   800   400   200 rpm
- In back gear   160   80   40 rpm

The actual range of speeds, particularly the maximum speed, is very much dependent on the design of the mandrel and its bearings. More speeds can be provided, by fitting in a four-step pulley for example, or by arranging a two-speed drive between the motor and the countershaft.

The back gears themselves need not be arranged at the back of the mandrel but can instead be arranged as a close-coupled pair of gears brought into engagement by a lever-operated eccentric mounting. Such an arrangement is shown in Figure 12.9. In this lathe, a sliding block attached to the bull wheel allows for engagement of the drive from the three-step pulley cluster when direct drive is used.

The provision of a back gear is important since it allows large-diameter workpieces to be rotated slowly enough to maintain the peripheral speed of the work past the tool (the cutting speed) within the normal range for the material being turned.

It is obvious that, if the back gear is engaged without disconnecting the direct drive to the bull wheel, the mandrel will be locked. Since many accessories are mounted onto a screw thread on the end of the mandrel, the ability to lock it when removing or replacing these is very

the three-step pulley, the whole assembly being free to rotate on the mandrel itself. The larger gear on the right-hand side, known as the bull wheel, is keyed or otherwise locked to the mandrel and provides the actual drive.

Lying behind the mandrel is a second shaft which has two gears locked to it. The shaft has two alternative positions, being latched into each by a simple ball detent. As shown in the figure, the gears on the rear shaft (known as the back gear shaft) are engaged with the two mandrel gears and drive to the mandrel is from the pulley cluster gear to the back gear shaft and forward once more to the bull wheel. There is, therefore, a two-stage reduction in speed from the pulley cluster to the mandrel and the lathe is said to be in back gear.

Direct drive is arranged by disengaging the back gear and driving the bull wheel instead from the pulley cluster. For this purpose, the bull wheel is provided with a captive pin which engages with a hole in the pulley cluster but which can be withdrawn when back gear drive is required. The head of this pin can be seen projecting from the right-hand side of the bull wheel in Figure 12.8. The arrangement provides for six mandrel speeds to be available, in two ranges of three, a typical set of speeds for a lathe of this type being:

Figure 12.9 A back gear cluster mounted below the mandrel.

convenient. This thread may be seen in Figure 12.6 while Figure 12.5 shows the catchplate fitted to this thread and a male centre pressed into the tapered socket in the front of the mandrel.

The mandrel

As noted above, the nose of the mandrel normally carries a screw thread to allow accessories to be mounted on it. Since, by their nature, screw threads do not provide accurate location, a simple threaded nose is not normally used. Instead, the mandrel incorporates a plain, parallel register, as shown in Figure 12.10. The register is concentric with the mandrel axis and behind the register is an abutment face. Attachments screwed on to the mandrel have an accurately sized bore which fits the register, but have a relatively sloppy thread so that the register and the abutment actually locate the accessory and hold it aligned with the mandrel axis. The accessories are most commonly workholding devices such as chucks, faceplates and collets, although they may also be used to hold such things as milling cutters. Workholding devices are described in Chapter 13.

Figure 12.10 This mandrel nose has a thread for attachment of workholding devices, but is also provided with a parallel register which locates items mounted on it.

Immediately behind the abutment face which locates accessories on the mandrel nose, the mandrel is supported in the headstock front bearing. To resist the axial pressure exerted on the mandrel by the turning process, the front bearing incorporates a thrust bearing. This may take the form of a thrust washer or a ball bearing specially designed to resist thrust, or thrust may be resisted by making the front bearing in the form of a taper thereby providing support and resisting the end thrust.

The mandrel is also supported in the headstock at its left-hand end but this bearing normally takes a cylindrical form which does not contribute to the resistance to end thrust. A means to adjust the longitudinal clearance (end float) of the mandrel in the headstock bearings is usually provided so that there is also a restraint to prevent the mandrel being drawn out of the headstock in the direction of the tailstock. This would otherwise happen, for example, when drilling an axial hole in work that is supported only at the mandrel end (and this is in practice the most common method of operating the machine). In any event, the driving pulleys may exert some side thrust on the mandrel and its positive location in the headstock bearings is an essential part of the design.

To allow turning to be carried out on the end of relatively long bars, the mandrel is drilled right through. The size of this hole is dependent on the outside diameter of the mandrel shaft, which is, in turn, determined by the general dimensions of the headstock assembly. The larger the lathe, the larger is this hole and the more versatile will the lathe be in dealing with large diameter bars. It is a fact of life that one's own lathe does not have a large enough through bore for all of the jobs which one wishes to undertake. There are fortunately other ways of dealing with longer lengths of large diameter stock.

Not only is the mandrel drilled through to provide clearance for long bars, but it is also normally bored to a shallow taper at the nose to allow centres to be fitted into the bore. Standard tapers for these sockets have long been agreed, the most common range for small machines being those known as Morse tapers.

The shallow taper which is used (an approximate 3-degree included angle for Morse tapers) means that the inserted centre jams itself into the socket, ensuring that it rotates with the mandrel. Other accessories may also be fitted in the same way, but if any great torque needs to be transmitted, positive steps have to be taken to ensure that the accessory is held into the taper because without some assistance the friction fit may not hold. When a centre rotates with the mandrel, this is not a problem, but a workholding device (or cutter) held in the taper socket does need some assistance to maintain the grip. However, once a tapered accessory has been inserted into the socket, it does need to be driven out fairly firmly.

Many items are available having taper shanks. They include centres for use in the lathe, drills, drill chucks and accessories such as boring heads. For one popular lathe, collets are available to fit the mandrel taper, together with an adaptor to close them up on to the workpiece. These are described in Chapter 13.

Headstock for screwcutting

A plain lathe is one on which it is not possible to perform screwcutting. That is, it is not possible to drive the saddle, and hence the tool, along the bed, automatically, at a rate that is related to the rotation of the workpiece. In truth, the design of the whole lathe is determined by the requirement to make it a screwcutting machine, but since the headstock needs to incorporate the required drive facility, the subject can be introduced here.

The basic requirement is to lead the saddle along the bed in sympathy with the rotation of the workpiece or mandrel. A vee-shaped tool is then fitted to the cross-slide and saddle and the tool cuts a vee-shaped spiral (a screw thread) in the work, as it rotates.

To lead the saddle along, a leadscrew is fitted along the length of the bed, supported in bearings at each end. At the headstock end, a key is fitted to the leadscrew to allow it to be driven by gearing from the mandrel. The end of the mandrel is permanently fitted with a drive gear so that further gears may be interposed between it and the leadscrew. The actual gearing is varied to suit the pitch of the screw thread to be cut. If a thread of 24 turns per inch (tpi) is to be cut, the gearing is arranged so that the saddle is driven along the bed by one inch as the workpiece makes 24 rotations, thus cutting the required thread.

The pitch of the leadscrew naturally affects the choice of gear ratio for a given pitch of thread, but since leadscrew pitches tend to be standardised it is possible to consult published tables for the arrangement of gears to be adopted for cutting common pitches of thread.

The gears which link the mandrel and leadscrew gears are known as changewheels and a screwcutting lathe might be supplied with a set ranging from 20 teeth to about 75 teeth, in steps of five teeth. This range allows all normal pitches to be cut but to allow flexibility, larger gears are also usually manufactured. For imperial lathes, a 'metric translator' gear is sometimes available which allows a close approximation to true metric pitches to be cut.

The changewheels are mounted on studs fitted into slots in a 'banjo' casting which clamps to the leadscrew bearing. Stud positions, and banjo position on the bearing housing, are adjustable to permit the full range of gear ratios to be set up and yet still allow correct meshing of the gears.

It is usual to arrange for simple reversal of the direction of rotation of the gear train by fitting a 'tumbler reverse' to the headstock. This is

Figure 12.11 Tumbler reverse, leadscrew drive and gearing at the headstock end of the lathe.

normal to describe it as a back-geared, screw-cutting lathe, this description frequently being abbreviated to BGSC.

The provision of a back gear is important for screwcutting since this usually cannot be done at a rush, especially for coarse-pitch threads, due to the need to disengage the drive at the point where the thread on the work is to end. The drive to the saddle is described below.

The gearing to implement the screwcutting facility may be contained within a separately mounted 'quick-change' gearbox which offers lever-operated selection of the gear ratio between the spindle and the carriage on which the screwcutting tool is mounted. Manual selection of the ratio by changewheels is a slower and dirtier process than the simple movement of an external lever on a quick-change gearbox, but changing this ratio is not undertaken too frequently. Lathes fitted with a quick-change gearbox are significantly more expensive than the changewheel type.

The saddle and its fittings

Cross-slide

The saddle of a small (3½in., 89mm) lathe is shown in Figure 12.12. The saddle can be seen to have a cross-slide mounted on it. This is provided with a feedscrew so that it may be traversed in a direction at right angles to the lathe axis (in and out relative to the workpiece). The feedscrew carries an engraved dial so that calibrated movement of the slide is possible, the dial being marked every .001in. or every .020mm or .025mm if the lathe is built to metric standards. On some machines the dial is loose on the feedscrew shaft, but driven by a friction-grip arrangement, so that it may be preset to some convenient reading to define, for example, the beginning or end of a cut.

shown in Figure 12.11. A three-position arrangement is provided which allows disengagement of the drive to the leadscrew, or selection of rotation in either direction.

If a lathe is of the screwcutting type, the gearing between the mandrel and the leadscrew may also be used to drive the saddle along the bed when doing plain turning. In this case the lathe automatically feeds the tool into the work and the rate at which this occurs is termed the feed per revolution. This use of the headstock gearing is generally described as the self act since the machine is 'looking after' itself, although the operator has to be present to disengage the drive to the saddle at the appropriate point, unless the machine is an automatic.

If the headstock is provided with both a back gear and an adaptation for screwcutting, it is

Figure 12.12 The saddle spans the bed of the lathe and has the cross-slide mounted on it.

Apron

The front of the saddle carries an apron that allows a mounting for the saddle traverse handle to be incorporated and also the mechanism that allows the saddle to be driven along the bed. Figure 12.13 shows the apron associated with the saddle and cross-slide of Figure 12.12. The large handwheel on the right is rotated to drive the saddle along the bed manually. This handwheel engages with a rack which is bolted to the bed of the lathe, immediately below the upper machined surface, suitable gearing being interposed between handwheel and rack to

Figure 12.13 The front of the saddle is closed by an apron which has the saddle traverse handwheel mounted at the right-hand side and the control for the half nuts on the left.

allow the handwheel to be used for 'putting on' the cut when bar turning is required. The gear ratio is a compromise, however, since it must be slow enough to permit a cut to be taken by use of the handwheel, yet fast enough to permit rapid movement of the saddle out of the way, for example when the tool must be withdrawn for the next cut, or to allow measuring or drilling operations to be performed.

Locking and driving the saddle

If the lathe is to be used for taking a facing cut (across the end of the workpiece) the saddle must be locked to the bed and the cut taken by use of the cross-slide traverse. Locking the saddle directly to the bed is not particularly convenient and since locking it to the leadscrew is required in any event, for screwcutting, this is usually the method adopted for the small lathe. Reliance is then placed on the thrust bearings at each end of the bed to locate the leadscrew without endfloat, and a definite arrangement for adjusting this should be incorporated into the machine.

The saddle is locked to the leadscrew by a pair of half nuts, or clasp nuts, mounted in slides inside the apron, and capable of engagement with, or disengagement from, the leadscrew. In Figure 12.13 the clasp nut lever is seen at the left-hand side of the apron. The lever rotates anti-clockwise to engage the nuts and clockwise to disengage them, these actions rotating the cup which engages with both half nuts. Anti-clockwise rotation lifts the lower half nut and pushes the upper one down so that the two halves engage the leadscrew, locking the saddle to it.

When screwcutting, the leadscrew is driven by the changewheel train at the headstock end and this leads the saddle along the bed. An alternative method of driving the saddle is to rotate the leadscrew manually and this is one

Figure 12.14 The leadscrew is sometimes fitted with a dial and handwheel so that the saddle can be moved by a known amount.

way of applying the cut when facing. It is also a valuable method of positioning the saddle when performing turning (or milling) operations and for this reason the leadscrew may be provided with its own handle and have an engraved dial. The dial-and-handwheel on my lathe is shown in Figure 12.14.

Topslide

It is perfectly feasible to mount the tool directly to the cross-slide and use the lathe for turning as described above. This can make for a very rigid, no-nonsense arrangement, but it is common practice to provide a further slide having screw-controlled movement and to mount this on the cross-slide. The resulting assembly is known as the topslide and it is this which normally carries the toolpost, as shown in Figure 12.15.

The need for the topslide is made more apparent by considering that, under some circumstances, the maximum diameter of work which can be accommodated on the machine is determined by the distance between the top surface of the cross-slide and the axis of rotation. It is desirable that this be made as large as possible, consistent with the other dimensions

of the machine, but in maximising this distance the designer places the support for the tool well below centre height and something quite substantial is required on which to mount the tool.

Put another way, this means that there is room for a two-part assembly; a base which bolts to the cross-slide and an upper slide, provided with a feedscrew and calibrated dial, so that controlled and precise movement is possible for taking a cut. To allow flexibility in use, the topslide is arranged to be readily removable so that the saddle and cross-slide may be used for mounting work when the lathe is not to be used for turning. The attachment of the topslide is arranged for rapid removal and replacement. A positive location for the topslide is frequently provided but there is not normally a register to guarantee that the topslide is at right angles to the saddle. Indeed, it is valuable to be able to set the topslide so that movement of the tool is not parallel to the axis of the mandrel.

When the tool is mounted to the topslide, the path taken by the tool is reproduced on the workpiece and unless the topslide is set so that it moves precisely parallel to the axis of the work, a tapered workpiece is produced. This ability to produce tapers is one advantage of the adjustable topslide. Tapered work is required

Figure 12.15 The saddle can have a topslide mounted on it. This has its own calibrated feedscrew and a mounting post for the tool holding arrangement.

fairly frequently, and rotation of the topslide to the required angle is one way to produce such work. The topslide base may be provided with a graduated angular scale to allow an approximately correct angle to be set relatively easily. Such graduations do only permit an approximate setting to be achieved and if a taper is required to fit an existing taper exactly, careful alignment or trial-and-error methods must be adopted to ensure a proper fit.

Toolpost

The toolpost is the name given to whatever device is provided for clamping a tool onto the machine. For most lathes this is mounted on the upper surface of the topslide. It may be a simple clamp arrangement for a single tool or may comprise an adaptor for mounting a number of tools. Some lathes are provided with an adaptor which accepts standard tool holders, permitting rapid interchange of tools while still guaranteeing that the top of the tool's cutting point is placed at centre height once the holder has been set up (an essential requirement, see Chapter 14).

The most common type of tool holder is the clamp-on-post type shown in Figure 12.16. This comprises a central pillar, threaded at the top for a clamp nut or ball-ended lever. Below the lever, a clamping plate is a loose fit on the pillar and is provided with a jacking screw fitted into a threaded hole in one end. Downward pressure from the clamp nut or lever, together with upward pressure from the jacking screw, provide clamping for the tool, which is normally packed up to centre height by strips of material placed below it.

The above type of clamp is perfectly satisfactory for much of the work which the lathe is required to perform. The tool may readily be mounted in different plan positions to 'attack' the work from the required direction and can

Figure 12.16 The normal tool clamp is used to clamp a single tool to the topslide. This clamp has a ball-ended clamping lever and a jacking screw which are adjusted in conjunction with one another to clamp a tool firmly into position.

be set to centre height readily as long as strips of material of various thicknesses are available. However, it frequently occurs that more than one tool is required in order to complete the turning operations on a particular item and the means to have several tools readily available is occasionally desirable. One way to arrange this is to purchase an interchangeable tool holder system, but this is a relatively expensive solution to the problem and a holder capable of taking more than one tool is a less-costly alternative.

Figure 12.17 shows a four-way toolpost mounted on the topslide of Figures 12.15 and 12.16. This fits onto the pillar and is provided with an indexing mechanism in its base that allows tools to be brought to eight positions, at 45-degree intervals, with reasonable repeatability. Tools may be mounted up for turning, chamfering, screwcutting and parting off, for example, and these operations performed in quick succession once the tools have been set up. This is especially valuable when several identical components are required, and a routine which uses several different tools needs to be repeated.

Figure 12.17 The alternative tool mounting is by use of a 4-tool turret. This one can be indexed round into eight positions to bring the tools to the working position.

The only serious disadvantages of the four-way turret are that the out-of-use tools do sometimes get in the way and they are also a potential safety hazard since their sharp ends are close to the operator when not in use. Nevertheless, it is a valuable means to satisfy the occasional multi-tool requirement.

A further method of making more than one tool available is to employ a second toolpost mounted directly on the cross-slide. This type is

most commonly employed for parting-off operations and Figure 12.18 shows a two-way toolpost which is used for this purpose. This indexes into two positions, 180 degrees apart, to bring one of the two tools into use. Parting off is described in Chapter 14 where a consideration of the advantages of using a rear toolpost for this operation will also be found.

The tailstock

The use of the tailstock for supporting work mounted between centres is described at the beginning of this chapter. For such service, a simple casting capable of being locked to the bed and having some means to mount the centre on it, is all that is required. However, other facilities are required, such as the capability to drill holes down the rotational axis of the work, and the tailstock has therefore developed into a more useful accessory.

A typical tailstock is shown in Figure 12.19. The main casting is bolted to a base which locates between the machined vertical faces of the inner U-section of the lathe bed. The base is held in alignment with the axis of the machine but the main (upper) casting may be moved relative to the base across the lathe axis. This

Figure 12.18 A rear toolpost has some advantages for mounting tools for parting off work from the parent stock. This one has a 2-position turret.

Figure 12.19 The tailstock is movable along the bed, but can be clamped to it in any position. It is used to hold drilling chucks and other accessories.

adjustment allows the axis of a tailstock-mounted centre to be moved relative to the mandrel axis so that it may be nearer to the operator, or farther away. This feature is a useful adjunct to the tailstock's other facilities, but for the moment it is convenient to consider that the tailstock centre is in perfect alignment with the mandrel axis.

The tailstock's main casting is bored through at centre height so that a substantial sleeve may be fitted. The sleeve is a good sliding fit in the main casting bore but is prevented from rotating by a screwed pin which locates in a longitudinal groove cut in the outside of the sleeve. The sleeve is threaded at its right-hand end (farthest from the headstock) and this thread engages the thread in a bored-and-screwcut handwheel which is captive, but free to rotate, in the end of the main casting. Rotation of the handwheel drives the sleeve in or out of the tailstock barrel.

The sleeve is itself sufficiently large to be bored through and is normally bored to one of the standard tapers, usually, but not always, matching that in the front of the mandrel. Accessories may be mounted into the sleeve provided that they are fitted with standard taper adaptors. Since adaptors are readily available, drill chucks, die and tap holders and multi-tool turrets may be mounted into the sleeve, in addition to the standard centres, allowing axial holes to be drilled and threaded 'from the tailstock', or work to be threaded externally by use of a die holder mounted in the tailstock. These operations are described in Chapter 15.

To permit these operations to be performed, the tailstock can be locked to the bed. This allows the thrust to be taken when the tailstock handwheel is used to advance a drill into the mandrel-mounted workpiece. The tailstock sleeve can be locked in the barrel, since this is required when supporting work between centres.

Mandrel and tailstock tapers

In considering the mandrel and tailstock bores, the normal practice is to bore these to a standard, shallow taper. The tapers used are described as self-holding tapers i.e. a tapered accessory such as a lathe centre may be jammed into the mating tapered socket where it will hold itself in. Provided that an excessive turning force (torque) is not applied, the accessory will be held in the bore until it is knocked out from the small end. The torque which can be resisted by the taper is certainly sufficient to resist the forces imposed when drilling a hole and the standard method of mounting a drill chuck in the lathe is by utilising a chuck with a tapered arbor and mounting it directly in the tailstock sleeve. A description of hole production methods on the lathe, and the use of the tailstock, is given in Chapter 15.

The most common tapers in use for the mandrel and tailstock of small lathes are the series known as Morse tapers. These range from No. 0 which is approximately ¼in. (6mm) diameter at the small end, to the massive No. 7 which is 2¾in. (70mm) diameter at the small end and some 10in. long. The most common sizes of taper socket for small lathes are the No. 1 and No. 2 Morse tapers, being roughly ⅜in. (9.5mm) and 9⁄16in. (14mm) in diameter at their smaller ends. Some larger machines utilise the No. 3. It is convenient if both mandrel bore and tailstock sleeve are bored to the same taper so this is usually the case.

Figure 12.20 shows the form of a standard Morse taper. The size of the small end of the

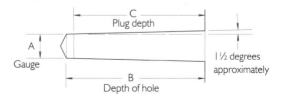

Figure 12.20 Standard forms of Morse taper sockets and arbors.

hole is specified by the gauge diameter, or plug diameter, A, and the length of usable taper by the plug depth, C. The hole depth, B, is usually the plug depth, C, plus $\frac{1}{16}$in. to allow for a reamer to be used to cut the taper and achieve the required plug depth. The standard dimensions for the four smallest sizes of Morse taper are shown in Table 12.1 below.

Table 12.1   Morse Tapers

No. of Taper	A Gauge Diam. (in.)	B Depth (in.)	C Plug Depth (in.)	Taper/in. on Diam. (in.)	Max. Diam. (in.)
0	0.252	2$\frac{1}{16}$	2	.05205	0.356
1	0.369	2$\frac{3}{16}$	2$\frac{1}{8}$	.04988	0.475
2	0.572	2$\frac{5}{8}$	2$\frac{9}{16}$	.04995	0.700
3	0.778	3$\frac{1}{4}$	3$\frac{3}{16}$	.05020	0.938

Notice, from the above table that the taper is not constant through the range of sizes but remains approximately .05in. per inch giving an included angle for the taper of about three degrees, or 1½ degrees per side, as shown in Figure 12.20. For the mandrel and tailstock bores, the Morse taper sockets take the form shown in Figure 12.20 except that the bores are not blind but continue as through bores, as described above.

Figure 12.21   A lathe centre, a drill chuck and some drills having Morse taper shanks or arbors.

Tapered arbors for lathe accessories are frequently of the plain type such as the centres shown in Figure 12.21, but it will be noticed that the drills and the arbor of the drill chuck in the figure are provided with a modified end in the form of a tang. This is designed to provide the drive for the drilling operation so that the taper provides the location and centring in the sleeve or spindle, but is not required to resist the torque which is exerted when drilling large holes. Provision for utilising the driving tang is not usually provided on a small lathe but is incorporated into larger drilling machines.

Clearances and adjustments

It will be obvious that the fits of the saddle to the bed, the cross-slide to the saddle and the upper part of the topslide to its base need to be maintained at what might be described as 'nice' clearances. That is, they should be free to slide under control from the associated feedscrews but should, as far as possible, be entirely free from shake or backlash. This means that the fit of the feedscrews should be good and they too need to be adjusted so that there is as little free play as possible. The different sliding members need to be maintained in correct alignment also, and so are arranged to move on slides which have adjustable clearances.

The mounting of the slides and the saddle usually takes the form of a pair of dovetail slides, as shown in the cross-section of Figure 12.22. Machined horizontal surfaces support the moving slide but it is located laterally by an upward projection with angled sides. The slide itself is machined with a mating cross-section but this is made wider than that on the lower member so that a filler strip , or gib strip, may be used to pack out the upper dovetail slot to adjust clearances. The gib is locked to the upper slide by screws which allow adjustment of the

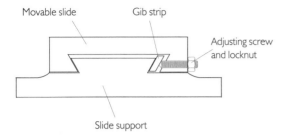

Figure 12.22   A cross-section of the type of mounting for the slides on a small lathe.

clearances. These screws locate in drillings or dimples in the gib ensuring that it moves with the slide. Adjustment of the screws removes any shake or play between slide and base and so allows the fit to be optimised. This adjustment should preferably be carried out with the feedscrew removed so that a proper feel of the fit can be obtained and the locking nuts (if provided) should be tightened and the fit finally assessed before the feedscrew is remounted.

Figure 12.23 shows the gib strip and adjusting screws, with lock nuts, on a slide of this type. This slide is also provided with two additional screws which can be used to lock it in a fixed position whenever movement is not required and the opportunity can be taken to provide the additional rigidity that the locking of the slide affords.

Figure 12.23   This underside view of a vertical slide shows the gib strip and the adjusting and locking screws.

The freedom, or otherwise, of the slides is one indicator of the condition of a secondhand machine. Any sensible owner in the habit of using his machine will have the slides adjusted as tightly as possible consistent with there being no actual binding. On a worn machine, this will mean that a good adjustment for places where there has been little or no wear will be sloppy at the most-worn spots. This effect will be noticed in the fit between the saddle and the bed, where sloppiness will generally be found at the headstock end, where most wear takes place, or binding at the tailstock end, according to the way the saddle clearances are adjusted.

It pays always to keep the slides clean and lubricated with a thin grease or thickish oil and they should be cleaned and lubricated regularly. It should be noted however, that cast iron is very dusty stuff to machine and it is best to keep the lathe as dry as possible while this metal is being machined, afterwards thoroughly cleaning the slides and re-lubricating them. In this respect, it is useful if the slides can be lubricated via an oil or grease nipple as introduction of the lubricant from the inside tends to push any debris out of the slideways.

Correct adjustment of the gib strips ensures that the slide fits without shake or looseness in one direction, but the attachment of the slide to its feedscrew, and the fit of the screw in its nut can still give rise to backlash or free play. The cause of this is explained by reference to Figure 12.24, which shows a section through a typical cross-slide, and the arrangement of the feedscrew and its nut.

The nut is naturally part of the saddle or is bolted to it, and the feedscrew engages with it. The feedscrew usually has a square-form thread (strictly, an Acme thread, which has a flat crest and root and tapered flanks) which engages the nut. At the outer end, the feedscrew is provided with a thrust collar and a reduced-diameter portion which passes through the endplate, which is attached by screws to the cross-slide.

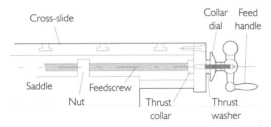

Figure 12.24 A longitudinal section of the cross-slide showing the means of adjusting the free play.

The outer end of the feedscrew is threaded with a conventional vee-form thread and the collar dial screws on to this and can be positioned so that it and the thrust collar embrace the endplate closely. The feed handle is also threaded to match the end of the feedscrew and it is screwed on tightly so that it acts as the lock nut for the collar dial.

There are two sources of free play in this arrangement. There must be clearance between the feedscrew and its nut, and there must be some clearance between the thrust collar and collar dial and the endplate. These clearances show themselves as lost motion or backlash when the rotation of the feedscrew is reversed. At each reversal, initial rotation of the feedscrew simply takes up the clearances.

For a given position of the cross-slide, there are two readings of the collar dial – one indication when driving the slide inwards and another when pulling the slide outwards. So, if the collar dial readings are to be used for positioning the slide, the slide's motion must always be in the same direction.

Careful adjustment of the components on the outer end of the feedscrew, and attention to any wear on the endplate will result in only small clearances at this point. There must naturally be clearance between the feedscrew and its nut, but even this can be adjusted on some machines which are provided with a split nut provided with a setscrew which can be used to squeeze the two halves together. On other machines, the nut is a low-cost, bolt-on item which can readily be replaced.

13

Holding work in the lathe

Introduction

Chapter 12 provides a description of the features of a typical small lathe, the mandrel of which is provided with a register and screw thread so that accessories may be mounted on it and remain aligned with the axis of rotation. Most of the items mounted on the mandrel nose are used for holding the workpiece. This is certainly true when the lathe is used for its intended purpose of turning, i.e. rotating the workpiece and bringing a cutting tool into contact with it by use of the saddle and cross-slide motions.

The faceplate

The manner of holding a shaft between centres and driving it by a catchplate-and-carrier arrangement is described in Chapter 12. This requires no accessories other than the simple ones described and is the manner in which early lathes were designed to be used, for example, for turning chair spindles and legs.

If the work to be turned is large in diameter but relatively short, a between-centres method of holding and driving is not appropriate, and a mounting which more resembles the shape of the workpiece is required. This is the faceplate. In essence, this is simply a catchplate without its driving peg, but when such a device is designed to be used as a faceplate it is normally provided with slots so that clamps to secure the work may be easily bolted on.

The faceplate is normally an iron casting bored and screwcut to suit the mandrel nose and turned on its front face and on the periphery, so that it runs true when mounted on the mandrel. The slots are usually cast-in and left unmachined, as is the rear surface. Figure 13.1 shows a pair of faceplates for a 3½in. (89mm) lathe, the smaller of which will be seen to have some extra holes in its face. These have been drilled and tapped to allow relatively accurate location of work on the face when several identical parts have been required and the faceplate has been used for holding the work.

Figure 13.2 shows a partly turned locomotive wheel bolted to the faceplate for the final facing back to thickness of the crank boss. A

Figure 13.1   A pair of faceplates.

Figure 13.2   A wheel casting secured to the small faceplate.

quently found on small lathes. This provides the capability to swing much larger diameters, just a few inches or centimetres long, at the headstock end of the lathe. In order to utilise this gap to the maximum effect, it is essential to use as slim a workholding device as possible therefore, generally, the faceplate will not be found to extend significantly beyond the end of the mandrel.

The larger of the two faceplates of Figure 13.1 is 9in. (229mm) in diameter yet it swings in the gap provided in the bed of the 3½in. (89mm) lathe for which it is an optional extra, as shown by Figure 13.3.

Figure 13.3   The large faceplate occupying part of the gap in the bed.

subsidiary mandrel, turned to be a good fit in the bore of the wheel, is fitted into the mandrel taper and locates the wheel relative to the centre of rotation. This alignment is not vital to this final operation but was used previously to ensure that the tread and flange were turned to be concentric with the bore.

The faceplate is adaptable to many turning operations, particularly when the work to be turned is large in diameter or is awkwardly shaped and unsuitable for holding by other means. One further advantage of the faceplate is provided by the gap in the bed which is fre-

Figure 13.4 Faceplate dogs used to hold a steel plate to the faceplate.

The method used for holding work on the faceplate is very much dependent on the shape of the item and the machining which is required. Figure 13.4 shows the use of slotted clamps for workholding, called faceplate dogs. They are used for clamping the work by use of bolts passing through the clamps (and faceplate), the toe of the clamp bearing against the work and the heel being supported on packing raised to a suitable height.

An alternative clamp is a section of angle, drilled for a clamp bolt and used in much the same way as a faceplate dog. Due to its shape it does not necessarily need packing at the heel and is considerably more convenient under

Figure 13.5 L-shaped clamps holding a boiler backhead to the faceplate.

some circumstances. Clamps of this type are shown in Figure 13.5. Both types of clamp are easily made and will repay the effort of their construction, being applicable also to the clamping of work to the cross-slide or vertical slide for boring or milling.

The disadvantage of the clamping methods described above is that the clamps themselves obscure the outer face of the work and prevent the whole of this surface being turned. These methods are most suitable where the work is provided with lugs which allow clamps to be applied outside the area to be machined, or where the work only requires holes to be drilled or bored.

A variation of the mounting method may be used when the complete face of the work needs to be machined. This utilises an angle plate bolted to the faceplate, the work being bolted to the angle plate rather than directly to the faceplate. This arrangement is shown in Figure 13.6 which also demonstrates an alternative method of securing the work by use of a long bolt through the angle plate, in this instance, in conjunction with a strap of mild steel, which forms the clamp. A thick strap of mild steel can be used, as shown here, but a length of steel angle adds some stiffness to the clamp, and might also be used, depending on the situation.

It frequently happens that work must be mounted out-of-balance on the faceplate (Figure 13.6, for example) and quite large out-of-balance forces can be generated when the mandrel starts rotating. This is potentially dangerous in that the friction clamps that are normally used may give way as the job tends to throw itself into orbit, but, in any event, it places an unnecessary strain on the mandrel bearings, particularly on the front bearing cap. Any out-of-balance mass naturally throws the mandrel about and since it has clearance inside its bearings, accuracy of the work will be affected and so a degree of balancing must be attempted. This is simply arranged, as shown in Figure 13.7.

Figure 13.6 If an angle plate is mounted to the faceplate it can provide the means to hold a cylinder block casting so that the portface can be turned. Lacking an angle plate, a length of square angle can be used in its stead.

Figure 13.7 Work may be mounted off-centre and need to be counterbalanced. A suitable source of weights is the lathe changewheel set but care must be taken that the tool does not touch the changewheel

Anything can be pressed into service for use as a balancing mass, but the basic requirement is satisfied by a few discs with holes through their centres so that they may be bolted into position. Lathe changewheels are ideal in this respect, but care must be taken not to allow them to strike the bed or cross-slide or to come into contact with the cutting tool.

The four-jaw independent chuck

A four-jaw independent chuck is the work-holding device which you must have – you cannot possibly manage without it. Not only is it capable of holding rectangular work but the independent movement of its jaws allows work to be accurately centred so that turned surfaces may be generated with reference to already turned diameters, or to reference points marked on the job. It is also suitable for turning circular-section bars, making it the most versatile of the available chucks.

This chuck developed from an adaptation of the faceplate which seems to have been in use in the early part of this century. As noted above, there is some disadvantage in the use of the faceplate since all work may not be suitable for mounting by clamps, or dogs, pressing onto the outer face of the work. Indeed, for most items, gripping by the outside edge is preferable since this naturally leaves the face clear for machining. Consequently, machine designers developed small, bolt-on assemblies comprising hollow rectangular channels in which a screw-controlled sliding jaw was located. These could be mounted into the faceplate slots so that the jaws could be positioned radially to clamp to the outside of the work. The four-jaw independent chuck is a development of this idea.

Figure 13.8 shows a 6in. (150mm) four-jaw chuck. It has a ground, cast iron body with four machined slides in which the jaws fit, each jaw being screwcut on the reverse to mate with a captive screw mounted in the body below each slot. The screws have square holes in their outer ends, and can be turned using a square-ended key.

The jaws and slots are matched during manufacture and each is numbered accordingly; jaws normally being hardened and ground and the slots also ground to size. Jaws should not be interchanged between slots. The body of the chuck carries a stamped or en-

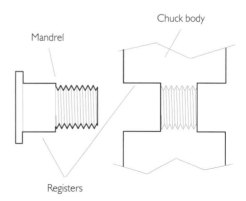

Figure 13.9 The normal mandrel fitting includes a screw thread and some form of location for the chucks etc. which will be mounted on it. On larger lathes, the chuck may be located by a short taper and be bolted into position.

shows one type of backplate-to-mandrel fitting.

All four jaws are reversible in their slides and are ground with two steps so providing flexibility for mounting different types of work. Figures 13.10 and 13.11 show work mounted in a four-jaw chuck.

One possible disadvantage of the four-jaw is that it is not so easy to cope with out-of-balance workpieces, but this can usually be resolved by use of the faceplate, as described above.

Figure 13.8 This four-jaw independent chuck is bolted to a backplate which is machined to fit the end of the mandrel.

graved serial number and each jaw is numbered to match the body. It goes without saying that jaws should not be interchanged between chucks.

Figure 13.9 shows that this particular chuck is mounted to a backplate which is bored and screwcut to fit the mandrel of the lathe. The chuck body is bored at the rear so that the backplate may have an integral, large-diameter spigot which locates the chuck accurately on it. There are two registers to hold the chuck body concentrically on the lathe mandrel. Figure 13.9

Figure 13.10 A casting for the rear toolpost mounted in the four-jaw chuck.

275

Figure 13.11 An awkward casting held in the four-jaw chuck for machining the outer face.

Self-centring chucks

Although relatively rare, self-centring four-jaw chucks are available and are naturally extremely convenient for holding both square and round work. The more normal self-centring chuck is the three-jaw, which is used for holding round and hexagonal workpieces.

In a self-centring chuck, all jaws are driven in or out together, coupled to a plate within the chuck body on which a spiral thread, or scroll,

is cut. The jaws carry mating sets of teeth for the scroll on their inner surfaces so that, when the scroll rotates, the jaws move in or out. Figure 13.12 shows a 4-in. (100mm) three-jaw chuck with jaw 3 removed and reversed so that the jaw and scroll machining may be seen. Note that the jaws and jaw slides are all numbered to match one another and the jaws also carry the serial number of the body to which they have been fitted.

The chuck illustrated in Figure 13.12 is of the type known as a geared scroll chuck since the scroll is driven through gears having a square socket in which the chuck key engages. An alternative type of three-jaw has the scroll fitted externally, at the rear of the chuck. In this type, the scroll is drilled radially to accept a tommy bar so that it may be levered round to open or close the jaws. This is called a lever scroll chuck and is normally less expensive than the geared scroll type.

Unlike the four-jaw of Figure 13.8, the three-jaw chuck has its body machined for fitting directly to the mandrel nose, and it is described as a recess fitting chuck. The aim here is to reduce the overhang and maintain the workpiece much closer to the mandrel than is possible if the chuck is fitted to a separate backplate, thus providing better support during machining.

Figure 13.12 A three-jaw chuck with one jaw removed and showing the internal scroll.

Figure 13.13 The chucking piece on a casting held in the three-jaw chuck.

Figure 13.14 A round steel bar in the three-jaw chuck.

Figure 13.15 The alternative pair of jaws holding a cast ring.

Due to the shape of the scroll, the jaws cannot be reversed as can those of the four-jaw chuck shown in Figure 13.8 so two sets of jaws are provided. These are normally described as inside and outside jaws. Together, the two sets provide for versatility in use and Figures 13.13 to 13.15 show different types of workpiece in the chuck for machining.

Accuracy and concentricity

In theory, a three-jaw chuck, if perfectly made and mounted, would hold round work axially in line with the mandrel and would do this for all diameters of work. It would be possible to mount a circular bar into the chuck, to reduce part of the rod to a smaller diameter and produce two diameters which were perfectly concentric with one another.

In practice, this is not possible. No self-centring chuck will hold all workpieces absolutely true, and a new chuck will have a nominal specified accuracy of .003in. or .004 in. (0.08mm or 0.10mm) total eccentricity. That is, if a perfectly circular rod were put into the chuck, the jaws tightened, and the mandrel rotated while measuring the position of a point on the surface of the bar relative to some fixed part of the machine, a total deflection of .003in. to .004 in. (.08mm to 0.10mm) will be measured. This eccentricity is called Total Indicated Run-out, or TIR.

It is likely that a new chuck, even of ordinary grade will actually be capable of better performance than the permissible tolerance, due to selective assembly of the parts by the manufacturer, but the three-jaw does not guarantee concentricity between the bar stock surfaces and newly turned diameters. If concentricity is required between the outside diameter of a round bar and further turned surfaces, either the bar must be set to the required degree of accuracy in the four-jaw independent chuck, or a larger stock bar must be used and both diameters must be turned at the same setting (as part of the same machining operation) without removing the work from the chuck.

The dial test indicator

A Dial Test Indicator, or DTI, is the instrument that permits the run-out of a circular bar mounted in a chuck to be measured. A DTI consists of a ball-ended lever or plunger connected to a pointer which moves over a scale as

Figure 13.16 A plunger-type dial test indicator.

the lever or plunger moves relative to the case of the instrument.

One type of DTI is illustrated in Figure 13.16. It has a circular case about 2in. (50mm) in diameter and is fitted with a plunger which passes through the case, along a diameter. The plunger is cut with a rack which engages a small gear set on a shaft, which has the pointer, or hand, on the front. As the plunger moves, the hand rotates over a printed scale, providing an indication of plunger movement. In order that the plunger can remain in contact with the object which is being tested, it is lightly spring-loaded outwards. For this DTI, the total plunger movement is ½in. (12.5mm).

The scale has major divisions of .001in. (one thou) and has minor marks showing the .0005in. positions. The gearing between the plunger and pointer is such that the hand rotates once for a plunger movement of .050in.

(1.27mm), and the number of rotations of the hand needs to be counted if large movements are to be measured. Some DTIs incorporate another hand for this purpose.

This is not a normal use for a DTI however, and to emphasise this, the scale is engraved with a zero and marked to indicate plus and minus .025in. about this position. The printed scale is attached to the bezel of the case and both can be rotated to bring the zero line to some convenient position when making a measurement of plunger movement.

DTIs having total plunger (or lever) movements of ½in. or 1in. are the normal types, usually having dials engraved in .001in. increments, but types with a smaller total movement are available, having engravings at .0005in. intervals. The corresponding main types of metric DTI provide 12.5mm or 25mm plunger or lever movement and have .01mm dial engravings.

The disadvantage of the type of DTI illustrated is that it cannot be used inside a hole, unless the hole is large enough to accept the complete instrument. A more versatile DTI is the lever type, an example of which is shown in Figure 13.17. In this DTI, a ball-ended lever is pivoted at one end of the case and is linked to the pointer such that .030in. movement of the ball end produces one rotation of the hand. The scale is marked with a zero line and shows plus and minus .015in. marked in .0005in. (half a thou) divisions.

The lever on this DTI is loaded by a double-acting spring to a central initial position. It needs to be brought up to the work and pressed against it to produce an initial deflection in order that it remains in contact with the surface which is being checked. The dial and bezel can again be rotated on this DTI to bring the zero indication to a convenient position. The movement of the ball is restricted to plus and minus .045in. from its central position.

Since the ball on the end of the lever is only

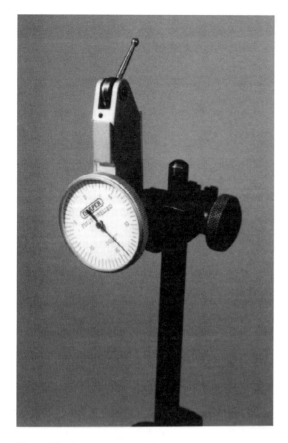

Figure 13.17   A lever-type dial test indicator.

$\frac{1}{16}$in. (1.5mm) in diameter, it will enter a small bore, which the plunger type will not. In addition, the lever is held to the internal levers by a stiff friction grip and can be positioned within an arc greater than 200 degrees, adding additional versatility.

Setting work in the four-jaw to run truly

Introduction

Initial turning operations in the four-jaw chuck may not actually require accurate setting up of the workpiece since the objective may simply be to clean up one face (of a casting, for example) to allow marking out and subsequent machining operations to be performed. As long as the job is securely held and reasonably balanced, this will suffice.

Frequently, however, the workpiece requires turning, or needs holes drilling or boring, with respect to existing turned surfaces or marked-out reference locations. Different situations can be envisaged, which require different approaches, and these are considered separately below.

Aligning to a cast-in feature

When dealing with cast parts which are circular, it frequently occurs that some surfaces cannot or will not be machined, but those that are turned should be reasonably true with the as-cast surfaces. The situation which comes to mind is when machining locomotive wheels or perhaps the spoked flywheel of a model traction or stationary engine. The inside of the flywheel rim may possibly not be machined since the prototype was not, and it is desired to follow prototype practice on the model. Similarly for a locomotive wheel. The heel of the crank boss cannot normally be machined and yet it looks so much better if the heel radius is reasonably concentric with the axle mounting hole. In both of these cases reasonably concentric is the description used and as long as the job is concentric by eye that is all that is required.

In these instances a 'sticky pin' is called for. This is nothing more elaborate than a dressmaking pin stuck into a piece of modelling clay pressed onto the end of a tool mounted in the toolpost. This provides a precise enough fixed point against which to judge any run-out of the cast surface and the job may be moved over by adjusting the jaws in pairs until the desired degree of truth is obtained. Figure 13.18 shows

Figure 13.18 A sticky pin in use when aligning the cast inner rim of a locomotive wheel.

a sticky pin in use to align the rim of a tender wheel casting.

Accurate alignment to a reference diameter

The situation in which a component must be remounted in the lathe and turned or bored to be concentric with an already finished diameter calls for the most accurate setting up. This can only be achieved by making an accurate measurement of the run out of the reference diameter using a DTI.

The technique simply requires an accurate DTI mounted in the toolpost or on a mounting base which can be stood on the bed or the cross-slide, so that run-out is measured directly on the dial. The DTI mounting must be rigid and free from shake, otherwise accurate setting is difficult or impossible to achieve. Figure 13.19 shows a very simple lever-type DTI on a stand with a magnetic base, in use for this operation.

The DTI is attached securely to the bed of the lathe by the stand's magnetic base with its lever resting on the parallel, circular part of the chimney which is mounted in a chuck having four independently adjustable jaws. The work can be moved in any direction relative to the centre of rotation of the chuck. The tubular centre part of the chimney has the base casting silver soldered to it, and the inside of the base has to be turned to mate with the inside diameter of the tube bore.

If the tube (assumed to be truly circular) is aligned with the axis of rotation, the DTI shows the same reading throughout a complete rotation of the work. If this is not the case, the DTI shows the relative movement of the surface of the tube as it rotates, showing that the centre of the work does not correspond with the axis of rotation. The DTI is said to show a run-out as the work is rotated. Jaws and work must be repositioned to bring the outer surface of the tube to the true-running condition.

Figure 13.19 A simple lever-type dial test indicator in use to set the run-out of the work to zero.

The way to do this is to rotate the work until one of the jaws is aligned with the DTI lever (or plunger, if the DTI is of this type) and note the reading, or set the dial to read zero. Rotating the chuck (and work) through 180 degrees then brings the opposite jaw into alignment with the DTI and the difference in the position of the two places on the work, aligned with the two jaws, can be calculated. It should be decided which jaw is 'up' and which 'down' and the pair of jaws moved in the required direction, ensuring that they still grip the work. If the work is already close to the true-running position, the jaws should be moved sufficiently to change the DTI reading by half the difference between the two readings.

If the work is still some way from the true condition, the DTI lever or plunger may lose contact with the work on one side. In these cases, again decide which jaw is up, and which down, and make a small adjustment to the two jaws in the required direction. The initial aim is to reach the condition in which the readings at the two positions are both on the scale of the DTI, after which, attention can be paid to the other pair of jaws, which are dealt with in exactly the same way.

Once the DTI reading is on the scale for the full revolution of the work, final adjustments can be made to both pairs of jaws, and a final check made through a full revolution.

If it is not possible to achieve satisfactory results, either the work is not truly circular, or the jaws are being tightened with variable pressure each time around, disturbing the readings, (or even squeezing the tube, in the example shown).

The particular example shown in Figure 13.19 is clearly not one in which great accuracy is required, but a DTI, even the simple one shown here, can be used to bring an already machined diameter or bore very accurately to a true-running condition before further machining is performed.

Boring or turning to a marked-out position

When marking out prior to machining, it is usual to mark the centre of any turned diameters or holes so that accurate location is possible. For holes to be drilled on the bench drill, centre punching is always used to allow the drill to find the centre. A centre punch mark is also convenient when marking out work which is to be turned on the lathe, since it is readily identifiable and easy to locate accurately on the job. The centre punch dimple also allows a dial test indicator and a simple accessory to be used to set the job so that the dimple runs truly.

Figure 13.20 shows an alternative way in which work can be set up in the four-jaw. A rectangular bar which is too long to hold between the jaws in the normal way has been set up cross-wise between the jaws. Notice also that two parallel blocks of mild steel have been placed behind the bar so that when it is tapped down to seat against them, it is parallel to the face of the chuck but its outer surface is clear of the jaws.

The bar has a section of mild steel soldered to one end and this is to be turned to a curved profile relative to a centre punch dimple roughly in the centre of the bar. The bar is aligned approximately before the chuck is mounted to the lathe mandrel (it is a lot easier to see when the chuck

Figure 13.20 A block mounted cross-wise in the four-jaw chuck. The frictional grip is quite adequate for safe use.

is horizontal) to minimise the time spent in achieving correct alignment.

Once the chuck is mounted on the mandrel, the tailstock sleeve is fitted with a centre. This is brought up towards the work and another centre interposed between it and the centre punch dimple in the bar as shown in Figure 13.21. If the chuck is rotated, the pointed end of the supplementary centre remains located in the dimple and if is this is not on the lathe axis, it naturally moves in a circle, taking the end of the supplementary centre with it. A DTI attached to a stand mounted on the lathe bed (or some other fixed part of the machine) can be used to meas-

ure the movement of the supplementary centre, and the position of the work adjusted in the chuck until the dimple runs truly. This is illustrated in Figure 13.22.

The DTI was more conveniently used with its dial facing the chuck, but the principle is nevertheless illustrated. Adjusting the work's position when it is mounted in this cross-wise manner is a little more difficult than for a conventional mounting, and it must be tapped through the jaws to adjust in one direction, and all four jaws adjusted together to move in the direction at right angles. It is nevertheless a quite legitimate way to hold the work, even though the grip relies solely on friction, rather than gripping on an abutment on the chuck jaws.

In cases where absolute accuracy of alignment is not necessary, the centralised position of the centre punch dimple may be found by simply bringing up the tailstock centre and adjusting the position of the work to bring the dimple into coincidence with the centre's point. Figure 13.23 shows how a half centre allows good visibility for this operation, demonstrated on the completed job illustrated earlier. This method of approximate alignment can, of course, be used as a preliminary to more accurate setting using the DTI.

Figure 13.21   A centre mounted between a dimple in the work and a centre mounted in the tailstock.

Figure 13.22   A dial test indicator can be used to monitor the loose centre when bringing the dimple in the work to the true-running condition.

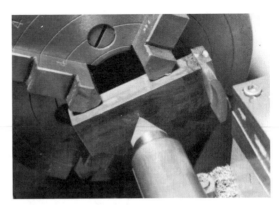

Figure 13.23   A dimple on the work can be aligned to a centre in the tailstock if a setting made by eye is adequate.

Figure 13.24 A hole in the work can similarly be aligned to a short stub of round material held in the tailstock chuck.

If the work has already been drilled to define a centre, alignment can be achieved by fitting a stub of suitable material into the tailstock chuck and adjusting the workpiece position to align the pilot hole with it. A tapered bar such as a scriber point often proves convenient but a parallel bar can be used, as shown in Figure 13.24, where a flanged copper boiler backhead, on the faceplate, is being aligned for boring the firehole.

Collets

One way to guarantee concentricity between the axis of rotation of the mandrel and the workpiece would be to machine the mandrel with a parallel bore (on the lathe itself) exactly the correct size to accept whatever (round) bar stock one wished to machine down to a smaller diameter. If the bar stock was perfectly circular, concentricity would then be automatically achieved. This is not exactly a practical solution but something very close to this idea is.

If the mandrel is bored concentrically at its nose, it can provide a mounting for a number of identically sized sleeves each of which is accurately (and concentrically) bored to a specific size. Each sleeve then provides accurate align-

ment of material of the particular size in the mandrel bore. A means must still be found to grip the material, but if the mandrel and sleeves are made as shown in Figure 13.25, and the sleeves are split symmetrically, a sleeve closes up on to the material when it is drawn or pressed into the mandrel bore, and therefore provides a grip. Correctly made, and kept clean, such sleeves offer the most accurate automatic alignment of round material with the mandrel axis.

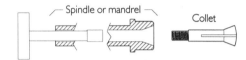

Figure 13.25 A typical fitting of a collet into the mandrel bore.

The sleeves are called collet chucks, or more simply collets. To be effective, a positive means must be provided to press or draw the collet into the mandrel bore. This may be a draw bar which engages a thread on the inside end of the collet and draws it into the mandrel bore, or may be an adaptor for the mandrel nose which presses the collet in. Figure 13.26 shows two collets, the smaller one being of the draw-in type, while the larger is of the press-in type.

This latter type is unusual in that it is designed to fit the No. 2 Morse taper bore in the mandrel of the lathe and is made having the relatively shallow 3-degree taper rather than the more usual 40-degree included angle of the draw-in type shown.

A further advantage of the collet is the very small overhang which it allows, permitting turning to be undertaken very close to the mandrel nose where the work is well supported. The set-up is therefore rigid and coupled with the accuracy which is afforded, is conducive to good work. Collets are, however, relatively expensive.

Figure 13.26 Draw-in and press-in collets and the larger example fitted with its nosepiece ready for installation into the mandrel bore.

Awkward shapes and second operations

Introduction

Some jobs do not really lend themselves to being mounted to the lathe mandrel by use of faceplate, chuck or collet and alternative means have to be found for holding the work. There is also the problem posed by the job which must be machined at both ends, often requiring a second setting up in the lathe, or a second operation.

The second-operation setting up may present difficulties if the chuck jaws grip a finished

(already machined) surface since, being hard, the chuck jaws bruise the work even when tightened to normal pressures. Gripping already machined surfaces directly in the jaws is not appropriate, and packing needs to be inserted between the jaws and the partly finished job.

Collets are especially useful for holding already machined diameters since they grip the work over larger areas and do not bruise the job as do chuck jaws. The only requirement is for the turned diameter to be correct for the available collet. If collets are not available, a homemade variety known as a split bush can be made up, as described below.

Some of the workholding methods which are appropriate to second-operation set-ups are also adaptable to awkwardly shaped workpieces so both problems can be considered together.

Problem pieces fall into several categories, including long work which is too large in diameter to pass through the mandrel, awkwardly shaped items and those second-operation set-ups which do not lend themselves to the use of chucks or collets for mounting the work. Some examples of each of these general types are considered below, but since the split bush is such a useful device, this is described first.

The split bush

Many small rods and pivots are made from materials such as silver steel and ground mild steel, which are already accurately sized on their outside diameters. If the lathe you have only has an old, worn three-jaw chuck which will not hold such material so that its outside diameter runs reasonably truly, you will find it a nuisance when the need is simply to reduce the material to a smaller diameter at one end, while maintaining the two diameters reasonably concentric.

In the absence of a collet to hold the work truly, the four-jaw chuck could be used, but there is then the nuisance of having to set the work running truly, and the possibility of bruising the work. As an alternative, a sort of temporary collet, known as a split bush can be made up. A stub of mild steel is cut off, placed in the chuck, faced off and chamfered at both ends. A piece about the same length as the chuck jaws is satisfactory and the diameter should be chosen to give a wall thickness of about ⅛in. (3mm) when the blank is bored to suit the workpiece. The position of number 1 jaw is marked on the outside diameter (centre punch mark) with the blank just protruding from the jaws and a bore of suitable size is made by centring, drilling and reaming or boring, this latter method producing the truest hole.

The marking of the position occupied by jaw 1 when the bush is made, means that it can always be replaced in the same orientation, and the bored hole should run truly each time the bush is used.

When it has been marked and bored, the bush is removed from the chuck and split by a single lengthways saw cut in a position which is central between two chuck jaws (the position opposite jaw 1 is convenient, since this has been identified). Once the burrs have been removed from the inside of the bore, a collet-like sleeve has been produced, but with only one slit. The bush can be placed in the three-jaw and will grip a circular bar (of the size for which it is made) when the chuck jaws are tightened.

Figure 13.27 shows two split bushes which demonstrate the versatility of such devices. One has been reamed to grip ¼in. (6mm) diameter material, and the other has been turned and bored to hold short lengths of ⅞in. (22mm) diameter steel.

The bushes need to be carefully made, and should have only thin walls, otherwise they need large pressures from the chuck jaws in order to close up on the work. This bruises the bush and its potential accuracy is lost. Since the bush grips the work over a large area, as does a

Figure 13.27 Two split bushes for mounting in the three-jaw chuck.

collet, it naturally offers the advantage of not marking the work. The larger of the two bushes shown has rather a thick wall, which has been weakened by external saw cuts.

Long, large-diameter bars

Long, solid bars which are too large to pass through the mandrel bore can be held at one end in a three- or four-jaw chuck but must be supported at the other end in some way when being turned, otherwise a wobble may develop, and the end fly about, which is exceedingly dangerous.

If the end of the work can have a centre drilled in it, it can be supported on a centre fitted in the tailstock sleeve, the headstock end still being held in a chuck. If diameters at both ends of the bar need to be concentric, centres must be put in both ends, the bar supported between centres, and driven by a catchplate and driving dog.

The difficulty comes in putting in the centres in the first place, since the normal method of facing and centring the end of a bar (described in Chapter 15) is not appropriate if the bar will not pass through the mandrel. Attempting to turn or centre when the job has a large overhang from the chuck is a recipe for disaster.

However, if the bar is reasonably stiff, which means not too long in relation to its diameter, a centre may be carefully put in to the unsupported end. Since the centre drill has two cutting edges, it presents a symmetrical load on each side of the axis of rotation of the work and, if put in with care, will provide the simplest means of cutting a centre.

It is vital to consider the stiffness of the workpiece in relation to the overhang however, and no attempt should be made to face or turn the outer end until the tailstock centre can be used to provide support. A 1in. (25mm) diameter bar with up to 6in. (150mm) overhang outside the chuck jaws can safely be centred without support, and even a ½in. (12.5mm) bar may be centred with this overhang, provided that the end is running reasonably truly. The way to achieve this is described below.

If the overhang of the bar is too long for a centre to be put in with safety, the end can be held in a special support, called a fixed steady, attached to the bed. Figure 13.28 shows a 12-inch (300mm) long bar held in a three-jaw chuck, with its end adjacent to a steady. This has a base which is clamped to the bed, and which has two adjustable fingers which can be brought up to touch the bar and then locked in position. The steady has a hinged upper half which can be clamped to the base, and which has a single adjustable finger that can be brought into contact with the bar and locked in position.

The first thing, however, is to make the end of the bar run truly. To assist this, three turns of paper are wound around the chuck end of the bar, and the chuck jaws are not tightened fully, so that there is a little resilience in the grip provided by the jaws. A DTI is set up with its stylus on the surface of the bar, near the end, and the chuck turned slowly by hand to locate the point at which the end of the bar is highest. When this point is located, the stylus of the DTI is lifted carefully away from the bar and the end tapped downwards with a soft-faced hammer.

Figure 13.28 A 12in. (300mm) long bar held in the chuck, showing the steady which will be used to support the outer end.

Figure 13.29 Paper wrapped around the bar at the headstock end.

Figure 13.30 Tapping the end of the bar to bring it true.

The set-up is illustrated in Figures 13.29 and 13.30.

The process is repeated, tightening the chuck jaws progressively, until the grip is secure and the bar end running truly. The lower fingers of the steady are brought into contact with the bar and locked, while the DTI is still monitoring its position, so that any disturbance can be detected. After this, the steady is closed and the upper finger adjusted and locked so that the end of the bar is properly supported. The fingers are lubricated with a little oil and a centre drill used in the tailstock chuck to cut the centre, as shown in Figure 13.31.

It is naturally helpful if the end of the bar is reasonably square, but it is adequate if the bar is cut off using a saw and then filed. If the end needs to be faced, this can be done by supporting the work on a half centre, and facing the end as described below.

If a steady is not available, and the work is too long and flimsy for an unsupported centring operation, the centres must be put in off the lathe. The workpiece should be sawn off squarely at both ends and filed so that it is reasonably flat and free from burrs. If the bar is cut off about $\frac{1}{16}$in. (1.5mm) longer than required, according to how squarely you can cut the ends, this allows for the ends to be faced to length later.

Figure 13.31 The true-running bar, supported by the steady, being centre drilled.

Figure 13.32 A bell punch.

To centre the ends of the cut-off bar, the drilling machine can be utilised, but to allow the centre drill to start, a centre punch dimple must be put into each end. If nothing else is available, the end of the bar may be marked out using Jenny calipers or odd legs and the estimated centre of the bar marked with a centre punch. The 'posh' way to find the centre is to use a centre square, but either method is adequate as long as the bar is somewhat larger than the nominal finished size, to allow for any inaccuracies.

The correct tool for centre punching the end of the bar is a specially adapted punch known as a bell punch. This comprises a bell-shaped holder bored to a sliding fit for a cylindrical centre punch, as shown in Figure 13.32. If the bell is placed squarely over the end of a circular bar, the punch automatically aligns itself with the bar's centre and a sharp hammer blow is all that is required to mark the bar. Naturally, if the bar is not faced off precisely, the centre punch dot may have a small positional error, but this can be accommodated by choosing a slightly oversize bar, as noted above.

Hopefully, the bar will not be so long that the drilling machine cannot be used for centring the ends and it should be possible to arrange to hold it in the machine vice, or clamp it to an angle

plate so that there is adequate support. This operation is illustrated in Figure 13.33. The centres should be put in a little deeper than required so that the later facing of the ends will leave the bar with adequate centres.

With the ends of the bar centre-drilled, it may now be set up between centres for the turning operation. If the tailstock is of the type which can be set over for taper turning, it must be set up first, as described in Chapter 15. Thereafter, the actual turning is straightforward. There remains however, the facing of the ends. Naturally this must be carried out at the tailstock end but cannot be accomplished if a normal, fully-coned centre is used. A cut-away centre, known as a half centre is therefore used in the tailstock, as shown in Figure 13.34. This has almost one half of its thickness machined or ground away at the pointed end, leaving just a small, fully coned tip to support the work but still allowing the tool to face the end.

Once both ends of the bar have been faced to length, a normal, full centre should be mounted in the tailstock since this provides proper support.

Figure 13.34 A close-up view of a half centre.

Large tubes

Large-diameter tubes may be dealt with in much the same way as large bars if the ends are

Figure 13.33 Centring a long bar on the drilling machine.

first plugged so that they may be centre drilled. Since concentricity may not be so important (tubes frequently only require their ends to be faced off to length) it is sometimes possible to use wooden plugs for providing the centres, but the mandrel end of the tube can usually be held in, or on, the chuck jaws, thus simplifying the driving arrangement and requiring only one end to be plugged and centred. Mounting to the chuck also simplifies the method of driving the tube.

If several tubes need to be turned, it is worthwhile making up a simple holding fixture which can take the form shown in Figure 13.35. This comprises a length of studding, with centres drilled in both ends, on which are mounted two wooden discs, turned to a suitable diameter to be a hand press fit in the tube. The discs are secured to the studding by nuts and washers on each side and a lathe carrier can be mounted on the studding at the mandrel end to provide the drive. The friction grip between the discs and the tube is generally sufficient to allow facing to length without problems, but, as noted above, it may be possible to hold the mandrel end of the tube in the chuck and this then provides the drive.

Figure 13.35 A length of studding and some wooden discs used for mounting large tubes between centres.

The wooden discs can be roughly cut out, drilled through their approximate centres and then mounted on the studding for turning to size. The lathe's top speed can be used for this operation and the turning performed quite adequately using an ordinary knife tool. The only problem is clearing away the sawdust!

Thin discs

Discs cut from sheet materials are sometimes needed and can present difficulties. If relatively small discs are required, having a central hole, they can be produced on the drilling machine using a trepanning tool or hole-and-washer cutter. For larger diameters, the drilling machine may not have a sufficiently slow bottom speed to keep cutting rates within comfortable bounds but the lower speeds of the lathe's back gear range do allow larger diameters to be turned, and the only problem which remains is how to hold the piece of thin sheet to the mandrel during turning.

If a disc is required having a central hole, this should be drilled, reamed or bored first so that it may be used for holding the blank for turning the outside diameter. If the chuck jaws will enter the hole, and bruising of the work will not be a problem, the hole may be used directly for holding the work on the chuck.

If the central hole is too small for this, a simple stub mandrel can be made up. This can take one of the forms shown in Figure 13.36, the mandrel being quickly turned up from a stub of suitable material. If the disc is large in comparison with its bore, two large washers should be used to increase the area used to grip the work, and if the friction grip provided by these is not sufficient, paper discs should be interposed between the washers and the job. In extreme cases, the use of emery paper discs will increase the grip still further.

For stub mandrels and other chuck-mounted

Figure 13.36 A selection of stub mandrels for mounting in the three-jaw chuck.

adaptors, it is useful to mark the position taken by jaw no. 1 when the mandrel is turned to diameter so that it may always be placed in the chuck the same way, thus helping to maintain the truth of the turned diameter.

If a disc is required which has no central hole, or if the hole is impossibly small in relation to the outside diameter which is to be turned, a chucking piece must be mounted on the blank prior to turning. This is a technique which is frequently adapted for castings and some jobs that serve to illustrate the method are shown in Figure 13.37. These are all castings for a steam locomotive – the smokebox door and the front and rear cylinder covers. Although each is to be machined to produce a broadly disc-like part,

Figure 13.37 Castings with chucking pieces.

each casting is provided with a chucking piece which is ultimately sawn or machined off, but which serves to hold the item for the initial machining operations. The smokebox door serves to show how the chucking piece is used and also demonstrates one way in which a solid disc, or disc with very small central hole, may be dealt with.

If the casting is reasonably circular, it is necessary only to clean up the chucking piece with a file (to remove any minor surface imperfections) and to mount it in the three-jaw. This allows one side to be faced and the outside diameter to be cleaned up to size. If the casting shows too much run-out when set up in the three-jaw in this simple way, the four-jaw should be utilised instead, to allow any eccentricity to be corrected. If there is a large amount of wobble of the face of the disc, this may have to be removed before turning can begin in earnest.

If the thickness and diameter of the casting allow it to be held in the chuck with the chucking spigot outwards, the spigot can be straightened by taking a cleaning-up cut on its outside diameter. If this is not possible, the spigot must be corrected by hand filing until the job runs as true as necessary to face the back and turn the diameter. Figure 13.38 shows these operations in progress.

The smokebox door requires a small central hole to be drilled so this is put in from the tailstock after facing off a small diameter (about the same size as the chucking spigot) to provide a good surface for the centre drill to start. With the hole drilled, the casting is removed from the chuck and the chucking spigot sawn off.

The central hole and the faced area on the inside allow another chucking piece to be made up and soldered into position so that the casting can be held in the chuck for the shaping of the outside surface. The chucking piece can be made up from anything that is available,

Figure 13.38 Turning a smokebox door held by its chucking piece.

provided that it can be soldered – a stub of brass is convenient. Since the door is drilled centrally, a short spigot should be turned on the end of the stub to locate in the hole and the outside diameter of the chucking piece cleaned up so that it is concentric with the short spigot. The casting should then run reasonably truly for the final turning on the outside. Figure 13.39 shows the door and chucking piece (the end of a length of stock brass rod) ready for soldering together. Soft soldering is adequate to attach the chucking piece to the inside of the casting and it need not be any greater than ½in. (12.5mm) in diameter for a 3½in. (90mm)

Figure 13.39 A smokebox door and its temporary chucking piece.

door casting, although naturally, huge cuts at high speed are not possible.

As noted above, if discs without a central hole are required, the soldering-on of a chucking piece allows the outside diameter to be turned, and if soft solder is used the chucking piece can be melted off once turning is complete, and the disc cleaned up.

Bored work

The chimney top and base castings shown in Figure 13.40 both need to be machined all over. Both have awkward shapes, to say the least, and the bores have a rough, as-cast finish and usually a good draw or taper to ease the mouldmaker's task at the foundry. With a little care, even these awkward shapes can usually be held in one of the chucks and Figure 13.41 shows each of the castings mounted up for machining the bore.

Figure 13.40 Chimney top and base castings.

Once bored to size, a casting may be mounted on a stub arbor for machining the outside, as shown in Figure 13.42. If the arbor is turned to be a firm fit for the bore, the casting can be pushed on, especially if the arbor is given a slight taper at its outer end. The friction grip is again normally satisfactory provided that things are taken gently, but if slippage occurs, a scrap of paper can be used as packing

Figure 13.41 Chimney castings held in the chucks for the first stage of turning.

Figure 13.42 The chimney top mounted on a stub arbor.

to improve the grip. Failing this, one of the modern metal glues can be used to secure the item. If an adhesive is chosen which breaks down at elevated temperatures, the job can be released from the arbor by gently heating the pair until the bond breaks down.

For turning the curves on items such as the chimney top, it is usually best to use a normal approach to establish the major diameters of the item, but to turn the curves by use of hand tools. Using these it is possible to produce such shapes relatively easily provided that a template is made to act as a guide as the shape develops, but it will frequently be found beneficial to remove the bulk of the metal, and establish reasonably true surfaces using the topslide and cross-slide motions together to approximate the required shape before using hand tools.

Screwed work

Workpieces having screw threads on one end can often be held in a screwed bush for any second-operation work which is required. Male or female threads on the work can be utilised by making a suitable adaptor for mounting in the chuck and simply screwing the work into (or on) the adaptor.

This method of mounting does not necessarily produce concentricity between the turned surfaces at the two ends of the job, since the fit of the screw threads on the adaptor needs to be just right to achieve this. The method is suitable if the end faces are required to be parallel or whenever the actual alignment is unimportant.

Figure 13.43 shows an extension tube which may be interposed between a lens and a camera body as a means of altering the focusing range. The abutment faces at each end must be parallel or the image will not be correctly focused at all positions on the film. The male thread was, therefore, machined first and then a screwed

Figure 13.43 A screwed cylinder and the ring used to hold it in the chuck.

bush was made, being bored, screwcut and faced off in the lathe ensuring, when making both threads, that the partly machined tube would enter the screwed bush right up to the abutment face. This is generally arranged by making the threads a sloppy fit and undercutting the male thread at the shoulder (see Chapter 15).

Once mounted in the screwed bush, the tube was faced off at the second end (to bring it parallel with the abutment face already machined) and then bored and screwcut itself to accept the lens thread.

In this instance, the tube was made from aluminium alloy and the bush was turned up from a slice of scrap brass. As noted elsewhere, it pays always to mark adaptors such as these with the position occupied by number 1 jaw when the device is made, as this helps to ensure accuracy should the bush be needed again. It also helps if the steps of the chuck jaws can be used to locate the adaptor, since without such location, a large, short cylinder is difficult to mount in the chuck without wobble.

14

Principles of turning

Tool materials

Carbon steel

The basic requirements for a lathe tool are that it should be hard enough to cut the work yet strong enough to withstand the forces imposed upon it during cutting. A high-carbon steel such as silver steel or gauge plate is suitable since it may be hardened and tempered to obtain a reasonable compromise between hardness and strength. These materials have the added advantage that they may be annealed and can be worked in the soft condition to form the tool shapes which are required.

However, carbon steel cutting tools have the serious disadvantage that they are tempered at pale yellow (210°C). The cutting action of the tool is such that chips bear down on the top and slide along the surface. This causes a frictional load on the tool and both it and the workpiece are heated up in the process. Indeed, chips or turnings may be removed from the work in the blue temperature region (300°C) and the tool point can, therefore, heat up considerably. If heavy cuts are taken continuously for a long

enough period, the tool point can quite easily heat up to a temperature at which the temper of a carbon steel tool will be 'drawn' and it will be rendered too soft to be used as a cutter without annealing, rehardening and tempering once more.

Carbon steel tools are only suitable when light cuts will be taken and the tool will not be overheated and rendered soft. Many of the jobs which will be undertaken are, however, well within the capability of carbon steel tools – indeed they were the only type available at one time. It was only the desire to achieve faster rates of metal removal and the introduction of much harder grades of material which required machining, that caused the abandonment of such tools.

One other situation in which the tool may become overheated is during resharpening. This is normally carried out on a grinder and friction between the fast-rotating abrasive wheel and the tool causes considerable heating and this may similarly draw the temper and render the tool useless.

Although case-hardened tools are seldom used, it is perfectly possible to do so and is worth remembering should a cutter be inadvertently made from a non-carbon steel, by mistake. The same arguments regarding the drawing of the temper also apply to the case-hardened tool, and once again, it should not be overheated.

Alloy and high-speed steels

Alloy tool steels were developed to provide resistance to what is called heat softening. In these materials, various alloying elements are added to high-carbon steels to produce increased 'hot hardness'. Sometimes, a single alloying element is added to the steel, sometimes two or three elements. Chromium, silicon, tungsten and manganese are among the elements used, various proportions conferring different tool hardness and different upper working temperatures.

The resulting alloy steels provide the capability to remain hard at 350 to 400°C, but another group of related materials known as high-speed steels remains hard up to 600 or 650°C. The name high-speed steel (HSS) is used as a general name when describing tool materials other than carbon steel and is sold generally under the HSS description which may be taken to describe a material with good hot hardness.

All high-speed steels undergo the appropriate heat treatment during manufacture and are supplied in the hardened condition. They do not require any further treatment other than grinding to the required form, and will virtually last until resharpened to such a small size as to be no longer useful.

Although extremely hard, they are nevertheless resilient and have good strength. They may also be ground on ordinary grinding wheels, are about half the cost of similar tipped tools (see below), and will perform all of the turning which is required.

High-speed steel is available in short lengths made especially for use as cutting tools, these normally being described as HSS tool bits. Square and circular sections are available, both having a ground finish, in increments of ¹⁄₁₆in. (1.5mm) from ³⁄₁₆in. to ¹⁄₂in. (4.8mm to 12.5mm). Figure 14.1 shows a group of tool bits from which it will be seen that they are ground-off squarely at the ends, or roughly cropped to length. Both types therefore need grinding to the appropriate shape before use.

Square bits are most commonly used for direct mounting in the lathe's toolpost, although they may equally be used in holders such as boring bars and fly cutters. For home-made versions of such tools it is usually easiest to utilise circular tool bits since the mounting hole may then simply be drilled and reamed to the appropriate size.

Figure 14.1   A selection of HSS tool bits.

Hard-alloy materials

In the search for even greater rates of metal removal, and the consequent need for materials with increased hot hardness, metallurgists were drawn to examine the properties of various carbides, based on the knowledge that it is iron carbide (cementite) which confers its hardness on steel. The carbides of tungsten, titanium, tantalum and cobalt were found to provide the desired improvement in hot hardness and various alloys of these carbides are now available, providing cutting materials which will operate at temperatures up to 1000°C, or well into the red hot range. The earliest of these alloys to come into general use was tungsten carbide and this name is used colloquially to describe tools having cutting points made in these alloys.

The materials themselves, known strictly as cemented carbides, are extremely brittle and are not suitable for use as lathe tools unless they are supported by a backing of ordinary steel to provide the strength. Cemented carbides are costly to produce and solid carbide tools would, in any case, be prohibitively expensive. Consequently it is normal for the carbide to be supplied as a small tip on a steel bar, and the name tipped tool is used to describe this type.

Two ways of providing the support are used. The first is to silver solder (or braze) a carbide tip onto a steel shank, thereby making what is commonly called a tipped tool. This is then ground to the required cutting angles and a strong, hard tool results. Since the carbide tip cannot be ground on an ordinary carborundum grinding wheel, commercially available tipped tools are supplied already ground to the required shape. They cost 20 or 30 per cent more than the same size HSS tool bit (which is not ground to shape, as purchased) but can be expected to give a long life between regrindings, especially for the amateur worker. Figure 14.2 shows tools with brazed-on carbide tips.

The second method of providing support for the cutter is to arrange that the carbide tip is mechanically clamped to a standard steel holder. The tips are manufactured to tight tolerances, in square, diamond and triangular shapes, and a variety of holders is available, allowing selection of facing, bar turning, boring or screwcutting tool shapes. The cutter tips are usually drilled centrally for the clamping or locating spigot and rapid removal and replace-

Figure 14.2 Tools with brazed-on tips.

ment of the tip is possible through the action of a wedging or locking screw. One form of holder and tip is shown in Figure 14.3.

The initial outlay for a holder and tip is relatively high, but replacement tips are approximately the same cost as a ¼in. (6mm) circular or square HSS tool bit. Considering that no resharpening is required, but the tool is nevertheless always sharp, the expenditure may be considered not too high a price to pay for the convenience.

The disadvantage of tipped tools, of whatever type, is the extreme brittleness of the tip itself. Any accidental heavy contact with the job, or a minor contact with the chuck jaws, is certain to knock a lump off the tip, thereby ruining it and requiring scrapping, or regrinding, if the tool is of the brazed-on type.

Figure 14.3 A tool fitted with a replaceable clamp-on tip.

For clamp-on tips, each corner of the tip is usable in turn, and accidental chipping, or final blunting of the tip, is rectified by bringing a new point into use. For the more common brazed-on tipped tools, a special grinding wheel is required, sufficiently hard to grind the cemented carbide tip. This is known as a green-grit wheel. These cost about 25 per cent more than normal grinding wheels, but once available, will allow brazed-on tipped tools to be reground without problem.

Industrially, the ability of tipped tools to operate at high tip temperatures is extremely important. In spite of the fact that such temperatures are unlikely to be reached in the amateur's workshop, the tools are nevertheless very useful. The cutting tip is hard and has a highly polished top surface. The chippings or turnings slide over the surface very easily, away from the cutting zone, and the hardness makes the tools useful for difficult situations such as the initial cuts on iron castings, or even those in gunmetal, which sometimes have a hard skin.

Theory of cutting

Cutting action

The basic process of cutting on the lathe is one of forcing a broadly wedge-shaped tool into the rotating workpiece. The action of cutting is one of tearing a 'chip' or 'turning' of metal out of the work. The chip bears down heavily on the tool, just behind the cutting point, and is severely compressed in the process. This is illustrated in Figure 14.4 which shows a section through the tool, looking straight on to the work. The process of removing metal continues so long as the workpiece is rotating and the tool continues to be fed into the cut.

The finish left on the work by the initial tearing process is extremely rough and it is left to

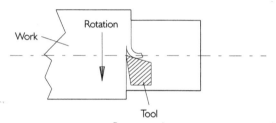

Figure 14.4   The basic cutting action.

the sharpened point of the tool to remove tiny chips from this surface and leave a smooth finish on the work.

The cutting action demands that the tool should have adequate strength and toughness, which in practical terms means that it should be as large as possible (particularly in the vertical direction) and it should be well supported in the toolpost.

Tool angles

The best slicing action for the tool of Figure 14.4 is provided by using as sharp a cutting angle as possible, but as Figure 14.5 shows, a narrow wedge angle results in a weak tool, making it unsuitable for cutting strong materials. Since materials vary in their composition and strength, the way in which the chip forms varies considerably as does the amount of power required to form a chip of a given size. Technically, different materials are said to have

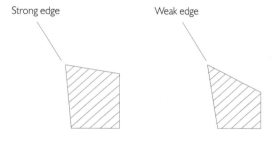

Figure 14.5   Weakness in a narrow-pointed tool.

different 'machinability'. Since the strength of the tool also needs to be considered, different tool materials can use different angles, even for cutting the same type of metal.

In practical terms, this means choosing the wedge angle of the tool to suit the material being turned. A quite different tool shape is used for brass from that used for aluminium, these being the two extremes.

Some compromises must be made in choosing the tool angles and it is possible to select for particular characteristics. For example, a tool may be ground to maximise metal removal without regard to surface finish, or to produce a good finish but not having the capacity to remove metal very rapidly. These compromises give rise to the descriptions roughing and finishing in relation to lathe tools and a two-stage process is sometimes adopted, using different types of tool, first roughing out and then finishing.

Another consideration also applies to selection of tool angles – the cutting point must be the only part of the tool that touches the work, and angles must be ground on the tool to ensure that this occurs. Below the tool point, these angles are called clearance angles, and in the plan view, they are known as relief angles, as shown in Figure 14.6.

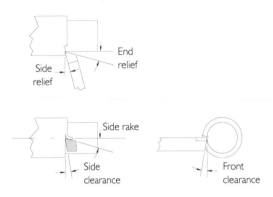

Figure 14.6   Clearance, relief and rake angles.

Clearance angles

Front and side clearance angles are essential to ensure that the part of the tool below the cutting point does not rub on the work. If these angles are made too large, the support for the point is weakened, as shown by Figure 14.5, and it is best to standardise on a value of 10 degrees for the clearance angles. If really tough materials need to be turned (stainless steel, or tool steels, for example) clearance angles of 5 degrees can be helpful in strengthening the tool point.

Relief angles

If the side and end relief angles are made too large, a very sharp point is presented to the work. This weakens the tool and also makes it difficult to achieve a good finish, so large relief angles are also best avoided, and 10 degrees is the usual value adopted. Side relief is provided by grinding the end of the tool at an angle of 20 degrees, in the plan view, and skewing the tool in the toolpost.

Right-hand and left-hand tools

Figure 14.7 shows plan views of the ways in which the tool can approach the work. At A, a tool is shown which performs a surfacing cut when fed towards the headstock. This tool is called a right-hand tool and requires side rake ground on the tool in the same direction as that shown in Figure 14.4. Figure 14.7B shows a left-hand tool (one which cuts when fed away from the headstock). This naturally requires side rake to be ground in the opposites direction to that required on a right-hand tool.

Notice from Figure 14.7C that, theoretically, a left-hand tool is required in order to make a facing cut with the tool fed towards the

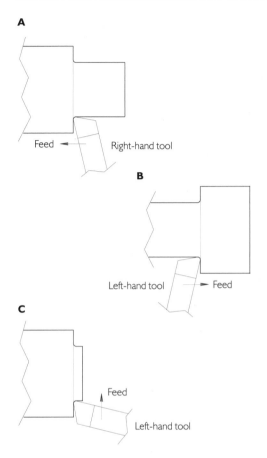

Figure 14.7 Right-hand and left-hand tools.

centre of the workpiece. In practice, this is seldom considered necessary, and it is common to use a right-hand tool for this purpose.

Rake for different materials

The clearance and relief angles, which exist largely to ensure that only the point of the tool touches the work, can be standardised at 10 degrees. With the side clearance, the side rake creates the wedge angle for the tool, which can have different values, depending upon the material being cut.

The accompanying Table 14.1 presents a list of side rake angles which can be used for turning common materials. As with many things, the angles selected represent a compromise between providing a sharp wedge angle and yet leaving a sufficiently strong point to withstand the cutting forces. Since the common tool material is high-speed steel, Table 14.1 has been compiled for tools made in this material.

Table 14.1 Recommended Cutting Angles For High-speed Steel Knife Tools

Material	Side Rake (degrees)
Free-cutting mild steel	20 to 25
BMS	15
Carbon steel (silver steel)	8 to 10
Cast iron	10 to 12
Hard brass and FC bronze	0 to 3
70/30 ductile brass	5
Copper, phosphor bronze and aluminiumbronze	20 to 25
Monel metal	10 to 14
Nickel silver	20 to 30
Aluminium and alloys	25 to 40*

* Aluminium alloys vary tremendously

Tools for bar work

Introduction

The basic shape of a right-hand tool for turning external surfaces is shown by the views of Figure 14.6. For want of a better phrase, this is described as bar work although it might also be described as outside turning. The principles of turning inside a hole in the workpiece (normally known as boring) remain the same but since there are several means of hole production available, as well as boring, these are all described together in Chapter 15 and this chapter is restricted to a description of bar work.

The knife tool

A tool which is provided with rake, clearance and relief angles shown in Figure 14.6 is known as a knife tool. It is by far the easiest tool shape to create from a standard tool blank and is extremely easy to resharpen, when this is required. A right-hand knife tool is illustrated in Figure 14.8.

The knife tool is satisfactory for both surfacing and facing cuts and will face the end of the work and cut up to and form a square shoulder on the work. If the end of the tool is ground back about 20 degrees in the plan view, this provides suitable relief angles, in the plan view, if the tool is skewed 10 degrees in the toolpost.

Figure 14.9   A right-hand knife tool in use.

Slightly modified tool shapes especially designed for the finishing operation are described below. These vary little from the ones already described and for most normal turning, knife tools having the side rake angles listed in Table 14.1, and front and side relief of 10 degrees, as shown in Figure 14.6, are the normal ones employed. Side and front clearance angles can be ground at 10 degrees, but if really tough materials need to be turned clearance angles of 5 degrees might be helpful.

Alternative tool shapes

Two modifications can be made to the tool in order to change its characteristics. The first of these is to grind the cutting point to a small

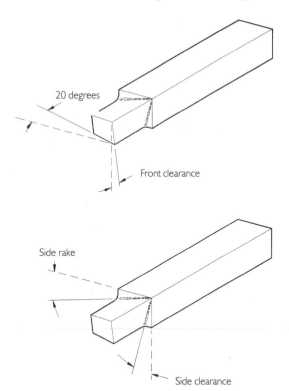

Figure 14.8   A right-hand knife tool.

radius, in the plan view. This has the double benefit of strengthening the point and of bringing a larger part of the tip into use for cutting, the finish is improved and so is the life of the tool.

Figure 14.10 Chatter on a locomotive wheel casting.

The radius must not be too large, however, since a larger contact between the tool and the work imposes a greater load on both and greater deflections occur. As a general guide, a radius of about ⅟₃₂ in. (1mm) should be supportable on a medium-size lathe, and is suitable for a tool ground from high-speed steel. Anything larger may cause problems, but it does depend on the situation.

As an example, the round-nose tool shown in Figure 14.10 has produced a good finish on the wheel tread, but it has also produced the uneven finish on the locomotive wheel flange. The tool has been used to form the side of the wheel flange. In this situation, the tool was cutting over a very large part of the tip. The work has consequently pressed down heavily on it, and it has deflected out of the way, reducing the load and causing the tool to deflect back up again.

This has induced a vibration of the tool which caused a varying cut to be applied and this spoiled the finish on the work. This type of rippled finish is called chatter, and it is, needless to say, unwanted. In this case the chatter was caused by too large a part of the tool being in contact with the work, but there are other causes, and these, and their cures, are considered in Chapter 15.

The large-radius, round-nose shape does have the advantage that it produces a very strong point, and it is frequently used to strengthen the end of a tool with a tungsten carbide tip, as is the example shown.

Tool shape for good finish

When turning, the tool must be traversed into the cut along the surface or face of the workpiece. The finish imparted to the work is a function of:
(1) The finish on the tool.
(2) The rate of traverse of the tool (if the tool

is traversed too quickly, the result will be a cut like a screw thread).

(3) The shape of the end (cutting point) of the tool.

The shape of the end of the tool is important in providing an overlap between cuts, or more correctly, in providing an overlap between successive turns in the spiral which the tool makes on the work. If a certain rate of feed is envisaged for a finishing cut, say .010 in. per rev (0.25mm per rev), an overlap between cuts (spirals) can be arranged by grinding a .020in. (0.5mm) flat on the end of the tool as shown in Figure 14.11. Such a flat is most easily put on by hand using a slip stone and can be done at the same time that the tool is being honed.

The flat must not be made too large however, since it represents a departure from the principle that only the point of the tool should touch the work and it may induce chatter, which is not what fine finish is about! To this end a proper clearance angle must be ensured below the flat, which therefore needs to be put on with care.

The best finish is obtained by using the 'self act' of the machine to provide the feed since this guarantees a steady traverse of the tool into the cut. However, it is a mistake to use too fine a rate of feed for the final cut in the belief that this will produce the best results. It does produce fine finish (initially), but it also blunts the tool, so if there is any significant length to turn, the tool may not be cutting to the same extent at the end of the cut and a size problem may result.

When considering the length of cut, it is necessary to consider the length of the spiral which the tool traverses and not just the straight length of the work. If the feed is set to .001in. per rev (.025mm per rev) the work will make 1000 turns for the tool to traverse a length of 1in. (25.4mm). On a workpiece 1.0in. in diameter, this means the tool travels 260ft (80m) along a 1.0in. length. Feeds of the order of .010in. per rev (0.25mm) will therefore be found much more satisfactory for finishing, due to the shorter distance travelled by the tool. This is especially true if the material is relatively hard on the tool and likely to cause some wear on one pass. Stainless steel, cast iron and perhaps carbon steel might come into this category.

There are other reasons for keeping the tool cutting, such as the avoidance of chatter, and a reasonable rate of feed and good depth of cut will generally be found preferable. Certainly, feeds less than .005in. per rev (0.125mm) should not be used.

Maximum-efficiency tool

The usual form of this tool associated with the list of rake and clearance angles (Table 14.2, p.304) is that illustrated in Figure 14.12. This tool provides the most efficient shape to maximise the rate of metal removal and is normally recommended where this is the prime consideration. It is provided with both side and top rake i.e. the rake is ground at a compound angle.

The tool is quite different from the conventional type. It is not provided with side clearance but is designed to cut along a significant portion of the side, this being arranged by grinding back the side to produce negative clearance. Consequently, the terminology relating to the tool point changes, and the angles in

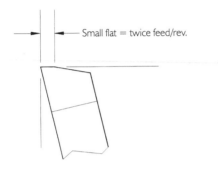

Small flat = twice feed/rev.

Figure 14.11   Creating a flat on the tool tip.

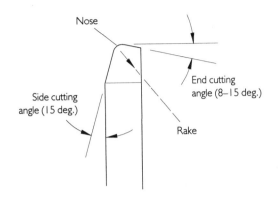

Figure 14.12 Maximum-efficiency tool.

the plan view are called the side and end cutting edge angles. The nose of the tool is radiused, essentially to strengthen the tip, since the tool is specifically designed for taking large cuts and a weaker, sharp point cannot be used.

Figure 14.13 shows the reason for the adoption of this tool shape. Two tools are shown, one of which has zero side clearance while the other has a side cutting edge angle of 60 degrees. A certain depth of cut is envisaged, together with a certain feed per rev, and the material which is removed from the workpiece is shown for each tool. Angling the side of the tool produces two effects – the chip which is being removed from the work is thinner, and a greater length of the tool is being used to remove the chip. Both effects naturally increase the life of the tool, hence its adoption when material removal is the prime consideration.

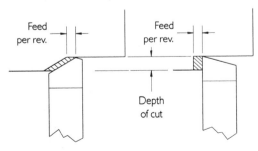

Figure 14.13 The effect of negative side clearance.

The tool is essentially for removing material quickly and although it will not form a square shoulder, it is ideal for roughing out as a first stage when turning a shaft with large integral flanges, when much metal must be removed. Afterwards, the shaft can be turned to size, and a fine finish, as a second operation.

In considering tool shapes and rake angles, it should be remembered that industrial processes are concerned with maximum economy. Consequently, tool angles are chosen to achieve a compromise between speed of metal removal, surface finish and tool life before regrinding becomes necessary. There is also much more power available on a professional lathe and this also influences the choice of speed and cutting angles.

The amateur is not usually so interested in maximising metal removal and will frequently not even want the bother of changing from a roughing tool to a finishing tool. Consequently, the preferred tool shape is the knife tool which is shown in Figure 14.8. Although designed for surfacing, it does cut well for facing the end of the work, or the abutment to a shoulder, and no other general-purpose tool is required for bar work.

Compound-rake tools

The knife tool is easy to grind consistently to the required angles and is equally easy to resharpen. It is consequently recommended for all external turning, or bar work. It is however, a simplified version of the generally accepted industrial form of tool which is described in most textbooks.

It is generally argued that the tool shown in Figure 14.6 is in reality cutting on the front and the side, and it consequently needs to be ground with both side and top rake. The rake angles are thus visible when the tool is viewed from the side and from the front.

Compound rake leads to better efficiency, both in the amount of material which can be removed, and in the life of the tool. So, compound rake can be beneficial and Table 14.2 shows the range of side and top rake angles which are generally adopted. A range of values for both rake angles is given, since the angles adopted depend upon how one defines maximum efficiency, and also since tool life, rate of metal removal and surface finish on the work, all need to be considered. Figure 14.14 shows how the top surface of the tool is ground sloping down away from the cutting point, for right- and left-hand tools with positive rake.

Parting tools

Once the initial turning operations have been completed, the workpiece will require parting off from the parent bar. This simply means

Table 14.2 Recommended Cutting Angles For Compound Rake High-speed Steel Tools

Material	Operation	Top Rake (degrees)	Side Rake (degrees)
Free-cutting mild steel	Roughing	5 to 10	15
	Finishing	20 to 25	0
BMS	Roughing	5 to 10	15
Carbon steel	Any	4 to 6	8 to 10
Cast iron	Roughing	8	10 to 12
	Finishing	6 to 10	0
Hard brass and FC bronze	Roughing	0	0 to 3
	Finishing	0	0
70/30 ductile brass	Roughing	5	5
	Finishing	2 to 6	2 to 6
Copper, phosphor bronze, aluminium bronze	Roughing	8 to 20	15 to 25
	Finishing	10 to 20	15 to 25
Monel metal	Roughing	4 to 8	10 to 14
	Finishing	15 to 20	0
Nickel silver	Any	10 to 20	20 to 30
Aluminium and alloys	Any	10	25 to 40*

* Aluminium alloys vary tremendously and it is difficult to give specific recommendations.

using a narrow tool that cuts on the front and plunging it into the bar, removing a slice of material and severing the partly made component from the bar. Figure 14.15 shows an embryo bolt being parted from the parent stock.

It is clear that the parting tool cuts on the end rather than on the side, and the wedge angle must, therefore, be ground on the end (usually called the front). The tool is then said to have top rake, as shown in Figure 14.16, which represents the normal form of parting tool. The sides are ground to give a small clearance of 1 degree on each side to reduce the friction and the tendency to jam. Front clearance is required but should not be excessive since this weakens the tool point. Zero top rake is sometimes used to give good strength to the point and this is also beneficial when it comes to regrinding as only the front needs to be ground,

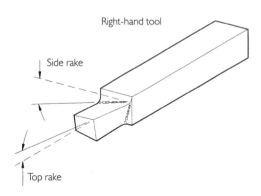

Right-hand tool

Side rake

Top rake

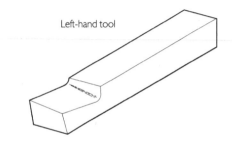

Left-hand tool

Figure 14.14 Compound rake angles.

Figure 14.15 Parting off a small bolt.

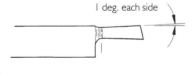

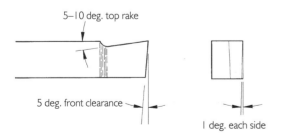

Figure 14.16 The parting-off tool.

but the material being worked may cut better if some top rake is given, but more than 10 degrees will weaken the point. Five degrees represents a good working compromise

To assist the workpiece to be severed without breaking off from the parent bar, and consequently still having a 'pip' attached to it, the front of the tool is sometimes ground at an angle so that the right-hand side leads the left and the job falls away just before the tool completes the removal of material from the bar.

Parting off is an operation which causes a great many problems, one reason being that the requirements for the tool are in conflict. Firstly, it is desirable that the tool is as thin as possible, since the metal it removes is just waste, but it is designed to cut over the full width of the face and this puts a strain on a narrow tool. To preserve strength, the tool has parallel or near-parallel sides so it tends to rub on the sides of the groove it is cutting. If the tool is not absolutely square to the work, it tends to bend and may jam in the groove and/or break off. The problems are magnified if a large diameter is involved since this requires a long tool with consequent aggravation of the strength problem.

Grinding a parting tool from a standard high-speed steel blank is somewhat tedious, due to the amount of metal which must be removed. Figure 14.16 shows generally what is required and it is obvious that quite large amounts of metal need to be ground away. To overcome this nuisance, special parting-off blades are available commercially. These take the form shown in Figure 14.17.

A narrow, hollow-ground blade, ½in. (12.5mm) deep and 4in. (100mm) long is ground at top and bottom to fit a dovetail slot. The blade is ground to this cross-section throughout its length and may be resharpened by grinding only at the end. Provided with a suitable holder, which may be home-made or purchased from a commercial source, a blade will last a lifetime. If the holder can be held in a

Figure 14.17 The end view of a commercial parting-off blade, mounted inverted in a rear toolpost.

rear-mounted toolpost, the best of both worlds is available.

Figure 14.18 shows an indexing, two-way tool holder specially constructed to hold two blades of different widths, in the inverted (rear) position on the cross-slide. This was made from a set of castings which is available commercially and remains on the cross-slide more or less permanently.

Figure 14.18 A rear toolpost fitted with two parting-off blades.

The blades described above do not seem to be available commercially in widths less than ¹⁄₁₆in. (1.5mm). If many small components are to be made in relatively expensive materials, the waste represented by the use of such a wide parting-off tool may be unacceptable. This problem can be solved by the adoption of a discarded hacksaw blade as a parting-off tool for small jobs. Where the diameters are small and the tool extension consequently short, a blade mounted in a simple holder works very well and probably represents one of the narrowest parting-off tools which can be used.

Of course, if all else fails, the component can always be cut off the bar using the hacksaw, and the job remounted in the lathe for facing off as a second operation. But please don't saw it off in the lathe!

Grinding and sharpening lathe tools

The bench grinder

If there is one general piece of advice which can be given, it is that one should never attempt to cut anything with a blunt tool. Some cutting tools cannot easily be resharpened without specialised equipment, but lathe tools and drills (probably the cutters used most frequently) may readily be reground to create new cutting edges, provided that a bench grinder is available. A grinder is also required to shape standard tool bits into usable lathe tools, and it is thus an essential in the workshop.

The normal type of grinder is illustrated in Figure 14.19. The one shown has 5in. (127mm) diameter wheels, and this size, or one with 6in. (150mm) wheels is necessary for the heavier work of grinding up lathe tools from standard blanks. Since this operation requires large amounts of material to be ground away, one of the wheels should have a coarse, open

Figure 14.19 A 5 inch bench grinder.

structure to allow material to be removed without generating too much heat. Nevertheless, heat will certainly be generated, and a bowl of cold water in which to cool the tool is an essential accessory when grinding up lathe tools.

For grinding the final cutting edge on a lathe tool, or sharpening drills, a wheel having a fine structure is required so that a fine, ridge-free finish is imparted to the tool. The grinder is normally double-ended and fitted with two differently graded wheels.

Considerable power is absorbed in removing material rapidly from the hard steel of a lathe tool blank, and the machine may stall when pressure is applied if the motor is not 'man enough' for the task. A grinder with 5-inch wheels thus needs a motor rating of ¼ hp, or about 200 watts, and 6-inch wheels require something larger, say ½ hp or 350 to 400 watts. Anything less than these power ratings may make bulk removal of material a slow business.

The wheels supplied with a new grinder will be suitable for grinding high-speed steel lathe tools and drills. If tipped tools need to be ground, a green-grit wheel is required. If one of these types of wheel is purchased, or indeed, any other wheel, it must match the machine for which it is intended. This means that its diameter and thickness should be correct, and its mounting hole diameter, but above all, it must

be suitable for the rotational speed of the machine.

Grinding wheels are designed to be used below a specified rotational speed which must not be exceeded or the material will be overstressed and the wheel will burst apart. **EXCEEDING THE DESIGNED SPEED IS VERY DANGEROUS.** All wheels must be marked with their maximum permissible speed, and any wheel which is not marked, or has lost its label, should be thrown away.

Wheel diameter is important since the rests on the machine must always be set close to the wheel to minimise the likelihood of anything becoming jammed between the rest and the wheel, which could also cause a wheel breakage. Therefore, smaller wheels cannot usually be fitted, and neither can larger ones, due to the presence of the wheel guard.

On the subject of tool rests, these are sometimes too flimsy to be of much use, or are attached to flimsy guards, and improvements can usually be made to improve the rigidity and the utility of the rests.

When creating a lathe tool from a high-speed steel blank, considerable amounts of material need to be ground away, and the blank must be pressed against the wheel with firm pressure. This pressure can only be resisted safely if the tool blank is pressed into the edge of the wheel, and the side of the wheel should never be used for the task of sharpening or grinding a lathe tool.

Wheels are made by a moulding process, and the periphery is then 'dressed' to render it smooth, with the abrasive particles projecting from the surface of the matrix in which they are embedded. Due to the fundamental weakness of the wheel, the sides are not dressed, but are left in the as-moulded state and may be quite rough, even on a fine-grit wheel, and are doubly unsuitable for grinding lathe tools.

Grinding wheels do wear in service, and the surface becomes clogged with abraded particles and the wheel ceases to cut cleanly. They occasionally need dressing to restore the surface to flatness and dislodge the abraded particles. This is usually done by using an industrial diamond which is sufficiently hard to cut the backing matrix of the wheel and dislodge the abrasive particles. The diamond is cemented into a holder which acts as a handle, and can be supported on the grinding rest. The dressing of a wheel naturally produces a large amount of highly abrasive dust and the workshop's machinery must be covered completely before the operation commences if the grinder cannot be moved to a less sensitive location. A face mask and eye protection are essential wear during wheel dressing and during tool grinding.

No attempt should be made to dress the side of a wheel since it is not strong enough to withstand the required pressure.

The wheels ordinarily fitted to grinders are intended to be used on hard materials such as those used for lathe tools and drills. Grinding soft materials such as mild steel, unhardened silver steel, and others, even softer, rapidly clogs the surface and renders the wheel useless. If you must grind soft materials, and it is strongly NOT recommended, fit an open-structured wheel such as one which uses silicon carbide as the abrasive.

Tool grinding

Given that the wedge angle is important in promoting good, clean cutting, it is clearly necessary to maintain the point angles on the tool within the normal range for the material being machined, during the grinding and resharpening processes. Referring to Table 14.1 it appears that it is necessary to grind specific angles (accurate to the nearest degree) in creating the tool point. As noted above however, the angles listed refer generally to the maximum efficiency type of tool and the recommendations may therefore be varied in practice. As noted elsewhere, the chief concern is finally whether the tool cuts and produces the required finish, and sharpness and lack of rubbing are the vital factors, rather than any strict adherence to published values for the various angles. It is, however, helpful to use a broad point to the tool for cutting tough materials, reserving tools with sharper points for lighter work, otherwise the point may break down very frequently

If a lathe tool is to be ground from a high-speed steel tool blank, the coarse wheel of the grinder is used to remove the bulk of the material on the three faces, following which, the ground faces are rendered smooth by grinding them lightly on the fine wheel. The periphery of the wheel must be used for the severe task of grinding high-speed steel. Even the fine wheel may not produce the finish on the tool which is desirable, and the faces should afterwards be honed by hand, using a fine carborundum slip stone to produce a polished surface on the cutting edges. Honing must be carried out carefully so that the angles ground on the tool are maintained and the edges do not become rounded.

Establishing the basic ground angles initially is not easy, and practice will be required if the periphery of the wheel on an ordinary bench grinder is used for tool grinding. If a tool blank is applied to the wheel by hand and its position is controlled by eye, the angle at which the face is ground is not precisely determined. During the grinding process, the tool, or tool blank, must be withdrawn from the wheel repeatedly to determine how the grinding is proceeding. The repeated applications of the tool to the wheel may mean that the surface may be ground as a series of disorganised flats which leave the surface immediately adjacent to the cutting edge – and that is where it matters – not at the required angle. This means that the cutting edge is not presented to the work in the ideal condition, and a clean cut and good surface finish may not be achieved. In extreme circumstances, the face of the tool adjacent to the cutting edge may be ground back too far, resulting in the tool simply rubbing and generating heat. However, many workers find little difficulty in developing the necessary skill and find the bench grinder entirely satisfactory for the preparation of lathe tools.

Without a proper guide for the tool, it is only the skill of the operator that determines the shape of the tool point. What is preferable is a wheel with a flat surface, which is provided with an adjustable rest so that the tool, or tool blank, can be presented to the wheel at the correct angle, repeatedly, thus ensuring that the ground faces are flat and lying at the required angles. The best form of wheel is the shape called a cup wheel, which is illustrated in Figure 14.20. A cup wheel is broadly cylindrical in form, but may be formed as part of a truncated (cut off) cone. It is moulded with one end closed and pierced for a mounting hole. The other end is open, and the edge of the cylinder is dressed squarely. Typically, a cup wheel presents an edge ½in. (12.5mm) wide which is intended to be used for grinding. This surface is flat, and can be maintained so by dressing, and will thus grind a flat surface on a tool blank.

If an adjustable rest is provided which allows the tool to be presented easily at a fixed angle to the wheel, the surfaces can readily be ground to create the tool angles. The range of angles which is listed in Tables 14.1 and 14.2 appears

at first sight to be very wide, but it will be found that just a few values (only) can be used to produce the whole range of tools likely to be required. These angles are 5, 10, 15, 20 and 25 degrees, and if means are available to set the adjustable rest accurately at these angles, the faces of any tool can readily be ground.

An adjustable rest is suggested in the view of a cup wheel shown in Figure 14.20. Such a rest can be set readily if a series of templates is made, say about 1½×1in. (40×25mm) in 20 swg (1mm) material each having two corners sawn and filed to one of the angles. With the rest loosely held by its clamp, a template can be stood on the rest and the rest adjusted so that the template determines the angle which the rest makes with the wheel's grinding face. The rest can then be clamped to its stand.

Figure 14.21 illustrates how an adjustable rest is used to grind a knife tool from a high-speed tool bit. In Figure 14.21A, the rest has been set using a 10-degree template and the cup wheel is used to grind the front clearance and end relief angles for a right-hand tool. Front clearance is determined by the setting of the rest, and the relief angles are created by pushing the tool towards the wheel at an angle of about 20 degrees. This is shown in the plan view of Figure 14.21A.

Note that the direction of wheel rotation is downwards past the rest, and the wheel is grinding the tool downwards from the cutting edge. This ensures that any burr left by the grinding process is on the unused (trailing) side of the tool. Cutting edges should always be ground in this way.

Once the end of the tool blank has been ground, the side clearance is ground to the same 10-degree angle, as shown in Figure 14.21B. Note, from the plan view in Figure 14.21B that the tool is maintained in line with the grinding edge of the cup during this operation.

Finally, the rest is reset and the tool is turned to the position shown in Figure 14.21C and its associated plan, for grinding the side rake.

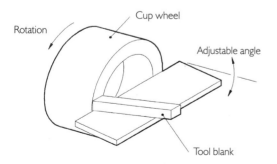

Figure 14.20 A cup-shaped grinding wheel and adjustable rest.

A Grinding front clearance and side and front relief

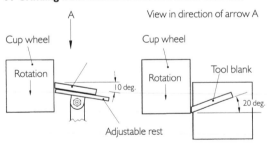

B Grinding side clearance

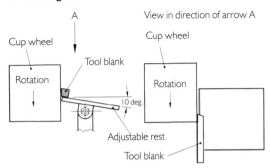

C Grinding side rake

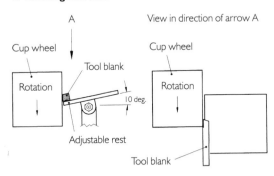

Figure 14.21 The procedure for grinding a tool using the adjustable rest.

When grinding up tools, note particularly that brass requires little or no side rake and the tool can therefore be flat-topped, it being necessary to grind only the clearance angles and ensure that there is side and front relief (of about 15 degrees). Some brasses may be found to machine more readily if negative top rake is given. That is, the top of the tool is ground away so that the point of the tool is lower than the tool body. Up to 10 degrees of negative rake is sometimes used.

Surface finish on the tool is important in promoting good finish on the work and although a coarse grinding wheel may be used to create the basic tool shape from an HSS tool bit, the final process must aim to produce as good a finish as possible on the tool. This must at least be performed on a fine wheel, but even this may not be entirely satisfactory in some instances. Honing of the surface of the tool by hand, using a fine slip stone is the only satisfactory means to create the required finish. This process should not destroy the carefully ground rake and clearance angles and a jig is frequently recommended for hand use to ensure that correct angles are maintained and sharp edges do not become rounded.

Resharpening

In use, wear occurs on the top of the tool, due to abrasion in the area on which the compressed chips bear down. It also occurs on the front and side of the tool, in the areas just below the point, as shown in Figure 14.22. Sharpening of a knife tool is quickly carried out by grinding back the front face of the tool to remove the worn point (on all three faces simultaneously) and thus create a new cutting point without grinding the top surface. If the top surface is truly flat (has zero top rake) there is no change of tool height after resharpening and the tool may readily be mounted back in the toolpost and turning continued.

If the tool has top rake, there is naturally a small loss of height at the point and this must be allowed for when remounting the tool. Nevertheless, resharpening should always be done on the end, since it is in its length that the tool has its longest life. It is also best performed little

310

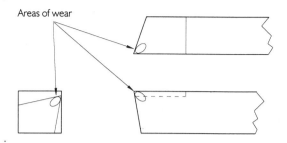

Figure 14.22 Areas of wear which develop on the tool.

and often since there is then little wear to take out. The job is then quick and easy and the tool is never allowed to become really blunt.

If tungsten carbide tools with brazed-on tips need to be reground, this must be done on a green-grit wheel. The need for regrinding means that the tip is chipped and it is usually on the top surface that this is evident. Since the remainder of the top surface is smooth and highly polished, and already established at the correct rake angles, such tools should also only be ground on the front and the side to establish a good cutting edge.

Height of the tool

Since the rake angles are important in determining how well the tool cuts, and also what quality of finish is achieved, it is important that the correct angle is created and maintained. In order to maintain the rake angles as ground on the tool, it is essential that it should be mounted at exactly the centre height of the machine.

Figure 14.23A shows a tool with both side and top rake in the correct position in relation to the workpiece. The point is at centre height and this establishes the top rake angle correctly. Correct front clearance is also achieved by this setting.

If the tool is too high (Figure 14.23B) the top rake is increased, but more seriously, the front clearance angle is reduced and the tool

point may not be touching the job and the tool will then merely rub on the work.

As Figure 14.23C shows, incorrectly setting the tool too low seriously reduces the rake angle and although the front clearance is increased, and the tool will not rub, poor performance and poor surface finish may result. Additionally, when taking a facing cut, the tool will not reach the centre of rotation of the work and so will leave a pip in the centre.

Attempting to turn with the tool too low is a recipe for disaster. Top rake is seriously affected, so the ability of the tool to cut the work is reduced and it may therefore dig in rather than dislodge a chip. If this happens, the

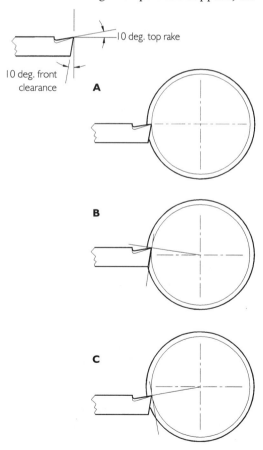

Figure 14.23 The effect of incorrect tool height.

workpiece will not necessarily stop rotating (it has a ⅓ horse power motor behind it) and it therefore climbs up over the tool. If it is small in diameter, it will bend and then jam up. If it is relatively strong (large in diameter) the jamming may occur instantly, but whichever way it is, the lathe is now stalled with the job pushed up above the tool and great strain is therefore placed on the chuck jaws and chuck body, the mandrel and front bearing cap and also on the tool, toolpost and topslide. Dig-ins should always be avoided.

Taking a facing cut is quite often the means by which the operator determines whether the tool is at centre height, the too-high or too-low condition being decided by determining whether the tool does reach the centre of the work, or passes below or above. This cannot be done if the work is hollow or is mounted between centres, and a simple height gauge may be made to suit the centre height of the machine.

It is frequently recommended that the tool should be set slightly high to compensate for the relative movement between tool and workpiece when cutting takes place as there is always bound to be some deflection of both items. The amount of recommended offset is small, however, the usual figure being between 1 and 4 per cent of the diameter of the workpiece. It is doubtful if any real benefit is to be gained from this type of setting. The final test is always an empirical one, "Does the tool cut and produce the required finish?". In any event, there are other factors which are probably more important and, in any case, proper support should always be given to both the tool and the work in order to minimise any deflections which occur.

When surfacing other than purely cylindrical work i.e. when turning tapers or using form tools, or when screwcutting, tool height affects the final shape produced and correct tool height is vital for such operations.

Cutting speeds

The importance of maintaining the correct wedge angle on the tool to ensure clean cutting and a good finish is emphasised above. A further factor affecting the cutting action is the speed of the workpiece past the tool.

It will be appreciated that some materials are naturally easier to machine than others, the softer or more brittle materials being easier on the tool which is therefore capable of removing material more rapidly. Different materials are said to have different machinability, four or five groupings being used by way of classification.

Speed is important because different materials cut best at different speeds and unless cutting takes place at approximately the correct rate, the surface finish may be much poorer than could be achieved and the power required to remove material may increase. This is sometimes noticeable if facing a large diameter, when it may be found that cutting is good at larger diameters but as the centre is approached the cut becomes ragged and uneven and markedly poorer results are achieved. This is because the cutting speed towards the centre is too low.

Cutting speed is the peripheral speed of the workpiece at the diameter being cut, measured in ft/min. i.e. the circumference, in feet, multiplied by the rotational speed (rpm). If you are used to working with metric units, the working circumference must be calculated in metres to yield a cutting speed in metres/min.

The problem with attempting to recommend cutting speeds is that several factors are involved. These include the physical and mechanical properties of the workpiece material, the rate of feed of the tool into the cut, the depth of cut, the tool size and point angles, and the tool wear which can be allowed before regrinding is required. Nevertheless, as described above there is a good cutting speed for each material and it is therefore useful to provide a guide. Due to the number of variables, different authorities tend to list different speeds depending upon the characteristics considered to be of prime importance. However, the cutting speeds shown in Table 14.3 are either within the range of speeds recommended by the references consulted, or are very close to the figures, in cases where a very narrow range of speeds is specified.

Although as amateurs we are not that interested in maximising tool life or production rates, it is wise to adopt cutting speeds which are within the reasonable capacity of the tool since in any event, time spent resharpening is also non-productive for us.

The table contains recommended speeds for turning when using high-speed steel (HSS) tools. If tipped (tungsten carbide) tools are used, speeds may generally be increased, a factor of two or three times being adopted commercially. This is possible due to the harder cutting point which the tip possesses and to the different wedge angle ground on the tool.

Table 14.3 should also be taken to refer to plain turning or boring. When parting off (or using form tools) speeds must be reduced due to the width of cut being used, a reduction to 50 per cent or even 25 per cent of normal cutting speed being desirable, depending upon the material and the actual width of cut.

As an indication of the mandrel speeds (rpm) which might be used in practice, a simple table can be constructed to allow selection of basic speeds. This is shown in Table 14.4.

Table 14.3 Recommended Cutting Speeds (HSS Tools)

Material	Cutting Speed
Bright mild steel Copper Gunmetal Phosphor bronze	85 to 100 ft/min (26 to 30 m/min)
Stainless steel Silver steel	60 ft/min (18 m/min)
Cast iron	70 ft/min (21 m/min)
Brass Nickel silver (90 m/min) Aluminium alloys	300 ft/min

Table 14.4 Cutting Speed, Diameter and rpm

Diameter	Cutting Speed ft/min				
	300	100	85	70	60
⅛in. (3mm)	9170	3060	2600	2140	1835
¼in. (6mm)	4580	1530	1300	1070	920
⅜in. (9mm)	3060	1020	865	715	610
½in. (12mm)	2290	765	650	530	460
⅝in. (16mm)	1835	610	520	425	365
¼in. (19mm)	1530	510	430	360	305
1 in. (25mm)	1150	380	325	270	230
2 in. (50mm)	570	190	160	135	115
4 in. (100mm)	290	100	80	70	60
6 in. (150mm)	190	65	55	45	40

Cutting fluids

General

When material is cut, there is naturally relative movement between the tool and the work. The resultant friction causes both the tool and the work to heat up, the heating effect increasing as the rate of metal removal increases. The temperature rise causes expansion of the work and the cutter, and this may lead to dimensional inaccuracy in situations in which the tool is cutting continuously. In an industrial production process, where the process is automated and continuous, the machine settings may need to be changed during any warm-up period.

Using modern tools, there may be no other effect which needs consideration, but in the days when tools were made from carbon steel, significant heating could not be allowed since it could cause the tool material to be tempered to a softer condition. Rapid tool wear occurred and machines had to be stopped frequently for the tool to be changed. This serious disadvantage of carbon steel tools naturally limited the rates of metal removal which could be achieved, but the problem was mitigated by spraying a liquid coolant onto the work continuously.

In spite of the adoption of the present-day types of high-speed steel tools, given this name simply because they do not lose their hardness at high temperatures, the practice of spraying a coolant onto the tool has developed, rather than diminished because cutting fluids lubricate, as well as cool, and this results in a better finish, longer tool life, lower temperatures, better accuracy, and so on.

If all this talk of industrial processes seems somewhat esoteric, put a piece of ⅝in. (16mm) silver steel into the lathe next time you have a spare moment, and set the lathe to run at 300 rpm. With a sharp knife tool in the toolpost, take a cut of .030in. (0.75mm) and force the tool into the cut firmly using the saddle handwheel. Watch the colour of the turning develop as the cut progresses, and you should see it turn gradually from silver to straw and finally to blue as the tool and the work heat up. An experiment such as this should convince you that coolants were a necessity in the days of the widespread use of carbon steel tools.

To be really effective, the cutting fluid needs to be present in large quantities, right at the point of cutting. This means that a pumped supply is necessary, and the machine has to be provided with a drip tray so that the overflow can be contained. Ideally the used fluid is collected, filtered and returned to the reservoir so that it may be reused. Thus, specialised equipment and/or adaptation of the machinery is necessary if the full benefits of using a cutting fluid are to be enjoyed. Some shielding of the machine is also necessary, since fluid can be thrown around in large quantities.

The use of cutting fluid in the sort of quantity necessary for effective cooling needs special provision, but in the amateur's workshop, continuous 'flat-out' cutting is unlikely to be required and so the need is not so much for cooling as for the improvement in surface finish which is achievable. It is not necessary to flood the work with coolant to achieve this, and some judiciously applied lubricant can be helpful, either applied as a mist spray, or as an occasional squirt from an oilcan, or applied by brush.

Types of fluid

Over the years, the development of cutting fluids has led to the use of a wide range of substances, from ordinary paraffin and other mineral oils, mineral and fatty oil mixtures and a range known as sulphurised oils. These latter fluids themselves comprise a wide selection with different uses, which vary according to the amount of sulphur they contain.

Due to its wide use for machining steel, the best-known cutting fluid is soluble oil. For use, this is mixed with water to produce a white fluid which is known universally as 'suds'. The oil content provides the lubrication and the water acts as the coolant, thereby satisfying both needs. The lubricating effect is naturally dependent upon the dilution, the traditional strength being determined by the relative costs of the two constituents – that is, mostly water!

The objections to soluble oil as a cutting fluid are its smell, which is distinctive, to say the least, and the fact that the water content tends to promote corrosion. There is also the problem that large quantities are required for effective cooling, and something a little more oily is preferable when the need is most likely for lubrication. There are, consequently, straight (not soluble) cutting oils which do not have an objectionable smell and which provide effective lubrication even when present in small quantities, and these are to be preferred if a full pump-and-recovery arrangement is not provided on the machine. The pumping equipment is generally known as a suds pump.

Choice of fluid

Since the purpose of a cutting fluid is to assist removal of material from the work, it follows that the way in which the chip forms influences the type of fluid which is used.

Some materials, such as cast iron and brass, machine very cleanly without the benefit of coolants. Most brasses cut very cleanly, and a good finish is produced, even at high speed. Cast iron provides a good finish due to the presence of graphite, which acts as a lubricant during machining.

Industrially, special types of soluble oil are used when machining cast iron, but they have to be present in large quantities to wash away all traces of swarf, which is highly abrasive and must not be allowed to enter the bearings or slideways of the machine since it will there do great damage.

As large quantities of coolant are not normally available in the amateur's workshop, and also since cast iron is only one of several materials which are dealt with, it is normally machined without the use of any coolant. Because the swarf is highly abrasive, it is usual to wipe any lubricant off the machine's slideways and keep the swarf away as far as possible, in order that the oil and cast iron dust do not mix to form a grinding compound. It is also useful to arrange a shield to collect and direct the turnings to where they can be better dealt with and a stiffish piece of plastic sheet can be taped to the cross-slide to achieve this.

Steels are very widely used and much machining of them is therefore undertaken. Steels are also tough and harder on the tool, tending to blunt it more rapidly and also heat it up. Industrially, a coolant is invariably used when machining steel, the most common type being soluble oil. If your machine is equipped to deliver the oil in sufficient quantities, and the smell poses no problem, then soluble oil should be used. If you can manage to cope with small quantities of the oil, then a drip feed or a mist spray can be beneficial in providing lubrication.

Aluminium and its alloys, although soft and easy to machine, have an annoying tendency to cause their rather 'stringy' turnings (swarf) to weld themselves to the tool's cutting edge or tip. This naturally masks the point and impairs cutting and hence spoils the surface finish. For this reason, paraffin is normally used as a lubricant when machining aluminium alloys.

When to use a coolant

Given that the workshop may not be equipped with suds equipment, coolants are unlikely to be used habitually. When they are, it is their lubricating properties which are usually desirable, and it is normally possible to apply sufficient fluid either by brush, oil can or drip feed.

Ordinary turning can be performed at speeds that are sufficiently low to provide the surface finish which is required. Paraffin may be used when turning aluminium alloys to help prevent the swarf building up on the tip of the tool. Here, the tool needs lubricating and the paraffin can be applied to it, rather than the work, otherwise the high speeds which are used ensure that there is a good distribution of the coolant around the workshop and this is obviously not desirable. Build-up also occurs on drill points and paraffin can also be used when drilling aluminium.

Another operation which benefits from the presence of a lubricant is the drilling of deep holes in phosphor bronze. This material is difficult to machine and a great deal of force must be applied before a chip will break away. There is much frictional heating as chips are deformed by the cutting edges and the resultant expansion can cause the drill to bind in the hole. A cutting fluid is useful both for its cooling and lubricating effect.

During turning, it is parting-off large-diameter work which leads to deep penetration of the tool into the work. Since the tool has very little clearance in the groove it is cutting, there is much friction and the operation once again benefits from the use of a cutting fluid.

15

Basic lathe practice

Introduction

General

The basics of what a lathe can do are described at the beginning of Chapter 12, together with a description of the major parts of a lathe and their principal features. The lathe is capable of carrying out both turning and milling operations although it is for the process of turning

Figure 15.1 The tool used for facing this gunmetal casting is set too low in the toolpost and has left a small pip as it has passed across the work below centre height.

that it is actually designed. Turning processes include plain turning, threading or screw-cutting and the production of form or taper surfaces. This chapter provides an introduction to those basic techniques of lathe utilisation which are common to all turning operations and the technique of screwcutting is among the topics described in Chapter 16.

Setting up and supporting the tool

The basic turning operations of facing and surfacing are described and illustrated at the beginning of Chapter 14, which also describes the basis for the grinding and sharpening of lathe tools to carry out these basic operations.

Before a tool can cut correctly, it must be mounted in the toolpost so that its cutting point is exactly at centre height i.e. it must be possible for the point to pass through the rotational centre of the work. To achieve this, the tool is supported in the toolpost on suitable packing which brings the tip to correct height. Several strips of differing thickness are used to adjust the tool's position, and a stock of tool-size

Figure 15.2 The packing below the tool must be arranged to give proper support, with all of the loose strips aligned carefully.

pieces in different thicknesses needs to be kept so that tools can be packed up correctly.

A quick way to set the tool is to mount it up in the toolpost, approximately at the correct height, and then to take a facing cut across the work. Figure 15.1 shows a square block which has been faced off using a tool set so that the point is just below centre height. The result is a small pip of unmachined material at the centre, indicating that the tool has passed across the face of the work just below centre height.

Figure 15.2 shows a correctly supported tool. Several metal strips have been used to

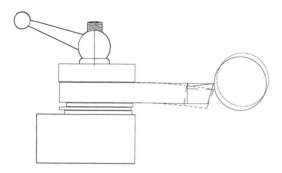

Figure 15.3 The load on tool and work, when cutting, causes the two to move away from one another, causing the cut to vary.

pack up the tool to the correct height. Their ends have been carefully aligned and the stack is positioned so that the full support offered by the toolpost base is utilised in supporting the tool. Provided that suitable strips are available, the packing is easy to arrange, but care must be taken to ensure that the tool is correctly supported.

Deflections of tool and work when cutting

The situation when turning is shown in Figure 15.3. When the tool is cutting, the uncut part of the work presses on the tool, causing it to bend down away from the work. However, from the work's point of view, the tool is pressing against it, and it, too, bends, but this time away from the tool. This means that in the initial stages of the cut, or when facing, both the tool and the work are being forced apart and experience a 'load' on one end. Both are supported only at the other end, and the amount of bending which occurs is related to the unsupported length, in such a way that the longer the overhang, the greater is the deflection. Put another way, this means that the least deflection occurs when the unsupported lengths are short. In practical terms, this means that the workpiece must not extend very far from the workholding device (chuck etc.) and the tool must be properly supported and not extend from the toolpost any further than is absolutely necessary. In other words, a short, stiff arrangement is required.

Due to the deflections which occur, the tool and work are pushed away from one another, when the cut is actually taking place, and a slightly smaller cut is taken than would be the case were there to be no deflection. If the tool is put into the job twice at the same setting, it removes more metal the second time around, the actual amount being dependent on the size of the initial cut, the sharpness of the tool, the

Figure 15.4 The deflections of tool and work may be sufficient to reduce the amount of material which is removed by the tool passing down the work. In these cases, a second pass of the tool along the work can result in a further cut being taken without advancing the tool any farther.

material being machined and also the rigidity of the set up.

The effects of tool deflection are illustrated in Figure 15.4 which shows the end of a long bar which is being turned, supported on the tailstock centre. Two effects are visible here, both resulting from passing the tool twice over the bar at the same setting, that is, without adjusting the position of the tool in relation to the work.

The end ½in. (12.5mm) of the work has been turned twice, once as part of a cut taken along the length of the bar, and again, by slowly traversing the tool over the end portion. The polish imparted by this second cut is evident, and so is a slow left-hand spiral caused by winding back the tool to the tailstock end without withdrawing it from the work.

The second cut taken by the tool is quite evident and this means that once the tool has traversed the turned surface in the correct direction, it must be retracted to clear the work before being returned to the start of the cut, when further feed may be applied and another

cut taken. This is the reason that feedscrew dials are so valuable – without them it is only guesswork that allows a cut to be put on, whereas their use means that precisely set cuts can be taken every time.

The load on tool and work is obviously dependent on the amount of material which is being cut, since a greater area of the work is pressing down on the tool when a greater cut is being taken. This area depends on both the depth of cut and the rate of feed of the tool into the work. The depth of cut is set by advancing one of the feedscrews by a known amount, and the rate of feed is determined by how fast the other feedscrew handle is turned. If the feed rate varies during the cut, the load and deflections also vary, and so, therefore, does the amount of material which is removed. An uneven feed rate results in a ridged and rough surface, so if a smooth finish is required a steady rate of feed must be used.

If the tool suffers a deflection when making the 'going in' cut, it is still deflected, and pressing in on the work, when it reaches the end of the cut. If it is held at this position for a moment or two, it cuts itself a groove as it tries to regain its natural shape. As the cut diminishes, the conditions for causing chatter are produced and if the tool is held too long at this position, the characteristic singing noise is likely to be heard, and the finish is spoiled. The problem of chatter is described below.

If the handwheel is used to traverse the tool from the end of the cut back to the beginning without retracting it from the work, it takes another cut 'coming out', as illustrated in Figure 15.4. If the work is reaching the finished size, this second cut may be a severe embarrassment, to say the least. The surface finish on the work may well leave a little to be desired since the tool is not cutting at its most efficient when coming out. The tool should, therefore, generally be retracted from the work before it is wound back to the beginning.

If graduated dials are fitted to the feed-screws, it is a simple matter to withdraw the tool before winding back to the start, and then to advance the tool precisely by a known amount for the next cut.

Retracting the tool before returning to the start of the cut may not always be absolutely necessary, since a roughing-out stage, often precedes the finishing stage, and it clearly doesn't matter if further material is removed when winding back – except, of course, that presenting the tool to the work going the 'wrong' way, might well blunt it more than taking a normal cut. If finishing is being carried out, it obviously does matter, since a good finish and an approach to the final size have been (or should have been) planned, and it is important that more or less the same tool deflection occurs on every cut.

Graduated dials are also helpful in allowing a definite approach to the finished size to be made without stopping to measure too often, since if the micrometer shows that the work is .050in. (1.26mm) over the required size, it is known immediately that the final cut will be taken with the tool advanced a further .025in. (0.63mm) into the work, give or take any differences in the tool deflections.

Depth of cut and feed rate

The rate at which material can be removed from the workpiece is dependent on the material which is being machined (some materials are softer, and easier to cut, than others) but the maximum rate is limited by the amount of power which is available to drive the work into the cut. If a deep cut is 'put on' and a high rate of feed is applied, the lathe mandrel may stall, since the power from the motor, transferred to the mandrel, is insufficient to remove material at the rate which the operator is trying to impose. Phrases such as "deep cut" and "high rate of feed" are exceedingly unhelpful in providing guidance, and it becomes a question of how deep is "deep"?.

In the descriptions of the basic turning processes presented in Figures 15.17 and 15.19, a cut of .010in. or .015in. is suggested for turning bright mild steel. Provided that the tool is sharp, a cut of .010in. should be supportable, even on a small lathe, and this can be considered as a suitable value from which to start. On a metric machine, this means taking a cut of 0.25mm.

Once the depth of cut has been set, the feed into the cut is applied using the other feed-screw. If this is turned by hand, the operator obtains 'feedback' to judge how the cut is going – the ears can detect a change of speed of the mandrel very easily, which is often the indication that the cut is too heavy.

The most important requirements are a firm feed and a steady rate, since it is only by maintaining the deflections of tool and work at the same values that an even cut and smooth finish can be obtained. The feed is always stated as the distance the tool travels into the cut during one revolution of the work i.e. the feed per rev. If .010in. is suitable as the depth of cut, it is also suitable as the feed per rev. Taking the example of the pivot bolt illustrated in the sequence of Figure 15.17, the bar is turning at about 300 rpm and .010in. per rev means 3 inches (75mm) in each minute, or 20 seconds to cut along 1 inch (25mm) of the work.

It is not necessary to use a stopwatch, but this figure does give a guide to the rate of feed which can be used on a bar turning at 300 rpm. If it is turning faster, the rate of traverse along the work can be faster to achieve the same rate of feed per rev.

The figure of .010in. per rev (0.25mm per rev) for the rate of feed can be considered as one which may be used for finishing the work (a finishing cut). If the object is simply to rough-out the shape of the work (a roughing cut) then the rate of feed may be greater, up to the maximum which the power available will allow. The depth of cut may also be greater than .010in. (0.25mm), and .020in. or 0.5mm can readily be used when the need is to remove metal more effectively. At the other extreme, cuts less than .005in. (0.125mm) should not generally be used since small cuts can give rise to chatter (see below). A large lathe can naturally support a deeper cut and 0.1in. or 0.25mm might well be used during a roughing-out operation.

Backlash and endfloat

When considering the feedscrews of the cross-slide and topslide, which are used to 'put on' the cut, it must be appreciated that there is inevitably some free play in the arrangement. This means that the cross-slide, for example, has a small amount of endfloat which may be large enough to be detected by pulling it backwards and forwards.

If the gib strips and screws are correctly adjusted, the endfloat should not be discernible simply by attempting to move the cross-slide directly, but if a to-and-fro motion is given to the feedscrew handle under these circumstances, a dead zone will be detected in which handle rotation does not result in motion of the slide.

This lost motion, or backlash, means that if the feedscrew is fitted with an engraved dial to indicate its setting, the dial reading for a given position of the tool relative to the work, will not be the same when moving the cross-slide in as when moving it out. When setting the cut or tool position using a feed handle and relying on the dial to indicate position, the feed must always be in the same direction. It is natural to take out the backlash before commencing the cut and it is usual to move the tool into the work to take the first cut, initially just advanc-

ing the tool to touch the work and then rotating the feedscrew handle further in the same direction each time a cut is taken, not forgetting the need to retract the tool before returning it to the start of the cut.

Chatter

Consideration of the deflections of tool and work, and the backlash which is always present in the feedscrews, naturally brings us to the turner's number one enemy – chatter. This is characterised by a singing noise from the tool when turning, and a ridged surface finish to the work on completion. Once the surface has become ridged, it is frequently very difficult to render it smooth once more, and if you are nearing its final size, the work may consequently be spoiled.

There are several root causes for chatter, but the ridged appearance of the work is caused by springiness in both the tool and the work. If the tool and the work are both large and substantial, the deflections which occur when the tool is cutting are small, and consequently cause no problem. However, any deflections do tend to push the work and the tool apart, which reduces the depth of cut and thus also reduces the force being applied to both items.

The scenario can go something like this: the cut starts and the tool and the work both bend under the load. As the two move apart, the cut is reduced slightly, reducing the load on work and tool, and allowing them to come closer together. This naturally increases the cut, which . . . and so on.

Since the cut is increasing and decreasing, the work naturally has a ridged surface. This produces a varying depth of cut on the next pass along the work, which varies the load on tool and work etc. The chatter thus tends to be self-perpetuating, since the conditions which first caused it are still likely to be present.

Figure 15.5 The situation to be avoided. Both tool and work are projecting too far from their holding devices, and large deflections will result if an attempt is made to take a cut on the bar.

The general cure for chatter – prevention might suggest the true requirement – is to reduce the deflections as far as possible. This means supporting the work, supporting the tool, and only using a correctly ground and sharp tool, properly aligned with the work.

Figure 15.5 illustrates a highly undesirable situation. There is a very long extension of the work outside the chuck and a long overhang of the tool outside the toolpost. Under conditions of large overhangs, the deflections caused are

Figure 15.6 The ideal situation. The projections of tool and work are the smallest possible.

significantly greater than for short overhangs, so the first requirement is to reduce any long extensions, or provide support when this is not possible. Figure 15.6 shows a vastly improved situation, which should be the one aimed for.

Extra support for the tool is not very conveniently arranged, so stiffness here must be by using a more robust tool. Support for the work can readily be provided either by using a centre in the tailstock, or by using a fixed or travelling steady.

In spite of achieving sensible support for the work, and using a stiff and short tool, chatter can still occur, which means that the forces acting on the tool and the work are still too high. This arises because there is too large a contact between the work and the tool, accounting for the chatter which is frequently produced when using a parting-off tool. This naturally has a large contact with the work since a groove of material is being removed.

An ordinary tool can, however, easily present a large contact with the work, principally if it is incorrectly ground, is blunt, or is incorrectly set in respect to the work. These conditions are illustrated in Figure 15.7. At A, the tool has been ground with insufficient end and side relief and as a consequence both front and side lie at too shallow an angle in respect to the work.

At B, the tool, although correctly ground, has become worn, producing a flat on the front. At C, although the tool is correctly ground, it is not set with proper end relief of between 5 and 10 degrees. All of these conditions increase the contact between the work and the tool, consequently increasing the load and the deflection and encouraging chatter to start. In these cases, the remedy is obvious; grind the tool properly and set it correctly in the toolpost.

If care has been taken in setting up the work and the tool to minimise overhangs and ensure correct clearance and relief angles for the cutting point, chatter arises most commonly from

A   Incorrectly ground

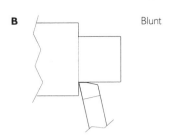

B   Blunt

C   Incorrectly set

Figure 15.7 Faults in the grinding and setting of the tool which may lead to chatter.

small changes in the overhangs of the tool and/ or the work, assuming these are possible.

The singing noise (chatter) which accompanies the production of the ridged surface is produced by vibration (oscillation) of the work and the tool. The frequency of the vibration is set by the length of the tool overhanging the support provided by the toolpost (or the similar overhang of the work outside the chuck)

Figure 15.8 This knife tool is blunt and is producing a significant burr while facing a gunmetal casting.

blunt tools. Bluntness can often be felt if the cut is being applied by hand, but it comes on gradually and the increased force which needs to be applied may not be noticed. In some situations the effect is noticed visually, rather than by feel, as shown in Figure 15.8. A knife tool is in use for facing a rectangular block, but its traverse towards the centre is producing a significant burr. That this is due to a blunt tool is without doubt, for the result of the same operation after the tool has been sharpened is shown in Figure 15.9.

If all of the above points (support, tool grinding and setting) appear to be in order, but chatter still occurs, it can often be cured by

Figure 15.9 Grinding the tool produces a lean cut, without a burr, on the same casting.

and this frequency determines the distance between successive peaks of the chatter marks on the work. The speed of the work is also a factor here, since, as the tool vibrates, it cuts less and more on the work and the pitch of the peaks and hollows is determined by how far the work rotates past the tool in each cycle of vibration.

So, if chatter occurs, it might be reduced on the next cut by changing the overhang of the tool and/or the work, or by changing the speed, preferably a reduction for both factors.

One other reason for chatter to develop is that you may be 'tickling' the work rather than taking a proper cut. The tool must cut at all times. If it is allowed to rub on the work it quickly becomes blunt, creating the conditions to allow tool vibration. If a larger cut is taken, and the tool pushed firmly into the cut, both tool and work are maintained in the deflected condition throughout the entire cut, thereby preventing the conditions for producing chatter developing in the first place.

So, try not to be forced into the position where you have to take a light cut when approaching the final size. Stop the lathe well before the final size is reached and measure the work so that the remaining cuts to the final size can be planned in advance, aiming to take a cut of at least .005in. (.125mm), i.e. a diameter reduction of .010in., as the finishing cut. If you are doubtful about this, take a practice finishing cut when still well above the final size to check that this will be satisfactory.

If the contact area between the tool and the work is very large, as for instance when using a form tool, the possibilities for producing chatter are much increased and the tool must be made as strong as possible. An example of this is shown in Figure 15.10. A tool with a large radius has been used for turning the tread of a cast iron wheel for a model locomotive. While turning the parallel portion of the tread, the radiused tip of the tool is in contact with the work, but as the flange is approached, the tool

Figure 15.10 Chatter can be produced if a large part of the tool is used for cutting, as here, where a round-nosed tool has been used to form the side of the flange and the root radius on a wheel casting.

cuts on a large part of the left-hand side, increasing the load on the tool dramatically and creating ideal conditions for the production of chatter, with the inevitable results shown in Figure 15.10. There is little which can be done in these circumstances, beyond employing a very large and stiff tool and reducing the speed as much as possible, even resorting to pulling the mandrel round by hand in extreme circumstances.

Using the tailstock

Introduction

The tailstock is most commonly used for centring and drilling, but tap and die holders are also used on the tailstock, and so are turrets which provide a mounting for several tools. The purpose of the tailstock (as a holder) is to bring tools up to the work in alignment with the lathe axis. The tailstock is also used to support the end of the work, particularly work which is mounted between centres. The principle of its use in this role is described in Chapter 13, and practical aspects of using the tailstock for support are considered below.

Since the sleeve of the tailstock may be advanced towards the headstock by the feed handwheel, a drill mounted in a tailstock chuck can be advanced for drilling an axial hole. To do this, the tailstock is brought towards the headstock and locked to the bed and the feedscrew handwheel used to advance the drill.

A turret holding several tools may also be advanced in the same way as for drilling. A turret may hold six tools in all and can be fitted with a set of items which is needed for a particular job. It might hold a centre drill, a pilot drill, a second drill, a reamer or D-bit, and so on, which are used in sequence, the turret being rotated and locked to bring the tools successively in line with the lathe axis.

If the turret is in use for a production run of similar items there is much slow-speed screwing of the sleeve in and out of the tailstock barrel, and for some lathes an adaptor can be fitted so that the sleeve may be advanced and retracted more rapidly by use of a lever.

Threading may be performed manually but tailstock-mounted tap and die holders greatly assist proper alignment of the cut threads. They are used with the tailstock locked to the lathe bed, but in these instances the tool holder is usually free to slide along a sleeve-mounted arbor and is prevented manually from rotating, or turned with respect to the stationary work.

The operation of putting on screw threads from the tailstock is so convenient that the use of a tailstock die holder is described later in order to establish the general principles of tailstock-mounted tool adaptors.

Centring and drilling from the tailstock

The tailstock sleeve is normally bored to a standard, shallow-angle taper so that accessories mounted on suitable tapered arbors may be mounted directly in it. The most common of these accessories is the drill chuck, and by far

the most common operations performed from the tailstock are centring and drilling.

Since the tailstock sleeve can accommodate a chuck, a workpiece revolving with the mandrel can be drilled. Just as with conventional drilling, a long and relatively flexible drill cannot be expected to produce a hole along the axis of rotation of the workpiece, unless it is provided with a centre into which it can commence drilling. For creating such centres, a short, stiff centre drill is used. This has a small diameter pilot, ground with straight flutes which extend away from the point into the body of the drill, the end of which is ground at 60 degrees. The drill therefore produces a 60-degree cone-shaped hole in the end of the work and the pilot provides clearance at the point of the cone. Centre drills are available in a range of sizes to suit large or small work and Figure 15.11 shows a set classified by the references BS1 to BS4, which range from ⅛in. to ⅜in. (3mm to 10mm) in diameter.

Centre drills are sufficiently stiff to start a hole without initial guidance, but to assist a true centre to be formed, the end of the work must be faced off, if this is possible, to present a smooth, true surface.

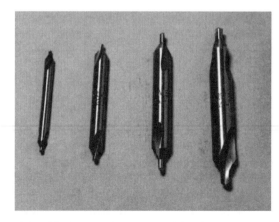

Figure 15.11 A set of centre drills, BS1 to BS4, ⅛in. to ³⁄₁₆in. (3mm to 8mm) in diameter.

Figure 15.12 Putting a centre drill into a large steel disc using the tailstock chuck.

Figure 15.13 Following up the centre drill with a ¼in. (6mm) drill.

Figure 15.14 Enlarging a drilled hole using a ½in. (12.5mm) drill.

Figure 15.12 shows a ⁵⁄₁₆in. (8mm) centre drill being used for putting in a centre, an operation usually simply called centring. It is not necessary to make the centre very deep, it simply needs a wide enough 'mouth' to accept the tip of the drill which is to be used, or to fit the centre, if the work is to be supported in this way. If a centre is being created in a part for a model, it may need to be made the correct scale size.

Once a centre has been made, a drill can readily be started in the work, and the hole opened out as required. Figures 15.13 and 15.14 show the initial drilling of a ¼in. (6mm) hole and its subsequent opening out to ½in. (12.5mm). With this larger drill, notice how it is possible to judge whether both lips of the drill are cutting equally.

When drilling from the tailstock it should be remembered that, on many lathes, the spindle speeds do not approach those available on the drilling machine, and much ordinary drilling can be carried out at the highest, or second-highest speed, particularly for drills below ¼in. (6mm) in diameter. It is usual to open out large holes progressively as for normal drilling, since this reduces the amount of metal to be removed with each cut.

One point must be made in relation to drilling – the drill does not necessarily remain exactly on centre and can be expected to wander from the truly axial path. The wander usually increases as the drilled hole becomes deeper and this should be borne in mind when deciding on the best method by which a hole should be produced. As a consequence, if a true, axial hole is required, it must be produced by drilling slightly smaller than required, followed by boring to the required size. Naturally, a sized finish can also be produced by boring to suit a reamer and then reaming from the tailstock. In both cases, the boring operation is vital since it is this which creates the truly axial hole.

Centre drills are naturally used to drill the

cone-shaped housings for the centres in the ends of work which is to be mounted between centres. The centre drill's pilot provides clearance for the point of the centre and since the body of the drill is ground to the same angle as the centre (60 degrees included angle) accurate location of the workpiece on the centre is assured.

Thread cutting from the tailstock

Turned work frequently requires internal or external threads to be cut and these are usually best done while the workpiece is still held in the chuck, or whatever, attached to the mandrel. If the thread is required to be truly concentric with other turned diameters, it must be cut using the screwcutting technique described in Chapter 16.

If the threads are required simply as fixings or fastenings, then the use of a tap or die is appropriate. Since a small amount of misalignment can usually be tolerated under these conditions, the tap or die can be applied by hand, but starting a die squarely is relatively awkward and some assistance in maintaining alignment is helpful. The time-honoured way to assist the alignment of a die when forming a thread in the lathe is to lock the tailstock to the bed and jam the die holder between the tailstock sleeve and the job. Then, with one hand pulling the chuck round and pressure maintained on the tailstock feed handle with the other, the die may be persuaded to start, after which it is just a question of whether you wind the mandrel round or turn the die holder, or some of each.

It has to be said that it does work, but it isn't really 'engineering' and a much better approach is to use a tailstock die holder. This consists of an arbor, tapered at one end to fit the tailstock sleeve and having a parallel portion on which a cylindrical die holder can slide, as shown in Figure 15.15. The die holder is

about 3in. (75mm) long and knurled over its central portion to allow a good grip. This particular holder accepts $^{13}/_{16}$in. diameter dies at one end and 1in. diameter at the other so it is suitable for the BA sizes and other small-diameter threads.

The holder is drilled to accept a tommy bar but this is only needed when cutting larger and coarser threads. Adequate torque can be applied manually to the body to cut threads up to $^3/_{16}$in. (5mm) or so in diameter, and the usual method of use is to cut the thread while the work is revolving, even at quite high speeds. All that is required is firm pressure into the job and

Figure 15.15 The components of a tailstock die holder and the sliding body fitted to its arbor which is mounted in the tailstock sleeve.

the die leads itself along the workpiece quite nicely, especially if opened up slightly. You should not be foolhardy in relation to speed since you need to let go of the knurled body soon enough or a nasty friction burn results as the die reaches the shoulder on the job (or the chuck jaws) and is forced to rotate. However, for small threads in brass, or such things as boiler stays in phosphor bronze, the one-pass, machine-running threading operation is far superior to the turn-chuck-by-hand and jam-in-the-die-holder approach.

The method naturally cannot be used if a tommy bar is needed to restrain the die holder body. In such cases, it is necessary to turn the die holder using the tommy bar, as far as is convenient, and then to rotate the chuck by hand to bring the holder and tommy bar into position to repeat the process.

For tapping axial holes in the work, it is possible to use an ordinary tap wrench provided that it has a short enough 'handle' to rotate over the cross-slide. If not, the tap wrench must be held and the work rotated. If the work is held in a chuck, the mandrel is most easily turned by inserting a chuck key in one of the opening/closing sockets and pulling the chuck round.

What are called piloted tap wrenches can be obtained for use in the lathe. These are of the type which has a cylindrical body fitted with a collet chuck, the body being drilled so that it can be located on a plain spindle which is held in the tailstock chuck. A wrench of this type is illustrated in Figure 15.16.

The tailstock chuck can be used to hold a tap but the tap cannot be expected to drag the whole tailstock along the bed, especially if it is small. However, the method can be used to start a tap, assisting the process by pushing the tailstock along the bed by hand. Once the tap has been started in the hole, it can be released from the chuck, a tap wrench put on the tap and the tapping operation completed by hand.

For much ordinary work it is usually satis-

Figure 15.16 A piloted tap wrench mounted on its arbor in the tailstock chuck.

factory to put the tap in by normal hand rotation all the way, but it does help alignment if the tap is held still and the chuck is rotated by hand, since it is then possible to see if the tap is developing a wobble.

Parallel turning

Turning without the topslide

The cutting tool on a lathe is mounted into a toolpost. This is attached to the cross-slide, which is in turn fitted to the saddle. Since there is the means of feeding both the saddle and the cross-slide independently, the tool can be moved both across and along the lathe bed. A tool attached to the cross-slide can thus be made to cut, both going into and going along the work. There is usually a topslide in addition, mounted to the cross-slide, but this is by no means essential, and much work may be done without using the topslide.

As a means of demonstrating the method of working without using the topslide, a sequence of photographs (Figure 15.17) illustrates the making of a substantial pivot bolt. Each stage of the process is illustrated and described, and the sequence can be read through as if it were part of the text.

Figure 15.17.1 Before starting, the chuck is cleaned to ensure that no swarf is present on the gripping faces of the jaws.

Figure 15.17.2 A final wipe over with the fingers is usually carried out since they are very sensitive and can detect any small particles which might still be present.

Figure 15.17.3 The cut-off bar is put into the chuck, ensuring that sufficient material projects to make the pivot and cut it off from the bar.

The sequence commences with the cleaning of the chuck jaws and assumes that a *sharp* tool is mounted in the toolpost at the correct centre height. Since the stock bar is circular (in cross-section), the three-jaw chuck is used for this item.

Figure 15.17.4 A right-hand knife tool is set up at centre height and adjusted in the toolpost so that it can reach the centre of the bar. The tool is ground with side rake of 15 degrees and is ground and set up so that it has side and front relief of 15 degrees. The tool has a well-radiused tip so that a good finish can be achieved. Since the bar is 1 in. (25mm) in diameter, the mandrel speed is set to about 300 rpm which gives a peripheral speed of 80 ft/min (24 metres/min).

Figure 15.17.5 The saddle is locked to the leadscrew and the leadscrew handwheel is used to advance the tool towards the end of the bar. The first turning operation is to take a facing cut across the end of the bar to provide a smooth surface. Although a right-hand knife tool is in use, it cuts well when fed towards the centre of the bar.

Figure 15.17.6 This first cut doesn't produce a fully turned end to the bar so the leadscrew handwheel is used to advance the tool again and another cut taken to bring the end to a smooth finish.

Figure 15.17.7 There is about 1¾in. (44.5mm) of the bar protruding from the chuck, so the facing operation is taken with care. Although the bar is substantial, subsequent operations benefit if support is provided at the outer end, so the end of the bar is centred using a centre drill mounted in the tailstock chuck.

Figure 15.17.8 The tailstock is locked to the bed and the centre drill is advanced, using the tailstock handwheel, until a centre about ³⁄₁₆in. (5mm) diameter (at the open end) is formed.

Figure 15.17.9 The tailstock chuck is knocked out and a hard centre is substituted. The tailstock is then pushed along the bed until the centre enters the bar, extending the tailstock sleeve, if necessary. Tailstock, cross-slide and topslide are adjusted so that the tool can cut the end of the bar without any of these major items clashing. The tailstock is locked to the bed and the centre withdrawn from the bar to clear the centre hole so that oil can be applied to the end of the bar.

Figure 15.17.10 The tailstock handwheel is used to bring up the centre into the bar so that it supports the work firmly and the tailstock sleeve is then locked. The angle of the tool in relation to the work is then checked and the saddle is moved down the bed to check that the tool can approach the chuck closely enough to turn the required length of the bar.

Figure 15.17.11 The tip of the tool is positioned about 1/32in. (1mm) away from the end of the bar by moving the saddle and the cross-slide. The leadscrew handwheel is then turned until the half nuts can be closed onto the leadscrew. The leadscrew is not turned again until the basic outside turning is finished.

Figure 15.17.12 The motor is now started and the saddle is disengaged from the leadscrew and moved towards the headstock using the saddle handwheel. The tool is advanced towards the work using the cross-slide feedscrew until it just touches the bar. The cross-slide feed dial reading is noted.

Figure 15.17.13 The tool is withdrawn and the saddle traversed back towards the tailstock until the tool is clear of the work. The cross-slide is advanced so that the tool is .015in. (0.4mm) further forward than the reading noted previously.

Figure 15.17.14 The first cut is taken by traversing the saddle firmly towards the headstock, again using the saddle handwheel.

Figure 15.17.15 Since the turned-down portion of the pivot is 1 3/8in. long, the saddle is advanced by this distance. The tool is then 1/32in. from the required position of the shoulder on the work. When the end of the cut is reached, the tool is withdrawn and the saddle returned to the starting position, ready for the next cut.

Figure 15.17.16 The saddle can be advanced into the cut by engaging the half nuts and rotating the leadscrew handwheel, in which case the handwheel is turned the requisite number of times to advance the tool by 1 3/8in. (35mm). If the lathe has a 1/8in. pitch leadscrew, this means 11 turns of the handwheel.

Figure 15.17.17 The saddle handwheel is frequently used since it moves the saddle more directly and the operator is more in touch with the cutting process and can feel what is going on. If the saddle is to be moved directly, the leadscrew must remain unmoved from its initial position and the saddle must be traversed along the bed until it is almost 1⅜in. along the work, and the half nuts can be engaged with the leadscrew.

Figure 15.17.18 Whichever means of saddle travel is used, the tool must be pressed firmly into the cut at all times, to keep it cutting. It must never be traversed slowly in the hope of achieving a better finish, since this may lead to chatter. A cut of .015in. (0.4mm) should be supportable on a 3- or 4-inch lathe (75mm or 100mm) with a ⅓ hp motor, provided that the tool is sharp and correctly set. Finish is unimportant at this stage and the first cut is usually uneven anyway, due to the fact that the outer surface of the bar does not run truly.

Figure 15.17.19 As soon as the outer surface of the bar is true, which should be after the first or second cut, the lathe is stopped and the outside diameter measured. The approach to the final size is now planned.

Present size   = 0.960 (24.4mm)
Required size   = 0.750 (19.0mm)
Still to remove = 0.210 (5.4mm)

It is planned to stop and measure again when the work is, say, .030 (0.75mm) larger than the required size, so 0.180 needs to be removed, or roughly 4.7mm if you are working in the metric sytsem. More cuts are taken, exactly as before, and the lathe is stopped again for measuring. How often this is done depends only on how well you are keeping track of where you have got to.

If you are working in the imperial system, advance the tool by 0.010in. (ten thou), say to yourself 'Twenty' and take a cut. Withdraw the tool at the end of the cut, return the saddle to the start and advance the tool to where it was, plus .010in. Say to yourself 'Forty' as you start the cut, do it all again, and say 'Sixty' and so on, until you feel that you should check the size once more.

If you are working on a metric machine, take 0.25mm cuts and say to yourself 'half, one, one-and-a-half' and so on as you take successive cuts.

When you think the size is nearing the first target you have set, stop the machine and measure again, and plan the approach to the finished size.

Figure 15.17.20 The final size for this pivot is to be such that it fits parts already made, so there is no absolute need to measure, but it is valuable in allowing the approach to the size to be judged. The bores to which the pivot will be fitted are measured, to check the sizes. Knowing the present size and the required size, a cut is taken which nominally leaves the pivot .005in. (0.125mm) larger than needed. After turning, there is always the likelihood that there is a burr on the end of the work, so a smooth file is used to remove this and put a small chamfer on the end. Although it doesn't look like it, the mandrel is turning and the fingers must be kept away from the chuck jaws. Always ensure that a handle is securely fitted to the file – a hang up could cause the bang to be driven into your palm and cause a lasting injury.

Figure 15.17.21 A final measurement is now taken and a cut is put on by advancing the tool HALF the amount still left to remove. A cut is then taken over the first ⅛in. (3mm) along the pivot and the saddle wound back quickly without withdrawing the tool.

Figure 15.17.23 The final cut is taken along the whole length of the pivot, right up to the present shoulder, and the fit of the mating part is tried, to confirm that all is correct. If the fit is too tight (and it had better not be too loose!) only a very small amount should be left to remove. If this is the case, the smooth file should be used to polish the surface and achieve the desired fit. Although the mating part appears to be different here, it has simply been fitted to the pivot the other way round. This will be its final assembled position on the pivot.

Figure 15.17.24 The body of the pivot is now finished except that the shoulder does not yet have a square corner where it meets the shank (due to the use of a radiused tool). The shank is also ⅟₃₂in. (1mm) too short. Since this lathe has a rear-mounted toolpost fitted with two parting-off tools, these are used to finish the corner of the shoulder. The wider of the two tools is brought up to the end of the pivot and the saddle position is adjusted so that it just touches the end. Note the use of the white card to improve visibility.

Figure 15.17.22 The tailstock is withdrawn and the mating part is tried on the end of the pivot and an assessment made of the fit which has been achieved. This allows the position of the cross-slide to be adjusted if the fit is not correct. Remember that it is not possible simply to wind back the cross-slide to a new setting on its dial, since any backlash comes into play when reversing the direction of rotation of the feedscrew. If the need is to retract the cross-slide, wind back the feedscrew far enough to take up the backlash (and some more) and then wind forward to the desired reading.

Figure 15.17.25 The saddle is advanced by 1⅜in. (35mm) and the square-fronted tool is used to cut the shoulder back.

Figure 15.17.26 To ensure that the corner where the head meets the shank is completely without any radius that would prevent correct seating, a narrow parting tool is used to undercut the shoulder.

Figure 15.17.27 The smooth file is used again to remove any burrs on the edge of the groove just turned that might interfere with the fit.

Figure 15.17.28 The head of the pivot is turned to the required diameter and then the rear parting-off tool is used to part off the pivot from the bar. Note that the end has been screwcut prior to parting off. Screwcutting is described in Chapter 16.

Figure 15.17.29 This operation is done at the slowest direct-drive speed, which on this lathe is roughly 200 rpm, and this cuts off the work without support from the tailstock, and without significant chatter, although the finish is not completely smooth and the almost-complete pivot is left with a large pip at the centre where it broke off from the bar. The amount of swarf produced by the operations carried out so far can be judged from this view of the cross-slide between the two toolposts.

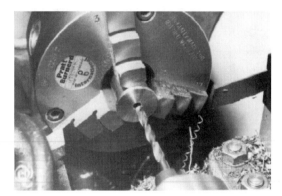

Figure 15.17.32 The end is then centre drilled and a drill put through to form a hole just large enough for the ¼in. reamer to enter.

Figure 15.17.30 Some of the powder swarf adhering to the chuck is evident in this view, and another cleaning up of the chuck is necessary before the final stages of finishing the pivot can be started.

Figure 15.17.33 The reamer is then put in the tailstock chuck and the tailstock sleeve advanced slowly and steadily with the mandrel turning at 200 rpm, until the reamer has been put completely into the hole. This creates a smooth, sized bore which should be a running fit on a shaft which is exactly .250in. in diameter, since if the reamer is new, or in its first 'life' it should cut just over the nominal size.

Turning using the topslide

The cross-slide and topslide are provided with feedscrews which have (or should have) dials so that a calibrated movement of the slides is possible. On an imperial lathe, the dials are calibrated in .001in. (one thou) steps and on a metric lathe might be .025mm or .05mm, roughly equating with one or two thou.

Figure 15.17.31 The pivot is mounted in the chuck with the top out so that it can be faced off to correct thickness. Since its body has been carefully turned to the correct diameter, with a smooth finish, it must be held very lightly, otherwise the hard chuck jaws will spoil the surface. In some cases it is necessary to place pieces of shim material between the jaws and the work to prevent marking. The head of the pivot is now faced off to the correct thickness, the same knife tool being used for this operation.

On a non-screwcutting lathe, the saddle is clamped to the bed when the lathe is used for turning and the cross-slide and topslide feedscrews provide the only means to advance the tool into the work, and it is these which are intended to be used when turning. Even on a screwcutting lathe, the leadscrew may not be provided with a calibrated dial, so the cross-slide and topslide again have the only calibrated motions.

The topslide is normally easily removed from the cross-slide and has a mounting arrangement that allows it to be attached so that the axis of traverse can be anywhere within 45 degrees on each side of its alignment with the lathe axis. On some lathes, the topslide can be attached with its axis of traverse at any angle relative to the mandrel axis.

The topslide can be set into a plan view such as that shown in Figure 15.22 and can turn a taper on the work. A taper is produced since the axis of traverse is not parallel to the axis of rotation of the workpiece. It follows that parallel (not tapered) work is produced when the topslide axis is parallel to the rotational axis, and if parallel work is required, care must be

taken to bring the topslide into correct alignment. The usual way to do this is an empirical one – a posh way of saying trial and error – which means turning the workpiece, or a stub of material put into the chuck, and measuring it very carefully to determine whether it has been turned to a parallel form.

On my lathe, the topslide is provided with an engraved angular scale, but the cross-slide which is now fitted is not provided with an index mark against which to set the scale. Figure 15.18 shows how a short rule is used to provide a temporary index to set the topslide to the approximately correct alignment.

The subsequent setting and checking can be carried out by mounting a short length of round bar, say, ½in. (12.5mm) in diameter into the chuck and taking a shallow cut along the bar to produce a smooth surface on which to measure, as illustrated in Figure 15.19.

Figure 15.19 With the topslide set approximately, a length of round bar is turned to create a smooth surface for measuring.

Figure 15.18 If the topslide feed is to be used for turning parallel work, it needs to be set accurately to the parallel condition. An approximate setting is being assisted here by using a rule as a temporary index for the initial setting.

The diameter of the turned length is then measured carefully at both ends, as illustrated in Figure 15.20. If the two diameters are different, the work is tapered and the topslide must be adjusted.

This is done by slackening the attaching bolts slightly and tapping the topslide (once)

Figure 15.20 Measurement of the two ends of the turned surface reveals whether the turned length is parallel.

Figure 15.21 If the work is not parallel, the topslide securing bolts are slackened and the topslide tapped in the required direction using a soft-faced hammer, and the process of turning and measuring repeated until the parallel condition is achieved.

with a soft-faced hammer, as illustrated in Figure 15.21. It is helpful to place a finger of the free hand so that it is touching both the topslide base and the cross-slide so that any movement can be felt.

When adjusting the topslide (or anything else) in this way, never hit it twice, or decide arbitrarily that it has moved too far and hit it in the opposite direction. If movement is felt, clamp up the fixings, take another cut of .005in. (0.125mm) and measure again. That is the only test of the magnitude of the adjustment.

The process of taking a cut and then measuring must be repeated until parallelism is achieved. But, be realistic about the aim. A good micrometer, properly used, can indicate a difference in the diameters of .0001in. (.0025mm) but a small amount of taper on ordinary work is unlikely to be a problem, and there may be no need to work to this degree of accuracy, and something better than .0005in. taper in 1 inch (.005mm per cm) may be acceptable. If there is taper, the outer end should be smaller.

If there is need for a long parallel length, or an item is being made for an interference fit, greater accuracy of parallelism may need to be achieved. In these cases, a longer test length can be used, making the test over a length of 2in. (50mm), say, rather than the 1in. which might otherwise suffice, if the requirement is less stringent.

For some jobs, it is possible to take the trial cuts and adjust the topslide position by turning the workpiece itself, particularly if the work requires significant stock to be removed, since this allows enough cuts to be taken to permit parallelism to be achieved.

Taper turning

Turning two identical tapers using the topslide

The normal use of tapers is for locating or holding items in axial alignment, and they are most commonly used for holding drill chucks to their arbors and holding arbors into the tailstock sleeve or mandrel taper sockets on the lathe. Other machine tools also use this method of location.

Since the topslide can be set up to turn parallel work, it can equally be set to turn tapered work. The problem lies not in the actual turning, but in setting the topslide to the correct angular position. The technique which is adopted differs according to whether the taper must match an existing one, or can have a free form.

If a matching pair of tapers (male and female) is required, but they do not need to match any others, it clearly doesn't matter exactly what the angle of the taper is, only that the socket matches the plug. In this case, both socket and plug must be machined at exactly the same angle, and once the topslide has been set, both tapers are turned without disturbing its setting. Figure 15.22 illustrates one way in which the two tapers can be turned.

With the topslide set over to the approximate angle, the plug is turned using a right-hand tool, having first turned the bar stock to the diameter required at the large end by using the saddle handwheel to provide the feed.

Turning the male taper

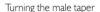

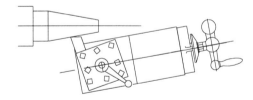

Turning the female taper
Boring tool turned upside down, cutting the far side of the bore

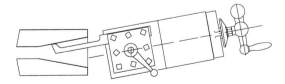

Figure 15.22 The method used for turning two matching tapers.

When the plug is finished, it is removed from the chuck and a bar suitable for making the socket is substituted. This is centred and drilled down, at least the length of the taper (plus a bit) about the same diameter as the small end of the plug.

A boring tool is set up on the toolpost, but upside down, so that it can cut on the far side of the bore, away from the operator. The socket is thus turned at the same angle as the plug. As turning proceeds, the plug needs to be tried in the socket to test the depth of entry.

This procedure produces two matching tapers if carefully carried out, and its only serious disadvantage is the difficulty of supporting the end of the plug on a tailstock centre, since the angled-over topslide is in the way. This can be quite an important disadvantage, since any narrow-angle tapers such as the Morse series, are rather long in relation to their diameters and tailstock support is often desirable.

Support might be achieved by extending the tool farther from the topslide, so enabling the cross-slide to be withdrawn farther, towards the operator. This works in some cases, but a long extension of the tool may be worse than not supporting the work, and may equally lead to chatter.

The alternative is to skew the topslide in the opposite direction, so that the plug is turned with its smaller end close to the chuck, as shown in Figure 15.23. This requires a groove to be turned as a preliminary, to provide clearance at the smaller end. This can be turned using a parting tool and should create a narrowed section, just smaller than the diameter of the small end of the taper. When the taper has been turned, the plug can be parted from the parent bar.

This method cannot readily be used if the small end of the plug needs to be drilled and tapped, as some tapers do, and the method of Figure 15.22 must be used in these cases.

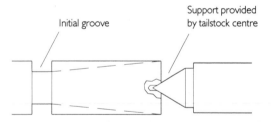

Initial groove

Support provided by tailstock centre

Figure 15.23 If a taper must be turned with the large end facing outwards, a groove should be turned at the inner end to provide clearance for the tool, and the work should be supported using the tailstock centre.

Turning a taper to match a standard

It sometimes happens that a taper must be turned to match an existing one, and it is then important that the angles match one another precisely. If the topslide is to be angled over to turn the taper, its position must be adjusted progressively by trying the fit of the two items repeatedly, after each cut, until a proper fit is achieved.

The process starts by setting the topslide to the approximate angle and taking enough cuts

Figure 15.24 If a chalk line is drawn on a male taper, and male and female brought together with a gentle rotary motion, the smudging of the line indicates the extent of the fit. Here, a commercial blank end arbor has been tested in a matching Morse taper adaptor sleeve.

to establish a reasonable length of taper to work with, say, about half the estimated length. The fit of the embryo taper has now to be tried with the existing component so that topslide angular position can be checked and then adjusted.

A marking medium is required so that the fit can be properly assessed. Chalk is the usual medium and it is applied by drawing a line down the plug. Socket and plug are then pressed gently into engagement, rotated through a small angle with respect to one another, and then pulled cleanly apart.

The result of this is that the chalk line is smudged where there has been contact between the two parts, and this shows in what direction the angle of the topslide needs to be adjusted. Figure 15.24 illustrates the results of testing a commercially made arbor in a Morse taper adaptor. The perfect fit is indicated by the smudged line all along the arbor.

If the fit is obviously not correct, the topslide is adjusted, and the test repeated after taking another cut. Adjustment must continue until the chalk line is smudging all along the plug.

Ordinary blackboard chalk can be used as the marking medium, but it tends to draw a heavy line with deep build-up of the chalk, which masks the true assessment of the fit. A stick of French chalk is better since it draws a thinner line.

Turning long tapers

The method of turning tapers by skewing the topslide has the limitation that the taper cannot reasonably be longer than the topslide travel. It is possible to use two positions of the saddle and try to join up the cuts at the two ends of the taper, but it is clearly not entirely satisfactory to rely on this method to produce a good fit.

An alternative method which does not place this limitation on length, is to utilise the setting-over capability of the tailstock and turn the taper between centres. This method is described below.

Mounting the work between centres

Tailstock alignment

The basic concept of mounting work between centres is described in Chapter 12. The method is appropriate where the work is too large to pass through the mandrel bore or whenever concentricity of turned diameters is required at both ends of the work and an alternative workholding method is not available. The workpiece must be faced off at each end and a centre drill used to create the 60-degree centres.

The basic procedure for doing this is described above. This supposes that the work will pass through the mandrel bore so that centring can be done conveniently in the lathe.

If the work is too large to allow this simple procedure to be used, the procedure for centring using the drilling machine and afterwards facing off using a half-centre must be used, as described in Chapter 13.

The tailstock is usually provided with a set-over arrangement so that the main casting may be moved relative to the base, across the lathe axis. A scale to show central alignment and displacement is not always provided and even if it is, should not be relied upon absolutely as the correct guide to the central position.

Since the tailstock is adjustable across the bed, a workpiece mounted between centres can either be turned parallel or tapered, as shown in Figure 15.25. If the tailstock and mandrel centres are aligned so that the axis of rotation of the work is parallel to the path of the tool as it traverses along the bed, the work is turned parallel. If the tailstock centre is displaced rearwards from the axis of the bed, a greater cut is initially taken at the mandrel end, and a taper is turned on the work.

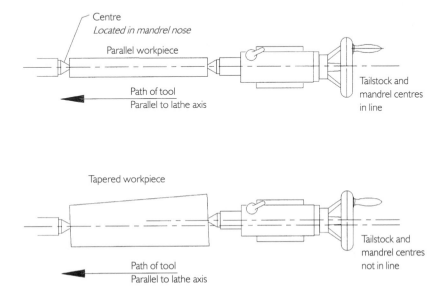

Figure 15.25 Tapers can be turned by mounting the work between centres and offsetting the tailstock by an appropriate amount.

A scale may serve as a good enough guide for ordinary centring and drilling, but better alignment is achieved by putting centres into the thoroughly cleaned mandrel and tailstock taper sockets and bringing the tailstock centre towards the mandrel centre. A careful check can then be made by eye to determine whether the two centres are in correct alignment, when viewed from above. The use of a watchmaker's glass allows quite small misalignments to be discerned.

Any required adjustment is made by slackening the tailstock body-to-base clamp and tapping the tailstock body gently with a soft-faced hammer to achieve the correct alignment. The centres used must be in good condition, having undamaged points which have not become rounded in use, and the taper sockets and centres must be cleaned thoroughly prior to the check.

Tailstock alignment for parallel work

Once the job has been centred it is ready for turning. However, before starting, the tailstock position may need to be adjusted in order to ensure that the job is turned parallel. The tailstock position should already be approximately correct. To achieve a precise setting, a test bar is required, centred at both ends, and about the same length as the work in hand to ensure that the tailstock is set at its working position on the lathe bed.

The test bar can take two forms, depending on the measuring method which is available, or is preferred. The ideal test bar consists of a hardened steel rod, centred at both ends, and ground on its outside diameter to give a fine surface finish which is concentric with the centres. In other words, a test bar which has been ground between centres. If such a bar is mounted between the lathe centres, parallelism with the longitudinal axis of the lathe bed may

be checked by use of a dial test indicator (DTI) mounted rigidly to the cross-slide, topslide or toolpost. A traverse of the saddle along the bed with the indicator plunger in contact with the front side of the bar, then reveals any lack of parallelism and the tailstock body may be adjusted by slackening the body-to-base clamp and using a soft-faced hammer to tap the tailstock body into the correct position. It must be borne in mind that any endfloat in the cross-slide feedscrew may affect the DTI reading and the cross-slide should be locked to the saddle if this is possible. If not, care must be taken not to disturb the cross-slide during the traverse along the bar.

The bar-and-DTI method is probably the quickest way to achieve the correct tailstock alignment, but if a properly made bar or DTI is not available, a second method can be used. This requires a test bar of the type shown in Figure 15.26. This is made from a length of bright mild steel which is centred at both ends, and the complete process of its manufacture can be part of the procedure for setting the tailstock.

Having aligned the tailstock centre to the mandrel centre by eye, a bar about the same length as the job is faced and centred at both ends. If possible, the chuck should be used to do this on a reasonably substantial bar, say, about ⅝in. (16mm) in diameter.

The bar is withdrawn from the chuck so that all but 1in. (25mm) can be turned, and the

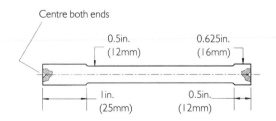

Centre both ends

0.5in. (12mm)   0.625in. (16mm)

1in. (25mm)   0.5in. (12mm)

Figure 15.26 If the tailstock needs to be positioned for parallel turning, a test bar can be made up and small cuts taken over the ends with the bar mounted between centres.

outer end supported on the tailstock centre. Using a round-nosed tool, the centre of the bar is turned away, reducing it, say, by ⅟₁₆in. (1.5mm) in diameter, leaving a short unturned portion at both ends. At the headstock end this should be long enough to allow a carrier to be put on (to drive the bar when mounted between centres) plus ½in. (12.5mm) or so. At the tailstock end, ½in. is left unturned.

The test bar is mounted between centres and fitted with a carrier at the mandrel end. Having mounted up the test bar, a small cut is taken over the unturned diameters at both ends, turning both at one pass, without retracting the tool or adjusting cross-slide or topslide feeds in any way.

Having turned both ends, the two diameters are measured using a micrometer or vernier. The relative sizes of the two ends indicate whether the tailstock must be moved closer to, or farther away from, the operator and a slackening of the clamp and tapping with a soft-faced hammer allows the adjustment to be made relatively quickly. The process of taking a cut, measuring the diameters and adjusting is repeated until the alignment is satisfactory.

If the workpiece is initially sufficiently larger than the finished size, the above procedure can equally be carried out on the work itself, turning the complete length as far down as the carrier, and measuring the two ends.

When setting the tailstock for parallelism, do retain a sense of proportion. If you are turning a locomotive axle 6in. (150mm) long, comprising a plain axle turned to a smaller diameter at each end, a difference in diameter at the two ends of the test bar (assumed to be the same length) of .002in. means that the parallel wheel seatings at each end will be tapered only .000033in. (.0009mm) if they are 1in. (25mm) long, and correspondingly less for smaller lengths.

Of course, if you need to turn the full length of an axle or other long shaft, so that it is parallel, you must be more fussy, and it is sometimes helpful to turn the shaft from both ends since this halves the distance the tool has to travel, and thus halves the taper.

It is vital to test the free play of the work between the centres at all times. Ideally, there should be no discernible play, but if things are adjusted too tightly, friction at the tailstock centre may cause things to heat up. On the other hand, as turning progresses and the centre at the tailstock end beds in, the setting of the tailstock sleeve needs to be adjusted, otherwise, progressively more endfloat develops. This is usually characterised by a rattle from the driving dog when a cut is not being taken.

Concentricity of diameters

General considerations

An important (and frequent) requirement in lathe work is that several diameters on one workpiece should be concentric with one another. Typical of items for which this is a requirement are the locomotive wheel and axle shown in Figure 15.27. The wheel tread rolls

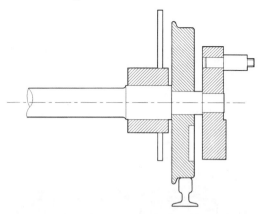

Figure 15.27 A cross-section through one end of a locomotive axle, illustrating the several diameters which need to be concentric if the wheel set is to be true-running.

along the rail and it is vital that the bore is exactly at the centre of the wheel otherwise the end of the axle rises and falls as the wheel rolls.

The axle has three diameters; one running in the axlebox bearing, one on which the wheel is seated and the third on which the outside crank is fitted. These three must share a common axis (they must be concentric) otherwise rotation may cause the centre of rotation of the crank to rise and fall, or the axlebox to be forced up and down.

The usual way to achieve concentricity is to turn all of the diameters without removing the work from the machine between stages. The axle of Figure 15.27 might therefore be turned between centres, turning it overall, followed by the two smaller diameters at the tailstock end, afterwards reversing the axle end-for-end and turning the smaller diameters at the tailstock end once more. Since all diameters are referenced to the two centres, concentricity is achieved provided that care is taken to adjust the tailstock sleeve correctly and remove all free play.

The wheel presents a different problem. It is usually recommended that the wheel is drilled and reamed with it held in the three-jaw chuck by its tread. At this stage of the process, the tread has not been finish turned, and being

Figure 15.29 When mounted in the mandrel taper, the arbor can be turned to provide a true-running diameter which can be used for locating a bored component.

inside the jaws, cannot be used as a reference. The problem of achieving concentricity is overcome by using the reamed bore as a reference.

The method employs a stub mandrel of the type shown in Figure 15.28. Blank end arbors can be purchased to fit the range of Morse taper sockets (the usual ones used on machine tools). As purchased, the outer end is unmachined and soft and can be turned if the arbor is fitted into the mandrel socket. Turning the end then creates a true-running spigot which can be used to locate a bored item for further turning of diameters which are concentric with the bore. Figure 5.29 shows this stub mandrel mounted in the mandrel socket.

Figure 15.30 shows a wheel bolted to a faceplate but located by a similar stub mandrel

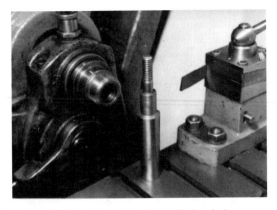

Figure 15.28 A mandrel made from a blank end arbor.

Figure 15.30 A bored locomotive wheel casting mounted to the faceplate and located by an arbor in the mandrel taper.

which lacks the threaded spigot of that shown in Figure 15.28.

Setting work running truly

It sometimes occurs that all of the diameters on a workpiece cannot be turned without removing the work from the machine between stages. If it is being turned between centres, this is of no consequence, and the same may be true if a collet is available to hold the already-turned diameter and is sufficiently accurate to provide the required concentricity.

If the work is being held in a chuck, it becomes necessary to set up the work for the second operation so that the existing diameter runs truly. The four-jaw chuck is used for this, and the existing diameter (cylindrical projection or bored hole) is monitored using a DTI which is mounted firmly to some fixed part of the machine.

The procedure for adjusting the chuck jaws to achieve the correct position is described in Chapter 13, where the use of the four-jaw chuck is considered in detail.

Hole production during turning

What type of hole?

The procedure used for making axial holes in turned work depends very much on the type and size of hole which is required. Holes may pass right through the work, or be blind (blocked off at the end). They may be clearance holes for bolts or studs, in which case drilling is appropriate, or they may be sized holes for running fits or press fits, in which case, a better size and finish is required than can be achieved by using a drill. The diameter of the hole also influences the method, since drilling a ½in. (12.5mm) hole is usually possible, whereas a 2-inch (50mm) one is not.

So, various methods may be used for making holes in turned work in the lathe, including drilling and reaming and suchlike operations, in addition to the use of a lathe tool inside a hole for the purpose of enlarging or sizing the bore. This latter method is known as boring and although the principles of cutting remain the same as for bar work (surfacing and facing) there are special problems associated with working inside a hole that demand separate consideration.

Since boring is not particularly appropriate to small holes, ways of dealing with these when accuracy is required are described first.

Sizing small holes

If a small hole is required to be a particular size, it can of course be bored using a very small boring tool, but if the hole is long in relation to its diameter it requires a long, thin boring tool, which is not a good combination. If the hole passes through the work, the most obvious choice of tool for sizing it is the reamer. Often, this will mean simply drilling a hole suitable for the reamer to enter, changing down to a lower speed and putting the reamer through from the tailstock.

The reamer is mounted in the tailstock chuck and may be put through using the normal feed with the tailstock locked to the bed. Alternatively, the tailstock body can be pushed along the bed by hand with the mandrel running at slow speed, when it is possible to feel the rate of cut and adjust the pressure accordingly. The reamer should not be allowed to run in the bore of the hole too long after it has sized the hole, or it may remove more material than is required.

If the hole is blind, a reamer cannot be used since it needs to be put right through the work, but a tool which cuts on the end can perfectly well be used in a blind hole. This end-cutting tool is the D-bit which is illustrated in Figure 15.31.

D-bits are available commercially but may perfectly well be home-made. A silver steel rod of the required diameter is reduced in width at

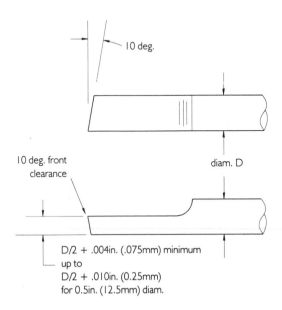

Figure 15.31 The usual form of D-bit. These are available commercially, but can easily be made from silver steel, and hardened and tempered.

one end until it is a few thou (say, .005in. or 0.127mm) over half the diameter. The end of the flat is then machined or filed away at an angle of 10 degrees or so, in plan, and to provide a front clearance of 10 degrees below the cutting edge. The end is then polished, taking care to maintain nice square corners, and the tool hardened and afterwards tempered to dark straw, as described in Chapter 3.

To use the D-bit, the work is drilled to depth, just below the D-bit size, and the D-bit put in from the tailstock to bring the hole to the correct diameter with a more-or-less flat bottom.

The D-bit may chatter as the width of cut increases at the bottom of the hole and it is best to keep the cut going right up to the point where the hole meets the full depth reached by the pilot drill point. Again, it is helpful to make the cut by pushing the tailstock bodily along the bed.

Chatter may produce a relatively poor finish at the bottom of a blind hole, and this may be important, when a D-bit is used to open out a hole to form a square-bottomed ball valve seating. Chatter can be avoided by reducing the width of cut by drilling or reaming the small-diameter passage which the ball will seal, before opening out the body with the D-bit and keeping the D-bit cutting right up to the final depth.

If a drill of the correct size is not available, the D-bit may of course be used to open out a through hole and will be found to give reasonable finish when used to remove the last few thou from the diameter. If the hole is deep, it is helpful to add a drop or two of oil to the D-bit shank, to reduce friction.

Drill wander may mean that a plain drilled hole, or one which has been sized by a reamer or D-bit, is not truly axial and may not be concentric with diameters turned on the work. If this is important, it is best to produce the sized hole first, and then use it to locate the work on a true-running spigot or stub mandrel, as described above.

Boring tools

There is no reason why small holes cannot be bored with a boring tool rather than being drilled-and-reamed or drilled-and-D-bitted. It is just that small tools are naturally weaker and more springy than larger ones and so are more liable to deflect, and spoil the job. Boring is therefore reserved for the larger holes, when a substantial tool can be used inside the bore.

The problem to be overcome is that of providing adequate clearance for the tool's cutting edge. For internal turning, or boring, the surface being turned is curving towards the tool and the underside of the tool must be ground away to provide clearance, as shown in Figure 15.32. Removal of this metal immediately below the point weakens it, and may mean that the point breaks down if a large cut is taken in tough material.

Rake and clearance angles for boring can be the same as for normal turning, although front clearance often needs to be greater and has to be increased until the tool cuts correctly within the bore being produced. In small holes, it is

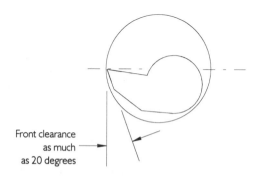

Figure 15.32 A cross-section through a boring tool illustrating the restrictions placed on the tool if it is necessary to bore a small hole.

Front clearance as much as 20 degrees

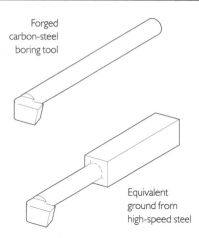

Forged carbon-steel boring tool

Equivalent ground from high-speed steel

Figure 15.33 An old-fashioned boring tool, and its equivalent ground from a square HSS tool bit.

sometimes helpful to set the tool a little above centre height to increase the front clearance.

To try to solve some of the clearance and strength problems, various traditional tool shapes are employed for boring. The one commonly shown in the text books takes the form shown in Figure 15.33. The tool has a cranked end which automatically provides good clearance behind the point and allows the cutting tip to be ground easily to the required shape. The tool shown is ideally produced by forging a high-carbon steel rod to the basic shape, filing up the cutting angles and then hardening, tempering and polishing. It is somewhat laborious to grind this shape out of a high-speed steel blank, especially if much length is required, but the shape is produced commercially by companies specialising in the supply of pre-formed tools.

The tool shape actually required depends upon the type of hole being produced, since if the bore is blind and must be turned across the internal end face, the tool tip must be angled (in plan) to allow the two cuts to take place.

The alternative is to use a tool which is generally called a boring bar. This is simply a

Figure 15.34 If space permits, the boring tool can be quite substantial.

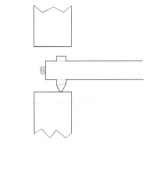

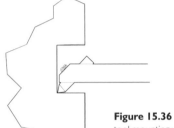

bar of material which may be clamped under the toolpost, and into which a tool bit can be mounted. Figure 15.34 shows a no-nonsense version made up hastily some years ago from a simple rectangular bar. This carries a length of 0.25in. (6mm) high-speed steel, ground to the required angles and although not very elegant does work well when there is sufficient clearance. The tool is fitted into a reamed hole and locked into position by a grub screw.

A commercially produced boring bar is shown in Figure 15.35. This is a ⅜in. (9.5mm)

Figure 15.35 A circular boring bar is more versatile and can be clamped into a split holder with the minimum extension of the bar to suit the job in hand.

Figure 15.36 The two types of tool mounting which can be adopted for through bores and blind bores.

steel rod, cross-drilled and reamed at right angles at one end and at 45 degrees at the other, so that two ³⁄₁₆in. (4.75mm) tool bits can be inserted. The bar is held in the toolpost by being clamped into a split sleeve which has a reamed hole which the bar fits closely, thus ensuring a firm mounting in the toolpost. In use, the toolpost clamp closes up the split sleeve and clamps the bar by friction. The sleeve mounting allows the tool overhang to be adjusted and the tool holder always provides the best possible strength.

Figure 15.36 shows the use of the two tool mountings which are provided. If the bore is a through hole, and there is sufficient clearance behind the workpiece, the right-angle mounting may be used for boring. If the bore is blind, or there is insufficient clearance, the 45-degree mounting allows the tip of the tool to be positioned ahead of the end of the bar, and the grub screw fixing, and permits the end of the bore to be machined.

Figure 15.37 An angled tool in use to counterbore the end of a through hole.

Figure 15.38 A commercial boring bar and another, made by me, which is suitable for very small bores.

Figure 15.37 shows a tool in the 45-degree mounting hole in use for counterboring the circular steel blank shown being drilled in Figures 15.12 to 15.14.

Figure 15.38 shows a home-produced boring bar which uses 0.125in. (3mm) diameter tool bits. The bar is ⅜in. (9.5mm) in diameter. One end is turned down to 5⁄16in. (8mm) diameter and will enter a bore of 11⁄32in. (8.75mm) diameter, provided that a short tool bit is fitted, so can be used for small holes. As shown, this bar is suitable only for through bores, but the idea can be adapted for a 45-degree mounting. Boring bars of this general type are available commercially, but often have a fixed, square central portion and the overhang cannot be minimised, as it can for the type in which a round bar is held in a split sleeve.

Boring and reaming an axial hole

A boring or counterboring operation such as that shown in Figure 15.37 produces a true-running bore in the work. If other faces or diameters have been machined on the work without disturbing its position in the chuck, the bore is on the axis of the diameters (and parallel to them) and is at right angles to the machined faces.

A sized hole can also be produced by reaming, but if it is to be truly axial, the initial hole must be bored as a preliminary to reaming. Figures 15.39 to 15.43 show the sequence which is used to produce an axial hole by reaming. The workpiece is an eccentric for a steam locomotive. It has already been finish-turned externally and has been mounted eccentrically in the three-jaw chuck for boring and reaming. Figure 15.39 shows the initial centre drilling and 15.40 the drilling of a ¼in. (6mm) diameter initial hole.

This hole is opened out using a ½in. (12.5mm) drill (Figure 15.41) and then bored

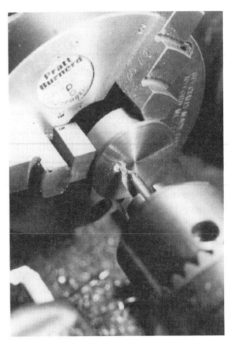

Figure 15.39 On the lathe, a centre drill is always used to create a positive location when drilling a hole from the tailstock.

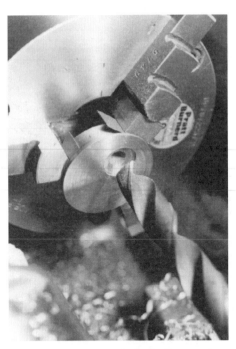

Figure 15.41 This permits a larger drill to be used to remove the bulk of the material, and bring the hole close to its final size.

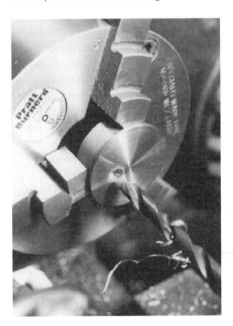

Figure 15.40 With an adequate centre to guide the drill, an initial hole can be drilled up to ¼in. (6mm) in diameter.

Figure 15.42 Simple enlargement of the pilot hole using a drill does not usually produce a true-running hole, and a boring tool must be used to enlarge the hole further and bring it to the true-running condition.

Figure 15.43 A reamer is then put into the tailstock chuck and the hole reamed to its final size.

using a tungsten carbide tipped tool (Figure 15.42) until the lead of a ⅝in. (16mm) reamer just enters the hole. This leaves just a small amount of material to be removed from what is now a truly axial hole. The reamer is fitted into the tailstock chuck and the tailstock pushed bodily along the bed, with the mandrel running at about 100 rpm, until the reamer stops cutting. Figure 15.43 shows the fine swarf on the end of the reamer following initial penetration of the bore.

The reamer should be passed through the work, and withdrawn, only once, otherwise there is a risk that the bore will be oversize. Reamers do cut slightly oversize in any event and can be expected to produce a clearance fit on the nominal size.

Boring a recess

Sometimes, a large-diameter shallow bore is required, not amounting to much more than a recess. It is necessary to drill a starting hole by entering the point of a large drill into a centre-drilled hole, but after this the operation is rather like facing, except that it is necessary to form the side and angle of the recess.

A short, stiff boring bar with the tool mounted at 45 degrees can also be used for this operation.

Turning things square

It may not be appreciated that things can be turned square in the lathe but this is an extremely valuable method of machining if full facilities for milling are not available, since turning away the unwanted material often results in a much faster rate of metal removal than when the lathe is adapted for milling. The method is particularly useful for cleaning up castings, but the problem is, it is very easy to get the job 'not square' and so a little forethought is necessary to promote success.

If a rectangular block is to be machined and it is already reasonably square, there should be few problems, but if an out-of-square work–piece is to be machined, there can be difficulties. Figure 15.44 shows a cross-section of a block mounted in the four-jaw, ready for one face to be machined. The out-of-square is perhaps exaggerated, but it at least illustrates the principle.

Facing off a casting mounted as shown produces one surface which is flat and machined smooth. This may now be used as a reference face from which to produce a second which is parallel to it. To do this, it is important that the next facing operation is carried out with the first face in contact with the ground surface of

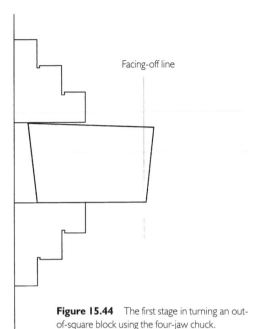

Figure 15.44 The first stage in turning an out-of-square block using the four-jaw chuck.

the four-jaw body, or one of the steps on the jaws. Figure 15.45 suggests that the block is large enough to be seated against the body. To assist maintenance of this position, narrow packing strips are used to clamp the work into the chuck, and before tightening the jaws finally, the job is tapped home against the body of the chuck to ensure that it seats properly. A hard-faced hammer can be used for this operation since the outer face is as yet not machined.

If the work is too small to span the hole in the front of the chuck, a reference can be provided by reversing the jaws, or using the other pair, and tapping the machined face down on to the step on the jaws. After seating the work correctly, a second facing cut produces another face parallel to the first.

If the chuck jaws are in good condition, it is now possible to produce a third side at right angles to the first two, by using the jaws to provide alignment, as shown in Figure 15.46. The

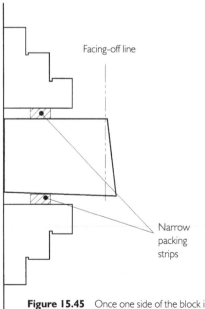

Figure 15.45 Once one side of the block is smooth and flat, it can be seated against the body of the chuck, or against the steps on the jaws, to permit the opposite face to be machined parallel to the first.

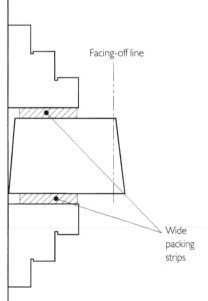

Figure 15.46 When two opposite sides are parallel, the jaw faces can be used to present a third side for facing, the block gradually being brought to squareness by turning.

process can then be repeated until a rectangular block with all sides machined has been produced.

If the block is very out-of-square to start with, it does sometimes prove difficult to get really square. It is always a good idea to remove any surface unevenness from a casting before starting, to give the chuck a chance to grip it properly and a little extra filing to produce two reasonably parallel faces before starting can also be beneficial.

It also sometimes pays to approach the task in two stages, facing off all over to produce an approximately square block and then going round again, producing a good surface finish and turning to size. Figure 15.47 shows two castings which have been brought to squareness by the method described, the large faces having been brought to parallelism within one or two thou in 2½in.

Once surfaces have been finish-turned, packing must be used to prevent the chuck jaws marking the job, but the packing should be used intelligently so that either the jaws or the chuck body (or the steps on the jaws) provide alignment and ensure squareness. Figure 15.8 shows a partly finished axlebox and its keep, in the four-jaw, protected in this way. This illus-tration shows a later stage in making an axlebox from the casting shown in Figure 15.47.

Knurling

Many small tools, or items intended for adjustment by hand, are provided with a ridged or diamond-pattern, external moulding which provides for a more comfortable and surer grip. Some items which have been finished in this way are shown in Figures 15.48 and 15.49. This type of surface deformation, for that is what it is, is known as knurling. It is produced in the lathe by pressing one or more hardened steel rollers, machined with the required pattern,

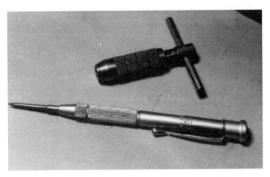

Figure 15.48 Straight and diamond knurling on small hand tools.

Figure 15.47 Two axlebox castings which have been turned square.

Figure 15.49 Coarse and fine diamond knurling on a tap wrench and an automatic centre punch.

351

against the work, while it is rotating. If sufficient pressure is applied, the hardened knurling wheel deforms the surface of the work, imposing on it a reverse facsimile of itself.

Knurling may take a straight form, as shown by the small pocket scriber of Figure 15.48, or may be a diamond knurl as seen on the other items. Knurling wheels are available to produce straight or diamond knurling, in coarse, medium and fine pitches, to suit the size of the work.

A knurling tool comprises a device carrying the wheel or wheels, which can be mounted into the toolpost of the lathe. The simplest form is that illustrated in Figure 15.50. This comprises a square-section bar, forked, and carrying a removable pivot screw, at one end, on which the single wheel is mounted. Although the simplest form of the tool, it has one serious disadvantage arising from the high pressure which is required for the knurling wheel to deform the surface of the work.

This places great pressure on the work which must be strong enough (large enough) to resist the force without bending. In turn, the pressure must be resisted by the workholding device and ultimately by the lathe's headstock bearing and great strain can be placed on the machine. The single-wheel tool is thus not ideal and certainly cannot be used for thin work.

An alternative form which overcomes this problem is the clasp-type knurling tool, a version of which is shown in Figure 15.51. This employs two wheels which oppose one another on the work, thus equalising the pressure and making the tool suitable for small-diameter work, or for use on a small, lightweight lathe. This particular tool uses commercial wheels but was otherwise made in the home workshop.

Figure 15.50   A single wheel knurling tool mounted in the toolpost.

Figure 15.51   A clasp-type knurling tool set up on a large-diameter steel bar.

The side arms were made much too thin, and are therefore a weakness in the design, but the tool has carried out all of the knurling needed in my workshop over the years, and has been borrowed by others. The tool represents a major landmark in that it was the first to be made after the initial establishment of the workshop, many years ago.

16

Adapting the lathe

Introduction

The lathe may be adapted in a number of ways to perform operations other than turning. These adaptations allow the machine to perform boring (or drilling) operations, or to allow milling to be carried out. Milling, drilling or boring are usually performed by using the mandrel drive to turn a cutting tool mounted into (or on) the mandrel nose and cutting on a workpiece which is held to the cross-slide, thus reversing the usual positions of tool and work. Since vertical adjustment of the workpiece position is possible only by use of suitable packing, it is normal to add a supplementary slide to the machine which allows controlled motion in a vertical plane. This attachment is known as a vertical slide.

The simplest adaptations are used for carrying out drilling or boring operations on work fixed to the cross-slide or to the vertical slide. This is a useful way to extend the workshop's drilling capability, since the lathe may well have much lower spindle speeds than the drilling machine and will drill or bore larger holes.

Accessories for the lathe

Morse taper shank drills

If the use of the lathe is to be extended in the ways suggested, some additional accessories will be required to provide better adaptations. The first extension of the lathe's use might be to use the capability which the mandrel has to turn at lower speeds than the workshop's drilling machine.

If the workshop has a drilling machine with a chuck having a only a ½in. (13mm) capacity, this naturally restricts the maximum size of hole that can be drilled. The machine will have a lowest speed which matches the maximum capacity of the chuck and the machine cannot really be used to produce larger holes.

Since the lathe is likely to have lower speeds than the drilling machine, it has the in-built capability to drill (or bore) larger holes, provided that a larger drill can be mounted in the lathe.

Twist drills having integral Morse taper shanks are available to fit the standard tapers used for the mandrel and the tailstock sleeve.

On a medium-sized lathe, a No. 2 Morse taper is frequently used for these tapers, which is the standard for taper shank drills between ⁹⁄₁₆in. and ²⁹⁄₃₂in. (14.5mm to 23mm), and although they are relatively expensive, it is useful to have one or two drills in this size range available, say, ⅝in. and ¾in. (16mm and 19mm) diameter, to provide the means to open out holes to sizes above ½in. (12.5mm).

If your lathe has only the smaller No. 1 Morse taper fitting, you will be restricted to a maximum size of ⁹⁄₁₆in. (14mm).

It is possible that the workshop is equipped with machines which have different taper fittings, or you have accessories which have tapers too small to fit the lathe. In these cases, you will need to obtain an adaptor sleeve, or Morse taper sleeve, to suit the two tapers. The sleeves are referenced by the sizes of taper which they will fit – the outside of a 1–2 sleeve fits a No. 2 Morse taper, while its inside accepts a No. 1 Morse taper.

Holding-down bolts

If access is available below a machine table, as on the drilling machine described in Chapter 10, common nuts and bolts may be used for holding down the work, or a machine vice. This access is not available on a lathe cross-slide, which is normally provided with tee-section slots (tee slots) to allow items to be secured to it. The correct items to use in the slots are specially shaped to fit the tee slot; either bolts (tee bolts) with special heads, or nuts (tee nuts), both being specially machined to fit the slot closely.

Although the slots have an inverted tee shape, what would be the head of the tee were it the correct way up, extends only a small distance from the side of the slot which would form the leg of the tee. The purpose of the special shape to the nut or bolt head which is fitted into the slot, is to maximise contact with the underside of the slot so that the upward pull which occurs when the fastening is tightened, is resisted over as large an area as possible. This means that the heads of the bolts need to be narrow enough to enter the slot, but long enough in the other direction to provide a good area of contact with the slot.

It is very tempting to produce tee bolts by machining down (thinning) the head of a standard hexagon head bolt, and filing two opposing sides of the hexagon, so that it will enter the slot. A head made in this way provides only a very small contact area with the underside of the slot and produces a high load locally which can damage the underside of the slot's head.

There is also the question of preventing rotation of the bolt when tightening and releasing the fastening, and the hexagon shape is also not ideal in this respect. If the machine table is an iron casting, which is usually the case, the material is easily crushed by the localised loads and severe permanent damage can easily be caused. Figure 16.1 shows a group of correctly made tee Bolts, a short tee bar having two tapped holes and some turned-down, hexagon-head bolts which illustrate the good and bad points.

Figure 16.1 Good and bad examples of tee bolts and nuts for fitting to the cross-slide.

It is naturally important that the fastenings used to hold items to the cross-slide are tightened adequately to be effective. When using tee nuts or a tapped tee bar, care must be taken that the bolt engaging the tapped hole does not protrude through the nut or bar, otherwise it will not tighten adequately, and will damage the bottom of the slot.

Parallels and vee blocks

It is frequently necessary to rest the workpiece on some form of packing when performing a machining operation. If the work has a flat underside, the packing might simply be a sheet of material having an even thickness, but if the work is awkwardly shaped support must be provided by matching parallel strips of material which are positioned to suit the job, and the operation.

Packing for this type of duty can readily be provided by preparing a selection of short lengths of stock bar material and reserving them exclusively for use as parallels. If a straight piece of stock bar is selected which has parallel sides, it can be sawn into convenient lengths and the ends filed and edges chamfered to produce acceptable substitutes for commercially made items. Some parallels 4in. to 6in.

(100mm to 150mm) long, made from ½in. × 1in. (12.5mm × 25mm) mild steel will be found generally useful.

Parallels are frequently required to pack up work which is held in a machine vice, since it is sometimes essential that it protrudes above the vice jaws. Some smaller parallels may be required for this task, since the vice may be small, and short lengths of something like ⅜in. × ½in. (10mm × 12.5mm) mild steel will be more appropriate.

When work is held in a vice, it is essential to make sure that it is firmly in contact with the base of the vice. To achieve this, the vice is partially tightened and the work tapped down on to the base with a small hammer (soft-faced if the work is already finished on this face). The vice can then be tightened fully. If there is packing below the work, this should be held firmly by the work if there is proper contact between it, the packing and the base of the vice. If the packing will move, the work needs to be tapped down once more to trap the packing, if necessary, using a larger hammer. If the outer face of the work is susceptible to damage, some soft material must be placed on it to protect it from the hammer.

Hardened steel parallels are available commercially. They are supplied in pairs and ground accurately to known dimensions. Half of a set of parallels comprising four pairs, in two widths and four heights, is shown in Figure 16.2.

The use of vee blocks for holding round bars is described in Chapter 10, but larger vee blocks can be used as parallel packing since they are made in pairs which are machined together and so have identical dimensions.

Angle plates

Angle plates are useful as work-holding devices both on machine tables and when marking out.

Figure 16.2 A half set of ground parallels.

Figure 16.3   A 4-inch (100mm) angle plate.

Angle plates are usually iron castings, machined on all of the external faces but not normally on the inside of the angle. They are generally slotted so that holding-down bolts may be fitted to clamp the work. A small plate, some 4in. (100mm) long, is shown in Figure 16.3. Larger angle plates may be made with stiffening webs at each end, and perhaps intermediate ones also, and even larger plates are made in box form and known as box angle plates.

Using the boring table

In the context of adapting the lathe, the cross-slide is frequently referred to as the boring table, since work may be clamped to it and drilled or bored by a tool driven by the lathe mandrel. These operations may naturally be performed when the work is mounted to the vertical slide (see below) but the method is suitable for larger workpieces for which mounting to the vertical slide is not appropriate. Without the benefit of the vertical motion which the vertical slide provides, the work must naturally be packed up to the correct height and then clamped, with its packing, directly to the cross-slide.

Figure 16.4 shows a boring operation being carried out on a model locomotive smokebox. An adjustable boring head is mounted into the mandrel taper and is in use to bore the top of a locomotive smokebox for the chimney's petti-coat pipe. The front of the smokebox is already bored out to a large diameter and clamping to the cross-slide is arranged by use of short tee bolts and a clamping plate. The required position of the bored hole is such that two pieces of standard-thickness plate, one on each side, have brought the smokebox to correct height.

To ensure a positive drive for the boring head, its arbor is tapped so that a drawbar can be fitted to hold it into the mandrel taper. Being a No. 2 Morse taper, the arbor is tapped with the standard ⅜ BSW thread, and the drawbar engages with this. Figure 16.5 shows the outer end of the drawbar. The bar passes through the mandrel and is fitted with a spacer sleeve and lock nut so that it may clamped firmly to the mandrel. The end of the bar is fitted with a knob which allows it to be screwed into the

Figure 16.4   Boring a smokebox for the chimney using a boring head mounted in the mandrel taper. The smokebox is clamped to the cross-slide.

356

Figure 16.5 The outer end of the drawbar used to hold the boring head into the mandrel socket. Without it, the force of cutting may loosen the grip provided by the Morse taper.

boring head arbor. The sleeve is then located in the end of the mandrel bore and the lock nut tightened to secure the boring head. The spacer sleeve needs to be quite long in this instance to allow the lock nut to lie outside the change-wheel guard, and thus be accessible. The guard has been removed in Figure 16.5 to allow the arrangement to be photographed.

Clearly, ordinary drilling may be carried out in the same manner as the boring operation, either by fitting a drill chuck into the mandrel, or by using a chuck or collet to hold the drill or centre drill. Taper-shank drills may, of course, be mounted directly into the mandrel taper, if they are available.

Another way in which work mounted to the cross-slide may be bored is illustrated in Figure 16.6. Here, a boring bar is mounted between centres to bore a cylinder block. This boring bar is a relatively straightforward device in which a circular, high-speed tool bit is fitted into a cross-drilled hole, roughly in the centre of a bar of BMS. A grub-screw clamp for the cutter is provided, and the head of a fine-pitch (40 tpi) screw bears on the back of the cutter bit, to allow the cutter to be advanced by a controlled amount.

If the work to be bored is relatively long, and particularly if the bore is long in relation to its diameter, there is advantage in adopting this method of boring. If the bore is turned, a long and relatively flexible tool must be used. In turning, this is naturally supported only at one end and is consequently much weaker than a bar of similar size which is supported at both ends. In practice, a between-centres bar can be made which is about ¾ of the diameter of the hole being bored, a condition which is not generally possible when boring holes greater than ½in. (12mm) diameter, by turning, due to the impossibility of mounting such a large tool in the toolpost.

The boring head shown in Figure 16.4 may equally be used for facing the work, rather than boring. For facing a workpiece of large surface area, it is normal to use the lathe in its turning mode, if this is possible, the work being held in one of the chucks or clamped to the faceplate. This is certainly the recommended method, hence the description of the method of producing square items, by turning, which is given in Chapter 15.

If the edges of a large plate need to be cleaned up, holding the work to the mandrel is naturally difficult due to the large overhangs which are inevitable, but the boring table may be used for such operations, by clamping the

Figure 16.6 A cast gunmetal cylinder block being bored using a between-centres boring bar.

work, with suitable packing, to the cross-slide. Figure 16.7 shows a wide, mild steel plate mounted on packing and clamped to the cross-slide. This has a finished size of 4in. (100mm) square yet is only ½in. (12mm) thick. Its edges are being cleaned up using a fly cutter.

Figure 16.8 shows how a square, in conjunction with the faceplate, may be used to set up the plate to the correct orientation by using an edge which has already been finished as a reference on which to place the square.

There is a theoretical objection to the use of a fly cutter in this fashion, which arises from the fact that the cross-slide is arranged to traverse in such a direction that a slightly concave (hollow) surface is produced when facing by normal turning.

If the work is positioned so that the fly cutter can cut on the upswing, when it is furthest from the operator, as well as on the downswing, when it is nearest to the operator, it takes a greater cut at the rear, thus creating a step in the work, unless the work can be traversed far enough to clear the cutter completely.

Alternatively, if the cutter swings on a sufficiently large radius to permit all cutting to be performed on the 'down' swing of the tool, the method may yield acceptable results. A further consideration of surfacing is given below.

Figure 16.7 A fly cutter in use to machine the edge of a steel plate clamped onto the cross-slide with packing below.

Figure 16.8 Setting a steel plate up relative to the faceplate, using a square seated on a machined edge.

The vertical slide

For milling operations, the work is most conveniently mounted on a vertical slide which is fixed to the cross-slide by tee bolts and nuts. A vertical slide is shown in Figure 16.9, set up facing the headstock end of the machine. The vertical slide is effectively an angle plate which incorporates slideways to locate a vertically mounted slide. The slide has a feedscrew carrying an engraved dial similar to that provided for the cross-slide. The saddle, cross-slide and

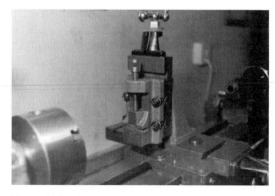

Figure 16.9 A vertical slide set up facing the headstock.

vertical slide motions thus provide the capability to position the workpiece in three axes.

In order to generate parallel work, it is essential that the vertical slide is set up correctly relative to the mandrel, and its face needs to be set up squarely across the bed of the lathe, or be aligned with it, along its length. If a machine vice is attached to the slide, this can be used to secure a parallel bar which is held tightly in contact with the base of the vice. Parallelism can be confirmed by standing a DTI on the bed, and monitoring the edge of the parallel as the cross-slide is traversed, as shown in Figure 16.10.

When a machine vice is attached to the vertical slide, it may not automatically be brought to a position with its fixed jaw parallel to the bed, or at right angles to it, and its position must be checked and adjusted using the DTI, by monitoring the top surface of the parallel bar.

Short workpieces are generally the case in modelmaking and a machine vice designed to fit the vertical slide may be rather small. If longer work does need to be machined, such as locomotive connecting or coupling rods, a small vice provides insufficient holding power to allow the ends to be machined and these longer items can best be held by clamping to a length of angle which is itself bolted to the slide, as shown in Figure 16.11. When setting

Figure 16.11 Long workpieces are a problem to hold since the machine vice for the vertical slide is very narrow. A solution is to attach a length of steel angle to the slide and clamp or bolt the work to this.

up work on the slide, or bolting on the machine vice, it must be remembered that careful alignment of the work in relation to the cross-slide upper surface is also required i.e. the work must be level.

The vertical slide shown in Figures 16.9 and 16.10 is of the type generally known as fixed – it operates only in a direction at right angles to its base. An alternative is the swivelling type, provided with the capability to be rotated about both a vertical and a horizontal axis. The version shown in Figure 16.12 is provided with a graduated angular scale to both adjustments and may be set relatively easily to the approxi-

Figure 16.10 A vertical slide being aligned to the lathe axis by monitoring the edge of a parallel held in the machine vice.

Figure 16.12 A swivelling vertical slide.

359

mate angular position desired. How accurately this setting may be made depends on the accuracy envisaged by the maker and the state of wear (or otherwise) in the registers which locate the various parts.

The particular swivelling slide shown has only a single-bolt fixing to the cross-slide and its attachment is consequently less rigid than the fixed type. It is not the only way in that permits different angles to be set, but it is highly convenient, especially if the graduated scales provide the required degree of setting accuracy.

Milling in the lathe

Introduction

In milling, a multi-toothed cutter is rotated and brought into contact with a fixed workpiece in order to remove material. The basic elements for milling are available in the lathe although it does need some adaptation to render the process really useful.

Milling machines are of two basic types – one having a vertical spindle (a vertical mill) and the other having a horizontal one (horizontal mill). The two machines tend to be used with different basic cutter types. The vertical mill uses relatively long, small-diameter cutters, which cut on the end as well as on the outside diameter, whereas a horizontal mill uses cutters which are broadly like circular saws but sometimes having teeth on the sides as well as on the periphery.

Figures 16.13 and 16.14 show these milling operations in progress. In 16.13, an end mill is in use to machine the edge of a steel plate. Although it does have end cutting teeth, the end mill is intended basically for profiling and performs best in this role and is not ideally suited to surfacing. It is also not intended for use in the formation of slots equal in width to its

Figure 16.13 An end mill profiling a steel plate.

Figure 16.14 Slitting an eccentric strap casting using a slitting saw mounted to an arbor held in the chuck.

diameter, for which a modified version known as a slot drill should be used.

Since the lathe mandrel is horizontal, a vertical milling operation such as this is carried out on its side which does sometimes present difficulties of visibility, but it is a very useful adaptation of the lathe and will satisfy the requirements for milling. The three-dimensional motion of the work allowed by use of a vertical slide secured to the cross-slide is essential if much serious milling is envisaged on the lathe.

In Figure 16.14, a saw having teeth only on its periphery is in use for slitting an eccentric strap casting. Such a saw is a great time-saver when compared with hacksaw-and-file methods since the cutting process leaves a smooth, machined surface. It is extremely convenient when several small lengths of angle are required for fixings or attachments since the process leaves the angle ends smooth and the cut-off items are ready to use. The filing of angles which have been cut off using a hacksaw is an extremely awkward process if clean, square ends are to be produced and the alternative of using a slitting saw for this process comes well recommended.

A variation of the slitting saw is the Woodruff cutter. Like the slitting saw, it has teeth only on the periphery, but is intended for cutting slots in shafts to accept semi-circular keys for engaging a driving pulley or gear (see Chapter 8 for an example) and is made to a specific outside diameter and width. Unlike larger-diameter milling cutters, the Woodruff cutter is manufactured with its own integral parallel arbor. It is very convenient for mounting in the lathe chuck and consequently finds much use in the amateur's workshop. Two Woodruff cutters are shown in Figure 16.15.

If the cutter has teeth on its sides as well as on the periphery, it is known as a side-and-face cutter. This type gives better results than a slitting saw due to the clearance between the workpiece and the sides of the cutter that is provided by the side teeth. Since the cutting action takes place on the sides as well as on the diameter, the cutter produces slots which are nominally the same width as the cutter itself and this is the tool which is used to produce slots which pass through the workpiece, when

Figure 16.15 Two Woodruff cutters. These are intended for cutting keyways but are also used for machining flutes in locomotive coupling rods.

using a horizontal mill. Side-and-face cutters are available in a wide range of diameters and widths but it is only the smaller sizes which are of interest for use on the lathe, since the power required to utilise the larger types is well beyond that which is available. The limited clearances which the lathe provides are also restrictive in this context.

Side-and-face cutters are normally used on a horizontal mill but may be mounted on a between-centres arbor in the lathe, and the workpiece traversed below the cutter, or used on a stub arbor held in the chuck, broadly as shown in Figure 16.14. A pair of side-and-face cutters is shown in Figure 16.16, together with

Figure 16.16 Two thin side-and-face cutters and an arbor which is used for mounting them.

361

a short between-centres arbor on which they can be mounted.

The cutters shown have keyways cut in them, and they are intended to be mounted to special arbors which incorporate corresponding keys. It is not essential that these are provided for the sort of work which is performed on a lathe. The motor power does not approach that which is available on a milling machine, and a frictional grip, perhaps improved by a large washer placed each side of the cutter, is more than adequate for the duties involved.

Since power is limited on the small lathe, only relatively light work is normally undertaken, and for the most part this means the use of small end mills and slot drills, carrying out what would ordinarily be light operations performed on a vertical mill. The light nature of the work is also dictated by the less rigid nature of the machinery which naturally arises out of the attachment of the vertical slide to the cross-slide. On a vertical mill, the table is substantial and slides on an even more substantial base and is, therefore, extremely well supported.

End mills and slot drills

The essential and obvious difference between an end mill and a slot drill is that the slot drill has one cutting edge which passes through the centre of the end face. There is also usually a difference in the number of edges which are provided; end mills having typically 4, 6, or 8, while slot drills have only two or three. These differences are illustrated in Figure 16.17.

A slot drill is intended to be 'plunged' straight into the work which is then traversed so that the drill cuts a slot. End mills are not suitable for plunging straight into the work – they can only cut a slot by entering work from the side and they do not perform well in this role.

Figure 16.17 End views of an end mill and a slot drill. The slot drill has one end cutting edge which spans the centre.

As each cutting edge of an end mill contacts the end of the slot it is cutting, a small deflection occurs. This tends to push the cutter sideways, causing one of the other teeth to dig into one side of the slot and induce a chattering motion. The result is an oversize slot with a rippled surface finish. This can be avoided by using a slot drill of appropriate size or by using a small end mill to cut the initial slot, subsequently opening it out by using an end mill of the required size, or cutting along each side separately, using a smaller cutter.

The lesser number of cutting edges on a slot drill is designed to allow a smooth-sided slot to be cut, thus conferring dimensional accuracy on the cut work. Once the work is being traversed and the cut is established, a 2-edge slot drill cuts only with one edge. There are no cutting edges on two opposite sides, so cutter and work deflection do not cause additional cuts to be taken from the sides of the slot and a smooth-sided slot results. Even a 3-edge slot drill gives a good finish and provides dimensional accuracy in the slot.

The likelihood of chatter developing while milling on the small lathe is increased, since the addition of the vertical slide adds another source of backlash into the motion and places the work up above the cross-slide, naturally weakening its support and thus increasing the likelihood of deflection. For all lathe work, it is vital to remove any lost motions, as described elsewhere, and it is advantageous, when milling, to lock whichever motion is not required to apply the cut, locking the vertical slide when the cut is taken using the cross-slide feed, and vice versa.

Holding the cutter

One of the problems with the use of end mills and slot drills is the seemingly simple operation of holding the cutter to the lathe mandrel. The three-jaw chuck is the most obvious choice and does frequently serve the purpose, but unless the mill (or drill) runs absolutely truly, a clean cut is not achieved and any run-out naturally destroys the dimensional accuracy when cutting slots using a slot drill. The four-jaw chuck is frequently the correct choice, therefore, on the basis that true running of the cutter is required.

There is, however, another problem. The cutting edges, or flutes, of the end mill or slot drill are shaped in the form of a spiral. This bears a similarity to a screw thread, albeit rather a quick one, and there is consequently a tendency for the cutter to screw itself into the work, particularly when used for profiling, since the cutting edges are then predominantly those of screw form, on the periphery of the cutter. Unless the cutter is securely held, it can pull itself out of the chuck, increasing the depth of cut on the end and most likely spoiling the work if a rebate is being cut. This situation is generally described as the cutter 'walking into' the work.

Industrially, the problem is solved by utilising end mills or slot drills having standardised, screwed-end shanks and fitting these into a spe-

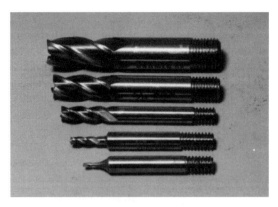

Figure 16.18 Screwed shank end mills.

cial holder or chuck of which the *Autolock* is one well-known type. This utilises mills and drills of the type illustrated in Figure 16.18. The cutter shanks are standardised in a limited range of sizes, allowing a holder with a small number of collet-like devices to hold a wide range of cutter sizes. The screwed end to the cutter engages with a thread at the inner end of the collet and the securing arrangement for the collet, together with an abutment face behind the cutter provides a positive lock against axial cutter movement in either direction.

Holders such as those described are available to fit the standard mandrel tapers on larger lathes now that vertical mills utilising these tapers have also become available. They are one possibility of solving the holding problem. The holder does need to be secured into the mandrel using a suitable drawbar, but this is not a problem. However, the holders are relatively expensive.

One possibility of overcoming the holding problem is the use of plain shank end mills or slot drills of the type illustrated in Figure 16.19. Ranges of this type of mill and drill are available with standard shanks, a ¼in. diameter shank being utilised for cutters from ¹⁄₁₆in. to ¼in. diameter. These cutters are provided with a flat ground on one side and may be mounted into a holder which incorporates a reamed hole

Figure 16.19 A holder for plain shank end mills and slot drills made from a blank end arbor.

and a grub screw fixing to hold the cutter. Figure 16.19 shows such a holder, machined from a commercially available No. 2 Morse taper arbor, known as a blank-end arbor. This has a fully machined taper but is unmachined (as purchased) at the outer end. It may be mounted into the mandrel and machined in situ which, with ordinary care, should provide a true-running axial hole in which the cutter can be mounted. The grub screw fastening provides the drive and prevents the cutter pulling out of the holder.

Blank-end arbors are available having an inner end tapped to accept a drawbar, and this is the type needed for an end mill holder. They are also available having a driving tang at the inner end. If your machinery is metric, or you prefer (or need) to use metric cutters, they are available having standard shanks, the 6mm shank size encompassing cutter diameters from 1.5mm to 6mm.

If collets are available for the lathe there is a temptation to use these for holding end mills and slot drills. This is not a viable alternative to the methods already described. Collets are themselves hardened and ground, as also are the cutters. Neither can deform readily to accommodate the other, as happens when a soft material is held in a hardened collet, and

the result is an extremely slippery hold. The cutter is thus highly likely to 'walk out' of the collet and into the job.

Direction of cut

Before going on to consider speeds, feeds and depth of cut, the matter of the direction of the cut must be considered. Put straightforwardly, the work must always be traversed into the cut, as illustrated in Figure 16.20, so that the work and the teeth which are cutting, approach one another.

If the work is traversed in the wrong direction i.e. such that the work and the cutting teeth are going in the same direction, the cutter drags the work in towards itself, tearing out large lumps as each edge contacts the work, and spoiling the finish due to the uneven rate of material removal. This grabbing action may be very severe.

These considerations apply to all milling operations, but are particularly important when milling on a small lathe where the arrangement may not be very rigid. Figure 16.21 emphasises the point in relation to the

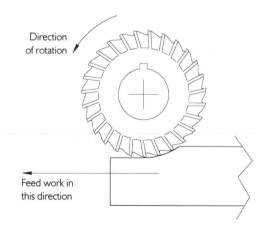

Figure 16.20 Direction of cut for a side-and-face cutter or slitting saw.

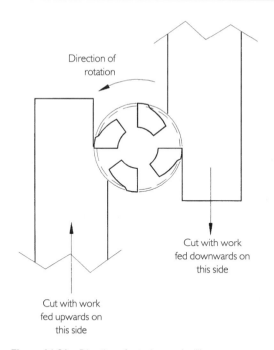

Direction of
rotation

Cut with work
fed downwards on
this side

Cut with work
fed upwards on
this side

Figure 16.21 Direction of cut when end milling.

use of an end mill, which can easily be broken by the grabbing action if the work is fed in the wrong direction. The position of the work relative to the cutter must be considered when arranging to hold and feed the work.

When cutting slots with a slot drill, the situation is not quite the same, since the cut is occurring at the end of the slot, rather than at the sides, and provided that the feed is steadily applied and the cutter is within the work, the feed can be in either direction.

However, if a slot drill or end mill cuts out through the side of the workpiece, the slow and steady feed of the work must continue until the mill or drill is completely clear of the work otherwise there is a high risk of the work moving too far between the arrival of successive teeth, and too heavy a cut being imposed. This is also likely to break off the cutting edge, or even break the end completely off the mill or drill.

Speed, cut and feed

As for other machining operations, the speed of the cutting edge over the work is dependent upon the material being cut (its machinability) and the material of the tool itself. There are recommended speeds for cutting different materials.

The speeds recommended by cutter manufacturers are based on conditions which are likely to be present in industry; very rigid machinery, good quantities of coolant present at the cutting edges, and a wish to maximise the rate of metal removal. In the amateur's workshop, neither of the first two conditions may be present, and it is likely that there is no wish to work in a 'flat-out' way. Somewhat lower cutting speeds are therefore appropriate and a list of those recommended for end mills and slot drills is given in Table 16.1.

As is usual, the theoretical speeds for small cutters, especially in the easier materials, are higher than are generally available, but a modern lathe will perhaps be capable of turning the mandrel at 1500 rpm, and this will prove entirely satisfactory for smaller cutters.

However, the other recommended speeds should be adhered to, if this is possible. The load on the individual teeth of a cutter depends on the rate at which the work is being fed into the cutter. The slower the speed of rotation, the slower must be the feed rate, in order to maintain the same size of cut. There is thus a law of diminishing returns if too slow a cutter speed is used. Any attempt to impose large cuts on individual teeth will surely break them off, or even break the end off the mill or drill.

Slot drills have only two or three teeth, or cutting edges, whereas an end mill might have six or eight. Its teeth therefore 'come round' much more frequently and an end mill might accept a somewhat higher feed rate than a slot drill.

Conversely, the recommended speed of rotation for a slot drill is often higher than the same size of end mill, on the basis that the same feed rate of the work is used in both cases, and the slot drill had better turn a little faster than the end mill, to equalise the load on the teeth in the two cases.

Feed rate is specified in 'inches per minute', typical figures being 3¾in./min. (95mm/min) for an end mill rotating at around 300 rpm, but less than 1½in./min (38mm/min) for a slot drill rotating at this speed.

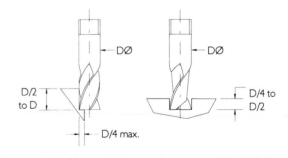

Figure 16.22 Depths of cut when end milling.

Table 16.1 Recommended Speeds for End Mills and Slot Drills (rpm)

Material	Stainless steel Monel metal	Cast iron Bronze Gunmetal Mild steel	Brass	Ali. alloy Tufnol
Cutter diam. in. (mm)	30 ft/min	45 ft/min	60 ft/min	120 ft/min
¹⁄₁₆ (1.5)	1835	2750	3665	7335
³⁄₃₂ (2.5)	1225	1835	2445	4890
⅛ (3)	920	1375	1835	3665
⁵⁄₃₂ (4)	735	1100	1465	2935
³⁄₁₆ (5)	610	929	1225	2450
¼ (6)	460	690	920	1835
⁵⁄₁₆ (8)	365	550	735	1465
⅜ (9)	310	460	610	1225
½ (12)	230	350	450	900

In addition to speed and feed rate, there are two other factors which need to be considered – both the depth and width of the cut which should be used.

It is recommended that the maximum cuts should not exceed those shown in Figure 16.22. The endwise depth of cut for an end mill should not be greater than the diameter of the mill, and the sideways cut should not exceed one quarter of the diameter. In practice, much depends on the rigidity of the set-up, but it should generally be possible to sustain a cut down the profile equal to one half of the cutter diameter, but it

may be necessary to adjust the sideways cut and rate of feed to values which appear comfortable. However, as for all machining operations, very light cuts bordering on no cut at all must be avoided since they simply generate heat and blunt the tool, which ultimately does very little for the finish produced.

When cutting a slot using a slot drill, the cutting depth can generally be up to one quarter of the drill diameter.

The figures given should be regarded as providing a starting point from which to work. Much depends on the rigidity of the machine and the workholding arrangements, and it may be necessary to reduce the depth of cut, in order to compensate for any shortcomings in this direction.

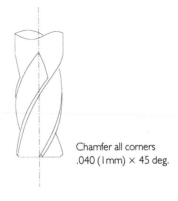

Chamfer all corners
.040 (1mm) × 45 deg.

Figure 16.23 The corners should be removed from an end mill which is to be used for surfacing using the end cutting edges.

Surfacing

Using end mills

Since end mills have cutting teeth on the end, it is very tempting to consider their use for milling large surfaces. They are not ideally suited to this type of work however, since taking shallow cuts on the end of the mill prevents it from cutting to its full potential because the principal cutting edges on the sides are not brought into use. The tool is therefore used inefficiently.

The corners of the cutting edges, where the side and bottom edges meet on the end of the mill, are very weak, being ground at an acute angle, and the effect of using an end mill for surfacing, using the end cutting edges, is to take off these corners. This results in the loss of the principal parts of the mill which are doing the work, and the mill has to be forced to cut (somehow) which results in the generation of much heat, and produces poor finish.

The way to alleviate the problem is to grind off the corners of the mill at 45 degrees, as illustrated in Figure 16.23, *before* using it for surfacing. This improves the end cutting action of the mill and allows it to produce a better surface finish. The corners can be removed by touching them on the side of the fine wheel on the grinder, presenting each corner in turn to the wheel, with the same pressure, and for the same amount of time. It is only necessary to touch the wheel very briefly with each corner of the mill, and once you have heard the 'zing' from the first, progress to the second, and so on, treating them all the same by listening to the zing.

A further point to consider is that end mills do not perform well if forced to cut to their full width, and considerable effort is required to force an end mill to cut a slot, even a shallow one. The first precaution in their use for surfacing is to restrict the width of the cut to something less than half the diameter of the mill. This means working around the outside of the face which is being machined, progressively advancing the cut towards the centre after each circuit of the face.

For some tasks, progressing around the outside is not immediately possible, for example when using the end cutting edges to reduce the width of the central portion of a stock bar to produce a slim rod with a boss at each end. In such cases, it is necessary to cut a slot at both ends of the central portion, using a slot drill, to produce a raised portion which can be approached from all around the edge.

Using a fly cutter

The most convenient tool for machining large surfaces is a fly cutter, which is both easy to make (or fairly cheap to buy) and easy to sharpen using the workshop's grinder. A simple commercially made fly cutter is illustrated in Figure 16.24. It consists of a turned steel body with an integral shank, the body being provided with a ³⁄₁₆in. (5mm) square through hole angled at 45 degrees to the axis. The hole accepts a standard tool bit, which is held in position by a screw.

The parallel shank can be held simply in the three-jaw chuck and the tool is readily ground

Figure 16.24 A small commercial fly cutter. Not very elegant, but effective nevertheless.

to the required shape – like a left-hand knife tool – and can be just as readily resharpened.

If making such a cutter in the home workshop, the angled hole should be drilled and reamed to a standard tool bit size (¼in. or ⁵⁄₁₆in., 6mm or 8mm say) so that a round tool bit can be used, and the body size chosen to swing the tool point on a radius of ¾in. to 1in. (20mm to 25mm). This will provide a tool which cuts over a 2in. (50mm) path, which is adequate for much of the work which is required. If a wider surface must be machined, a second pass can be taken.

Surfacing using a fly cutter held in a chuck, with the work mounted on the cross-slide, poses a potential problem if the work is longer than twice the swing of the cutter. The cross-slide is not set precisely at 90 degrees to the mandrel axis, in such a way that the part of the work which is beyond the mandrel axis is slightly nearer the cutter.

This means that the cut on the upswing of the tool is greater than that on the downswing, and the work will, therefore, have a shallow step in it unless it can be traversed completely through the upswing cutting position.

This presents difficulties however, due to the limitation of the maximum cross-slide movement. On my lathe, this is only 7in. (178mm) so the limitation is quite severe. Nevertheless, a fly cutter is cheap and convenient, is easily resharpened, and there are benefits from its use.

If the cutter is working on a large diameter, its rotational speed needs to be low to maintain the normal cutting speeds. Another factor which limits the speed is the fact that the material which is cut from the work is thrown out at the speed of the tool tip. This may pose a safety hazard apart from being objectionable, and it may also impose a constraint on the speed used. The direction in which the chips are ejected also needs consideration, and may dictate the way the work should approach the cutter.

To reduce the load on the machine, and the speed of ejection of the chippings, it is prudent to operate a fly cutter at a lower cutting speed than that used for turning. Recommendations in this respect are given in Table 16.2. The rotational speeds are calculated for three diameters of cutter swing, and relate to cutting speeds of roughly half of those used when turning.

When fly cutting, the depth of cut can be somewhat like that used for turning. There is considerable shock (and noise) when the cutter first strikes the edge of the work and this limits the depth of cut on a small lathe. It is sensible to adopt modest cuts, and .010in. (.25mm) will perhaps be the limit on a small machine. For the final, finishing cuts, the depth can be reduced to .005in. (.125mm). To produce a fine finish,

Table 16.2 Recommended Speeds for Fly Cutting (rpm)

Material	Cutting Speed*	Diameter of Cutter Swing		
		1in. (25mm)	2in. (50mm)	4in. (100mm)
	BMS			
Copper	40–50 ft/min	150–200	80–100	40–50
Gunmetal	(12–15 m/min)	rpm	rpm	rpm
Phos. bronze				
Stainless steel	30 ft/min	120 rpm	60 rpm	30 rpm
Silver steel	(9m/min)			
Cast iron	35 ft/min	135 rpm	70 rpm	35 rpm
	(10 m/min)			
Brass	150 ft/min	570 rpm	285 rpm	140 rpm
Nickel silver	(45 m/min)			
Ali. alloys				

* Approximately one half of normal values

the tip of the tool should be given a small radius, as described in Chapter 14, for use when turning.

Commercially made fly cutters can accept square tool bits and are sometimes very substantially made. Figure 16.25 shows a cutter which accepts a ⅜in. (9mm) square tool bit, which is held into a very substantial body by three Allen screws. The body is forged integrally with a No. 2 Morse taper shank, which is tapped for a drawbar so that it can be held securely in the lathe mandrel. As illustrated, the cutter is mounted into an adaptor which permits its use in a larger mandrel bore than that provided on my lathe

This cutter is shown in use in Figure 16.7.

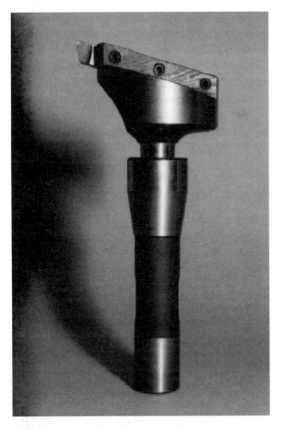

Figure 16.25 A large counterbalanced fly cutter.

Finding a reference position

Introduction

The milling process is generally concerned with machining steps, recesses and slots in the work. These features are often positioned relative to existing surfaces or faces, and there is a frequent need to find a starting position, on an edge or surface, from which to measure the feed of the cutter for subsequent operations. A similar operation is required, when it is necessary to locate an already-marked position in alignment with the axis of the cutter.

For some work, it is convenient to locate the position of holes during the milling operation. This can be done by aligning the axis of the cutter with an edge on the work, and then traversing the work to the required position.

Once the edge, or other reference position has been found, the feedscrews can be used to traverse the work the required distance from this position. This means locating the edge of the work, or the reference position, accurately as a preliminary operation.

Finding a surface or edge using an end mill or slot drill

The intuitive way to find an edge or surface using an end mill or slot drill is simply to bring the cutter towards the work until it is seen to touch it, note the feedscrew dial reading (or set it to zero) and use this to define the start. The basic problem with this method is that it is impossible to discern exactly when the cutter is just touching the surface. By the time that it is obvious, a small cut is already being taken so there is a positional error at the start. In any event, there may be no requirement to cut the surface which has just been cut, and the damage is therefore not required on this account.

The way to avoid this error is to stick a small piece of cigarette paper to the surface or edge using spittle, and then to feed the work towards the cutter until it cuts the paper. Since the thickness of the cigarette paper can be measured, the cutting edge of the mill or drill is a known distance from the work when it 'picks up' the paper, and the work can be advanced by this amount, using the feedscrew, to find a reference position. Typically, a cigarette paper is .001in. (.025mm) thick so the cutter is this far from the work when the pick-up occurs.

It is important that the paper is properly in contact with the work, without any edges or corners protruding, and it is best to cut a small patch from the whole paper, ¼in. (6mm) wide, thoroughly wet it with spittle by stroking it on the tongue, and then carefully press it down on to the surface of the work. The surface tension of the spittle keeps the thin paper in contact with the surface for quite long enough to use the mill or drill to find the surface, but the paper dries out and becomes detached after a few minutes.

When approaching the cigarette paper, the cutter must be clean and free from particles of swarf, otherwise a clean contact may not be made with the surface of the cutter. The contact between the cutter and the cigarette paper may simply cut the paper, but it is often picked off the work and rotates with the cutter. Either way, the position has been found.

If the side of an end mill or slot drill is approached by the work, and the cigarette paper is picked up, the edge of the work has not immediately been located below the axis of the cutter, only on its periphery. A further movement of the work equal to half the diameter of the cutter is needed to bring the axis into alignment with the edge, if this is required. This is illustrated in Figure 16.26.

Once the edge has been found, the work can be traversed to another position, using the feedscrew, and a cut taken from the work. It

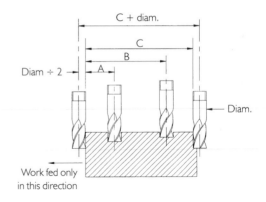

Figure 16.26 Finding a reference position.

must be borne in mind that there is sure to be backlash and lost motion in the feedscrew drive, and the position achieved is only accurately referenced to the edge if the feed continues in the same direction in which the edge was approached.

Distances A, B and C may be specified on the drawing. If the edge has been found by touching the side of the cutter on a cigarette paper stuck on the work, the position A is reached by moving the work by A plus the cutter diameter divided by two. A similar consideration applies to position B. To cut the edge defined by dimension C however, the work must be traversed by C plus the cutter diameter.

Centre finders, edge finders and wigglers

If a small, straight, pointed bar were to be mounted in a four-jaw chuck, its point could be set to rotate truly when the mandrel was turning. The feedscrews on the cross-slide and vertical slide could then be used to bring the intersection of two scribed lines on the workpiece into alignment with the mandrel axis. This would allow a hole to be drilled at the marked position, using a centre drill or stub drill, and further positions could be drilled simply by traversing the work through further increments, using the feedscrews, thus setting out the hole positions without further marking out.

This method of working relies on the fact that there are feedscrews which control the movement of the work in two directions at right angles to one another, and a position can be defined as being X in. (or mm) in one direction, and Y in. (or mm) in the other direction, from some reference point. A true-running pointed bar allows the reference point to be located in line with the mandrel axis permitting the method of traversing the work by the feedscrews to be adopted.

The essential need is for a device which can be mounted in a chuck and quickly brought to the true-running condition. The most convenient device is contained in what is called a wiggler and centre finder set. A typical set comprises the items illustrated in Figure 16.27.

The tool consists of a collet mounted on the end of a plain shaft which can be held in a chuck. The collet can accept one of several tools, each of which consists of a thin rod having a ball at one end. The ball can be gripped in the collet to provide a firm, but movable grip, so that the rod can take up any position on the end of the collet, within about 30 degrees of being axially in line with its shaft.

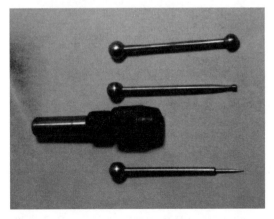

Figure 16.27 The components of a wiggler and centre finder set.

If the collet is mounted into a chuck, with one of the tools gripped by the collet head, the outer end of the tool will wobble, if it is not running absolutely truly, when the chuck rotates. Since the tool can be moved into alignment while it is gripped by the collet, it can be pushed towards the true-running position while the chuck is turning, simply by pushing on the end with a piece of wood, for example, or by pressing against it with the work.

Two basic tool types are provided with the wiggler – a rod with a point at the outer end, and rods with a either a small cylinder or a ball at the outer end. The pointed rod is the centre finder. Its end can be brought to the true-running condition by bringing a piece of wood slowly towards the end when the wiggler is turning.

If the end is rotating eccentrically, it begins to touch the wood once during each rotation as the wood approaches. These contacts progressively push the point towards the true-running position, and as the wood is pushed further the point will at some stage stop wobbling as it comes to the true axial position. This position is readily identified by eye, and the centre of rotation can be found easily.

If the contact point between the piece of wood and the centre finder is pushed beyond the true axial position, this forces the point to run eccentrically and the end of the rod 'walks' along the wood to a point at which it can swing freely, out of contact with the wood. There is thus a little danger from the sharp point at this stage, and the hands should be kept away from the working area. Once the centre is running truly, it can be used to locate a marked position on the work on the axis of rotation, and thus reference other positions, drill a hole, machine a slot centred on this point, or whatever.

A sticky pin will also serve for the purpose of finding a centre, provided that some means can be found to attach one end of it to the chuck jaws. Modelling clay is ideal for the purpose, provided that it has not been rendered too soft by excessive handling.

The other rods in the wiggler set have precisely sized outer ends, either cylindrical (or disc-shaped) or turned as a ball. They work in precisely the same manner as the centre finder, but the true-running position is generally found by approaching them with the work. They are described as edge finders, and they are used to bring the work to a position which is half the diameter of the ball or disc from the axis of the wiggler.

A ball-ended wiggler which is rotating eccentrically is shown in Figure 16.28. In Figure 16.29, a steel angle mounted on a vertical slide is being brought slowly upwards toward the ball, which is now rotating less eccentrically. The point is soon reached at which the ball appears to be running without eccentricity, Figure 16.30, and a further upward movement of the steel angle of .001in. (.025mm) causes the wiggler ball to roll along the angle to take up the position shown in Figure 16.31. It now has freedom to rotate past the edge of the steel angle to indicate that the angle has been raised just too far to permit the ball to rotate truly above it. The angle has, therefore, been raised to a position which is half the diameter of the ball away from the axial centre. Further upward movement of the steel angle can now be accurately made, controlled by the feedscrew on the vertical slide.

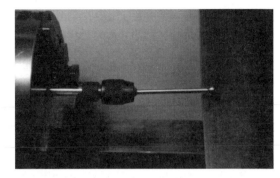

Figure 16.29   Wiggler ball touching angle.

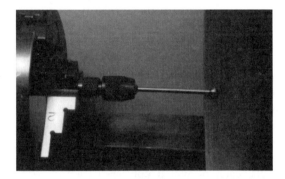

Figure 16.30   Wiggler ball running truly on the angle.

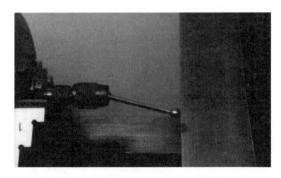

Figure 16.31   If the angle moves upwards by .001in. (.025mm) the ball flies away.

Indexing and dividing

The ability to divide circular work into a number of sectors is especially useful for a wide variety of engineering activities. Really accu-

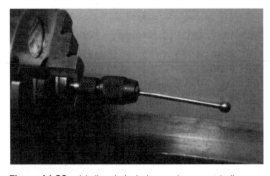

Figure 16.28   A ball-ended wiggler running eccentrically.

rate division requires the use of a dividing head, an expensive accessory which comprises a spindle, to which the work is attached, which may be driven by hand through precise angles under control of a handle-and-shaft which drives the spindle through a worm reduction gear. Ratios of 48:1 or 60:1 are commonly used in the drive, giving good control of the work, and indexing plates (steel discs with circles of precisely located holes) allow the handle to be turned through whole turns and/or fractions of a turn, to position the work with considerable precision. For gear cutting, or the cutting of clock wheels and pinions, a dividing head provides the appropriate degree of accuracy.

However, there are occasions when such accurate division is not required, but nevertheless several holes need to be equi-spaced (nominally) around the work, for example, when building spoked wheels or when drilling holes in cylinder covers. Simple dividing of this type might be carried out while the work is mounted in the chuck. Since chucks have three or four jaws, there are this number of reference points immediately available from which to mark off the work.

The pitch circle for the hole positions can also be marked, while the work is in the chuck, by using a sharp, vee-pointed tool held in the toolpost, and facing the work. Centre punching and drilling can then be undertaken in the usual way.

It is possible to remove the chuck from the lathe, with the work still held in the jaws, and scribe across the work against a rule which is aligned with the jaws by eye. With care, it is possible to scribe lines centrally between jaw positions, or to bisect the angles marked from alignment with the jaws, using dividers. Having three- and four-jaw chucks available, division of the work into 3, 4, 6 or 8 sectors is possible quite readily.

Great accuracy of division may not actually be essential, unless interchangeability is vital, since it is possible to drill, say, a cylinder cover after marking out on the lathe, subsequently drilling the cylinder block, for tapping, using the cover as a jig. Provided that a reference mark is made on both the cover and the block, to identify the fitted position of the cover (a small centre punch dot will suffice) it is of no consequent that the holes have only been spaced by eye.

Greater accuracy of division can be obtained by mounting the work to a commercially cut gear and then indexing by counting teeth on the gear. This often serves to provide the accuracy required. A suitable number of teeth on the gear provides the number of divisions required. For example, a 48-tooth gear can be used to divide the circle into 2, 3, 4, 6 or 8 parts, and so on, up to 48, simply by counting teeth and using a detent to lock the gear (and the work) at the required points.

The most obvious source of a suitable gear is the set of lathe changewheels, and the most obvious place to mount the work is to the lathe mandrel, consequently allowing the usual workholding devices to be used. This requires that the gear is also fitted to the mandrel and this can be achieved by using an adaptor which fits into the left-hand end of the mandrel bore, permitting a changewheel to be mounted to it. A spring-loaded detent fitted to the headstock then allows the mandrel to be indexed as required.

An alternative approach is to machine the work directly, by mounting a cutter or drill into a toolpost-mounted, motorised spindle and use this to machine the work. This type of supplementary drive, known as a milling and drilling spindle, may be self-contained or driven from a separate motor, perhaps with a countershaft, mounted behind, or above, the lathe bed. Numerous designs for auxiliary spindles of this sort have been published over the years, since they are extremely useful in extending the capability of the lathe, by allowing holes to be drilled in the face of work mounted to the mandrel in normal ways, or allowing radial slots or recesses to be machined.

The work required to make a suitable spindle and its drive is not difficult, especially if a published design is followed, but for occasional use it is possible to make up an adaptor to attach a DIY electric drill to the cross-slide, or perhaps more usefully, one of the small, high-speed hobby drills, which are now widely available. An adaptation of this sort may well satisfy the occasional need for straightforward drilling, or the spotting of hole positions.

Since much dividing, apart from that actually required for gear cutting, is required for drilling holes, it is convenient if the indexing operation can be carried out on the drill. A simple indexing device, mounted on the drilling machine table, is shown in Figure 16.32. This utilises a 48-tooth changewheel mounted on a spindle which accepts a lathe chuck on its upper end so that work may readily be mounted below the drill spindle and indexed into the required positions. For spotting the hole centres, it is usual to use a small centre drill which is sufficiently stiff to drill a locating dimple without undue wander. A stub drill can be used equally well.

Figure 16.32　An indexing head on the drilling machine.

This indexer can of course be used on the cross-slide of the lathe, or on a vertical slide, to permit flats to be machined on the work, using either an end mill mounted to the mandrel, or a side-and-face cutter mounted on a between-centres arbor.

Screwcutting

Introduction

It may seem odd to consider screwcutting to be an adaptation of the lathe, since it is most likely described as a screwcutting lathe at the outset. However, it does need adapting in order to cut screw threads.

It is possible that very few threads will need to be cut on the lathe, since the simplest way is to use taps and dies to cut threads. However, these may not be available for every thread which is required, since there are bound to be occasions when threads outside the normal range are needed.

Taps and dies for the larger threads in particular may not be available due to their high cost, especially if the size is not required very frequently. In some cases, it is economic to buy the tap of a particular size, and to cut the matching male thread on the lathe. In these cases, the female thread, or nut, should be made first and the male thread screwcut in the lathe until the one fits the other.

If the lathe must be used to cut both male and female threads, it is probably simpler to cut the male thread first, especially if you are new to screwcutting, since this gives better visibility for the first operation, which simplifies things considerably.

The basic principle

It is obvious that the lathe can be used to cut a screw thread, provided that it is fitted with gearing between the mandrel and the leadscrew. The gearing must be the correct ratio for the screw pitch which is to be cut, and the lathe is supplied with a set of changewheels which are used to set up the correct gear ratio.

The gear trains required for standard screw pitches may well be specified in the maker's handbook, or on a plate attached to the machine, if the outfit is new, but if you have a secondhand machine, such a table may not be available. Fortunately, it is a simple matter to calculate the set-up of changewheels which is required.

The principle is illustrated in Figure 16.33. A workpiece is mounted to the mandrel in some way and is to be cut with a screw thread. The pitch of the thread can be read from the tables in Chapter 11, or from an engineer's reference book, in terms of the number of threads per inch (tpi).

The thread is cut by mounting a vee-shaped tool, of the correct angle, in the toolpost, and allowing the leadscrew to drive the saddle, with the cross-slide, topslide and toolpost on it, along the bed, as the mandrel and workpiece turn.

On an imperial lathe, the leadscrew usually has a thread of 8 turns per inch (tpi) so, to cut a thread of known pitch on the workpiece, the speeds of rotation of the work (mandrel) and the leadscrew are related simply by the pitches of the two threads. Thus, if a 24 tpi thread is to be cut, the mandrel must rotate 24 times while the leadscrew rotates 8 times. To put it another way, 8 turns of the leadscrew move the saddle (and tool) 1 inch along the work. During this movement, the work rotates 24 times and has 24 turns of thread cut on it.

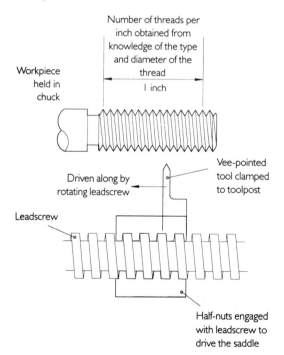

Number of threads per inch obtained from knowledge of the type and diameter of the thread

Workpiece held in chuck

I inch

Driven along by rotating leadscrew

Vee-pointed tool clamped to toolpost

Leadscrew

Half-nuts engaged with leadscrew to drive the saddle

Figure 16.33   The principle of screwcutting on a lathe.

Gearing

For the above example, the mandrel must rotate three times faster than the leadscrew. A gear ratio of 1:3 is therefore required and any combination of gears which gives this ratio may be used. In a standard set of changewheels, which normally range from 20 up to 80 or 100 teeth, the most convenient pair of gears to choose for a 1:3 ratio comprises 20-tooth and 60-tooth gears, as shown in Figure 16.34A. Since the mandrel naturally drives the leadscrew, the 20-tooth gear is called the driver, and the 60-tooth, the driven gear. The gear ratio is simply written down as:

$$\frac{\text{Driver}}{\text{Driven}} = \frac{20}{60} = \frac{1}{3}$$

which may equally be written down as:

$$\frac{\text{leadscrew tpi}}{\text{mandrel tpi}} = \frac{8}{24} = \frac{1}{3}$$

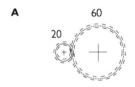

A

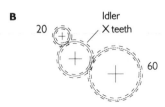

B

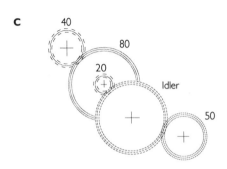

C

Figure 16.34 Gear ratios for cutting 24 and 40 tpi using an 8 tpi leadscrew.

In theoretical terms, the gear ratio is fine, but the two gears are not of such sizes that the one (20-tooth) may be mounted on the mandrel and the other (60-tooth) on the leadscrew, yet have them in mesh with one another. It will in general be necessary to use a third gear to couple them together, as shown in Figure 16.34B.

When considering any gear train, the gears are always considered in pairs, and the ratios are always written down as driver divided by driven, so for the case of Figure 16.34B, the overall ratio is:

$$\frac{20}{X} \quad \text{followed by} \quad \frac{X}{60}$$

Since gear ratios have a multiplying effect, the overall ratio may be written down as:

$$\frac{20}{X} \times \frac{X}{60}$$

The two values X cancel one another and the ratio remains as 20:60, or 1:3. The only effect of the intermediate gear is to reverse the direction of rotation of the 60-tooth gear, and the intermediate is therefore known as an idle gear, or simply as an idler.

Naturally, choosing a very convenient ratio of thread pitches such as 8:24 has resulted in an extremely simple determination of the ratio required. If the thread to be cut on the workpiece is, say, a less convenient 20 tpi, a little arithmetic manipulation is required to determine a suitable set of gears.

The required ratio is still written down in the same form, giving:

$$\frac{\text{leadscrew tpi}}{\text{mandrel tpi}} = \frac{\text{Driver}}{\text{Driven}} = \frac{8}{20}$$

It is now necessary to scale the 8 divided by 20 fraction until numbers are produced which allow a viable gear ratio to be discerned. Provided that the same arithmetic operation is carried out on both the top and the bottom of the fraction, it remains unaltered in value. The most useful first step is to multiply both the top and bottom of the fraction by 10 and then make further numerical adjustments in whatever direction looks hopeful. So, it might work out like this:

$$\frac{\text{Driver}}{\text{Driven}} = \frac{8}{20} = \frac{80}{200} = \frac{20}{50} \quad \text{OR} \quad \frac{40}{100}$$

Both of these last results look hopeful, since there are certain to be 20-, 40- and 50-tooth gears in the changewheel set, and quite possibly a 100-tooth as well. Remember that the mandrel has the finer thread and must rotate faster

376

than the leadscrew, therefore it needs the smaller gear.

Further ratios can be worked out to increase familiarity with the manner of adjusting the figures. For example:

For 11 tpi:

$$\frac{\text{Driver}}{\text{Driven}} = \frac{8}{11} = \frac{80}{110} = \frac{40}{55}$$

For 14 tpi:

$$\frac{\text{Driver}}{\text{Driven}} = \frac{8}{14} = \frac{80}{140} = \frac{40}{70}$$

For 26 tpi:

$$\frac{\text{Driver}}{\text{Driven}} = \frac{8}{26} = \frac{80}{260} = \frac{40}{130} = \frac{20}{65}$$

For 28 tpi:

$$\frac{\text{Driver}}{\text{Driven}} = \frac{8}{28} = \frac{80}{280} = \frac{40}{140} = \frac{20}{70}$$

For some thread pitches, the sums do not turn out to be as convenient as those above and this can be demonstrated by considering the requirement to cut a 40-tpi thread:

$$\frac{\text{leadscrew tpi}}{\text{mandrel tpi}} = \frac{8}{40} = \frac{80}{400}$$

This fraction doesn't look at all promising and it therefore needs to be adjusted in a different manner. The ratio required is too great to be accomplished by one pair of gears, assuming you are limited to the range from 20 to 80 or 100 teeth, which is generally the case. The need is to find multiplying factors for the top and bottom of the fraction which allow the ratio to be obtained from two pairs of gears. This adjustment might be:

$$\frac{80}{400} = \frac{4 \times 20}{8 \times 50} = \frac{4}{8} \times \frac{20}{50} = \frac{40}{80} \times \frac{20}{50} \quad \frac{\text{Driver}}{\text{Driven}}$$

Two pairs of gears can be used to provide the

required ratio; a 40-tooth gear driving an 80-tooth, and a 20-tooth, rotating with the 80-tooth, in turn driving a 50-tooth gear.

Figure 16.34C shows how the gears might be arranged. A 40-tooth gear on the mandrel drives an 80-tooth, to which is keyed (or pinned) a 20-tooth gear. This in turn drives the 50-tooth gear fitted to the leadscrew through an idler, this being necessary otherwise the 80-tooth gear would foul that on the leadscrew.

There are other ways to obtain this ratio, since the 40:80 could equally be 30:60 or some other 1:2 ratio, and the 20:50 might equally be 40:100. Also, it does not matter which driver gear drives which driven gear, since:

$$\frac{40}{50} \times \frac{20}{80} \text{ is the same as } \frac{40}{80} \times \frac{20}{50}$$

but the drivers must drive and the driven gears must follow, otherwise one of the fractions has been inverted.

Practical gearing

Figure 16.35 shows a view of a changewheel sector plate, or quadrant, with only some changewheels fitted. At the top is a 25-tooth gear fitted permanently to the outer end of the mandrel. Below this is a meshed pair of 20-tooth gears mounted on studs fitted to a bell crank which can be positioned so that a second 25-tooth gear can be driven in either direction from the mandrel gear. This arrangement is described in more detail in Chapter 12.

Whichever direction of rotation is chosen, the lower 25-tooth gear rotates at the same speed as the mandrel. This gear runs on a shaft fixed to the bell crank and has a sleeve projecting outwards which is fitted with a square key. Any changewheel, all of which are machined with keyways, can be mounted on this sleeve and rotates at the same speed as the mandrel. This is the first (or only) driver of the gear train.

Figure 16.35 The changewheel end of the lathe showing the driving keys.

Figure 16.36 A close-up view of a changewheel stud and its sleeve and driving key.

Figure 16.37 The changewheel set-up for cutting 24 tpi using an 8 tpi leadscrew.

The end of the leadscrew is also machined for a key so any changewheel can be mounted on the end of the leadscrew and drive it. This gear is the last (or only) driven gear of the gear train.

To allow for intermediate gears, the sector plate, or quadrant, has two slotted arms, each of which has a movable (but lockable) stud fitted to it. The studs are each provided with a sleeve which is free to rotate on the stud and is fitted with a square-section key. Each sleeve is long enough to mount two gears on it which can be locked together by inserting a key. In this case, the two gears on the sleeve must rotate at the same speed. Figure 16.36 shows a close-up of a sleeve mounted on its stud, with a 20-tooth gear fitted.

The quadrant is locked to a housing around the end of the leadscrew but can be rotated around the housing so that the upper arm can be positioned as required in relation to the 25-tooth gear. This movement, together with the adjustable positioning of the studs, allows idlers or intermediate gear clusters to be arranged to suit the gear ratio required.

Figure 16.37 shows a set-up to cut 24 tpi. The changewheels are set up so that the 20-tooth is mounted on the bell crank spindle, and therefore rotates at the same speed as the mandrel. Each stud carries an idle gear, and the leadscrew has the 60-tooth changewheel mounted on it. Since the sleeves on the studs are long enough to accept two changewheels, each is fitted with a small gear, as packing.

Figure 16.38 shows an arrangement for cutting 40 tpi. This is arranged with the 30-tooth gear on the bell crank spindle, and the 60-tooth on the leadscrew. The arrangement uses 30:60 as the alternative 1:2 ratio, which is suggested

Figure 16.38 The changewheel set-up for cutting 40 tpi using an 8 tpi leadscrew.

above. Between these gears, the first stud carries an idler, and the second carries a 50-tooth gear keyed to a 20-tooth. This arrangement is suggested by the lathe manufacturer as being easier to set up than the layout suggested in Figure 16.34C.

Engaging and disengaging the drive to the saddle

The screw thread on the work is cut by driving the saddle along the bed by engaging the half nuts with the leadscrew. Mandrel and leadscrew are naturally rotating and this leads the vee-shaped tool along the work. The tool is advanced so that a shallow cut is taken on the work, and the process is repeated, advancing the cut each time, until the full depth of thread has been cut.

Generally, the start point of a single-spiral thread is not important and so it does not matter when the half nuts are closed for the first cut. However, for the second, and subsequent, cuts, the tool must pass along the work down the same spiral and herein lies the problem.

Figure 16.39 shows this diagrammatically. The workpiece and leadscrew are frozen in some position in which the half nuts are engaged with the leadscrew and the tool is aligned with the partially cut thread.

If the half nuts are disengaged from the leadscrew, and the saddle (and tool) moved to the left, the half nuts will next engage with the leadscrew when the tool is ⅛in. farther down the workpiece. This position may, or may not, correspond exactly with the spiral of thread already cut, depending upon the pitch being cut, in relation to the leadscrew pitch. So, not every position at which the half nuts will close maintains the tool in synchronism with the partially cut thread on the work.

A similar effect may occur if the drive from the mandrel to the leadscrew is disconnected

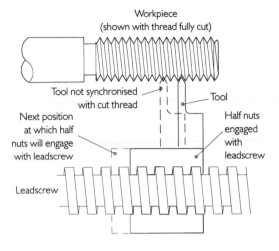

Workpiece
(shown with thread fully cut)

Tool not synchronised
with cut thread

Tool

Next position
at which half
nuts will engage
with leadscrew

Half nuts
engaged
with
leadscrew

Leadscrew

Figure 16.39 Synchronising the tool with the partly turned thread.

once a cut has been taken on the work. If the mandrel turns without turning the leadscrew, or vice versa, the thread being cut and the leadscrew thread become displaced with respect to one another and synchronism is lost. Thus, in the absence of any means to indicate the relative positions of the partially cut thread, the tool and the leadscrew, the half nuts must not be disengaged from the leadscrew, and the drive from the mandrel, through the change-wheels, must not be disconnected, once both are engaged for the first cut.

This means withdrawing the tool by the cross-slide feed, when the tool reaches the end of the thread, stopping the lathe (by switching off) and turning the mandrel backwards to bring the saddle back to the starting position. When the tool is back at the starting position, it is advanced to the reading used for the previous cut and is then advanced farther for the next cut. There is much winding in and out of the cross-slide feed and notes need to be kept of the feedscrew dial readings.

The aggravation of having to wind back manually is removed if the lathe's motor is

fitted with a reversing switch so that a return to the starting position can be made under power.

The cutting of the thread is simplified if the half nuts can be disengaged from the leadscrew at the end of each cut and then re-engaged in synchronism with the partly cut thread for the next cut. To allow this to be done, the lead-screw position is monitored, and shown on a dial attached to the saddle. This attachment is called a thread dial indicator, an example of which is illustrated in Figure 16.40. This small device comprises a housing bolted to the right-hand side of the saddle. The housing holds a shaft, on the bottom of which is a gear which engages with the leadscrew, and on the top of which is a dial engraved with 8 lines, alternate ones being identified as 1, 2, 3 and 4.

The position of the thread dial indicator is adjusted on the saddle (spacing washers are usually provided) so that the half nuts will close on to the leadscrew when the dial '1' engraving coincides with the index mark. The ratio between the leadscrew and the dial is 16:1, and the half nuts actually engage with the leadscrew at 16 positions of the dial, eight of which are identified by the engravings.

If the saddle is stationary, with the half nuts not engaged, but the leadscrew turning, the dial rotates to show the relative position of the leadscrew thread with respect to the position of the half nuts. When the first cut is about to commence, the saddle is brought up so that the tool stands just off the end of the workpiece and is set so that a shallow cut will be taken. With the machine turning, the thread dial indicator is rotating and the half nuts can be closed when the '1' engraving, say, is aligned with the index. The saddle now starts to move and the cut occurs, but since the saddle is travelling with, or being driven by, the leadscrew, the dial indicator does not rotate.

When the tool reaches the point at which the thread is to finish, the half nuts' control handle is lifted (smartly!) to stop the motion of the

Figure 16.40 A thread dial indicator.

saddle and prevent the tool over-running the end of the threaded portion. The tool is now stationary, and the work is turning, so a groove is cut around the work. The tool is now withdrawn using the cross-slide feed, and the saddle wound back manually to bring the tool clear of the work.

The mandrel and leadscrew are still turning and so is the thread dial indicator. So, the tool can be advanced into the cut and the half nuts closed again when the '1' on the dial is again aligned with the index on the indicator, ensuring that the tool advances in synchronism with the already-cut thread.

It doesn't matter which of the engraved lines on the indicator was used to engage the half nuts for the first cut – provided that the same engraved line is used for subsequent cuts, they will all be in synchronism. While this is true, it is not actually essential to use the same engraving each time, but it does depend upon the screw pitch being cut.

The ratio between the thread dial indicator and the leadscrew is 16:1, so that it needs 16 turns of the leadscrew to rotate the dial through one turn. Since screwcutting normally must be undertaken at low mandrel speeds, the leadscrew is rotating only slowly, and the indicator dial rotates only at $\frac{1}{16}$ of this speed and it takes an age for it to complete one revolution.

However, the 16:1 ratio is chosen for a very good reason. 16 turns of the leadscrew correspond with 2 inches of saddle movement, 8 turns with 1 inch, and 4 turns with half an inch. If the thread being cut has an even number of turns per inch, it naturally has a whole number in half an inch. So, if the saddle moves half an inch, or 1 inch, or 1½ inches, or 2 inches, from the position at which the half nuts were closed for the first cut, the tool will be in synchronism with the first thread and the thread dial indicator will show one of the remaining, numbered engraved lines aligned with its index. For an even number of turns per inch in the cut thread, the half nuts may be closed when the thread dial indicator shows any of the numbered lines aligned with its index.

If the cut thread has an odd number of turns per inch, the half nuts can be closed only when alternate numbers are aligned with the index, since it is only every inch on the work that the thread and saddle positions can again be in step, or synchronised. This is indicated by the thread dial indicator rotating half a turn. So, for an odd number of threads per inch, the half nuts may be closed either at 1 and 3, or 2 and 4, but you naturally cannot change halfway through.

If the thread pitch being cut includes half a turn per inch, say, 12½ tpi, synchronism between saddle and work can only be achieved every 2 inches, corresponding with 16 turns of the leadscrew, or one turn of the thread dial indicator, so, if you have to cut 12½ tpi (or any figure plus a half) the half nuts can be closed only every turn of the indicator dial.

Grinding the tool

To cut the screw thread, the tool point is ground to the correct angle for the thread being cut. In form, the tool can be something like a parting-off tool, with a narrow cutting point on the left-hand side of the tool blank. It must be flat-topped (have zero top and side rake) and it should have a front clearance of 5 to 10 degrees. There should be clearance below both cutting edges so that the tool will cut cleanly on both sides.

To assist tool grinding, a standard template can be purchased, into which are ground vee notches corresponding to the angles of 47½, 55 and 60 degrees which are used for the major thread systems. The grinder should be used to grind away the bulk of the material, on the right-hand side of the tool blank, after which the flanks of the tool can be ground carefully so that the point fits the gauge.

Preparing the workpiece

The work needs to be prepared correctly before thread cutting starts, in order that the process can be successful. First, the portion of the work

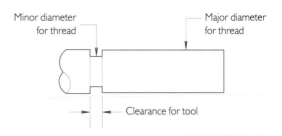

Figure 16.41 Work prepared for screwcutting.

which is to be threaded must be turned to the correct diameter. It is simplest to turn the work to the nominal diameter of the thread, and ignore any need for truncation or rounding of the crests, at this stage.

If possible, a shallow groove (an undercut) should to be turned at the point at which the thread will end, as shown in Figure 16.41. When cutting the thread, the drive needs to be disconnected when the tool reaches the end of the thread (assuming that a thread dial indicator is fitted) or the tool withdrawn, and the groove marks this position clearly. The groove allows for the operator's reaction time in disconnecting the drive or withdrawing the tool, allows a space in which the tool can lie when it is not being driven, and prevents the tool running into the shoulder, if there is one.

If there is a shoulder, the groove allows the mating item to abut properly when it is screwed on, this being otherwise difficult to arrange since the thread cannot be cut to its full form and then come abruptly to an end. The undercut is normally cut to the full depth of the thread so that it acts as a gauge while cutting is in progress; when the thread is cut to the full depth, the tip of the tool just cuts the bottom of the undercut.

Setting up the tool

The thread is cut by setting up the tool in the toolpost and then taking successive cuts by plunging the tool into the work, using the cross-slide feed. Before mounting in the toolpost, the tool should be stoned to a fine finish on the flanks, after which it needs to be set up with its top surface exactly at centre height, otherwise it will not cut a vee shape of the correct form. The gauge is used to establish the tool at the correct, square-on position, as shown in Figure 16.42. The gauge is held against the prepared work and the tool is advanced into one of the notches corresponding with the angle of the vee form for which it has been ground. The toolpost clamp is slackened to allow the tool to be brought into alignment with the notch in the gauge and the clamp is then tightened to hold the tool firmly in position.

Figure 16.42 Aligning the tool to the work.

Taking a trial cut

Now to the final task – the actual cutting of the thread. First of all, a comfortable mandrel speed needs to be selected so that there is time to open the saddle half nuts at the end of the cut, or to withdraw the tool. One of the back gear speeds should be selected, depending on the pitch of the thread to be cut, and the experience of the operator. The more time you have to see what is happening, the better, and something like 30 or 40 rpm will not seem too slow a speed to use.

With the half nuts not engaged with the leadscrew, the drive to the leadscrew should be engaged, and a little oil put on the change-wheels. The lathe can then be started and a check made that the leadscrew is turning in the direction to drive the saddle towards the headstock. If not, the motor should be switched off and the other direction of rotation of the leadscrew selected, using the tumbler reverse.

As a final check, the tool should be brought up to the work, with the mandrel not turning, and the saddle wound down manually towards the headstock to check that there is clearance all round. It is sensible to turn the mandrel by hand, with the tool at its closest approach to the headstock, to check that there is proper clearance for the chuck jaws to pass the saddle and the topslide.

Finally, the saddle can be wound back so that the tool is near the end of the work, but in a position to be advanced on to the diameter which is to be threaded. The tool should now be advanced until a cigarette paper is just gripped between the tool and the work. The tool is now a paper thickness away from the work, and if it is advanced by this amount, it will just touch the surface.

Having advanced the tool by the thickness of the paper, the cross-slide index reading is noted, or it is set to zero, if it is settable. The saddle is then wound back further until the tool is not touching the work, and the cross-slide is then advanced for the first cut. The lathe can then be switched on.

The first cut should be a very shallow one, say, .002in. (.05mm), so the cross-slide should be advanced by this amount and the half nuts closed at some convenient point, using the thread dial indicator, if there is one. Only a shallow cut is taken on this first pass and the operator should be ready to disengage the half nuts, or withdraw the tool by the cross-slide, according to whether a thread dial indicator is fitted, or not.

With the first cut taken, the lathe should be stopped and the actual pitch which has been cut should be checked, either using a thread pitch gauge, or by counting the turns of thread against a rule held on the work, counting over some convenient distance such as ¼in., ½in. or 1in.

Once you are sure that the pitch being cut is the correct one, the thread can be cut progressively to the full depth of the vee form. See Table 16.3. Individual cuts should not be too large, and although it might be acceptable to commence by advancing the tool by .010in. (0.25mm) for the first and second passes, it is better to adopt a standard in-feed of .005in. (0.125mm) since the tool is cutting on both sides and the total length of cut soon becomes significant, especially if the pitch is coarse. It pays to be patient. Clearly, it is no hardship to adopt .005in. as the standard cut for finer pitches, since the total in-feed is, in any case, small.

The cutting process benefits from the use of a coolant, which can be applied using a brush before each cut.

Depth of cut

None of the commonly used thread forms actually uses a full vee form for the thread. Instead, the crests of the thread are reduced in height by rounding or truncation, as described in Chapter 11. When grinding up a screwcutting tool by hand methods in the home workshop, it is ground to the simple vee form, and no account is taken of the need for truncation or other modification of the crests. The tool therefore has a sharp point, without flat or radius.

The full vee form of the thread is cut, and the total in-feed of the tool from the starting position is determined by the height of the full thread form. The adoption of this method permits the outside diameter of the blank on which

an external (male) thread is to be cut to be turned down to the nominal diameter of the thread.

Engineers' reference books do not always give a figure for the full depth of thread, but give instead the depth of the rounded or truncated form. Table 16.3 provides the required information for the 55-degree vee, Whitworth form which is used for the most commonly used threads.

The full depth of thread for the pitch being cut can be read from Table 16.3 and the tool fed in progressively until the in-feed has reached the required value. As this point is approached, the flat crest on the partially cut thread will just disappear as the sharp end to the vee is formed. The approach can therefore be observed by inspecting the thread as cutting progresses, using a watchmaker's eyeglass.

When the thread is theoretically fully cut, as indicated by the cross-slide feedscrew dial reading, it is usual to pass the tool once more along the thread using the same setting, to work out any spring in the tool, and ensure that the required depth has really been cut.

Table 16.3 Full Depth of Thread for 55-degree Vee Form Threads (BSW, BSF, BSB, BSP and Model Engineer Threads)

Bsaic pitch	Full depth of thread	Truncation for flat crests
40 tpi	.024 in.	.0037 in.
32 tpi	.030 in.	.0046 in.
28 tpi	.034 in.	.0053 in.
26 tpi	.037 in.	.0057 in.
20 tpi	.048 in.	.0074 in.
19 tpi	.050 in.	.0078 in.
18 tpi	.053 in.	.0082 in.
16 tpi	.060 in.	.0092 in.
14 tpi	.069 in.	.0106 in.
12 tpi	.080 in.	.0123 in.
11 tpi	.087 in.	.0134 in.
10 tpi	.096 in.	.0148 in.
9 tpi	.107 in.	.0164 in.
8 tpi	.120 in.	.0185 in.

Although fully cut, the thread is not quite correctly formed, since the crest needs to be removed to provide the truncation. From Table 16.3, it can be seen that a pitch as coarse as 8 tpi requires only .0185in. removed form the crest, and the necessary amount can usually be removed from the crests of a male thread by gently applying a fine file to the rotating work, a second-cut file being used for the coarser threads. Table 16.3 shows the truncation for the pitches which are likely to be required. The figure given is the amount to be removed from the crest all round the work, and there should therefore be a reduction in the measured diameter of the work of twice this amount.

After truncation, the work should be brushed thoroughly to remove any swarf or debris from the vee (a toothbrush is ideal) after which it can be tested for fit using a thread pitch gauge and a watchmaker's eyeglass. The gauge should fit comfortably against the flanks of the cut thread, with small gaps evident between the gauge and the cut thread, at the truncated crests and the root of the thread.

Off-setting the topslide

The method of cutting screw threads which is described above suggests that the tool should be fed into the work using the cross-slide feed. The in-feed of the tool is applied directly to the workpiece and the method has the merit that the work is cut directly to the depth required.

The disadvantage of the method is that the tool is cutting on both sides, and as it advances into the work, the length of cut increases, and can become quite significant when cutting a coarse thread. This places more strain on the tool and the conditions become progressively more like parting off, which tends to encourage chatter, as described in Chapter 15, resulting in noise and poor finish.

This can be avoided if the topslide is set over

to take up the position shown in Figure 16.43 and the tool advanced into the cut using the topslide feed. The topslide has been turned to lie roughly parallel to one side of the thread vee form. The tool is still aligned squarely, so that the correct vee form is cut, but the tool is advanced into the work using the topslide feed, rather than the cross-slide. This causes the tool to advance into the cut predominantly on one side, thereby reducing the strain on the tool and helping to avoid chatter.

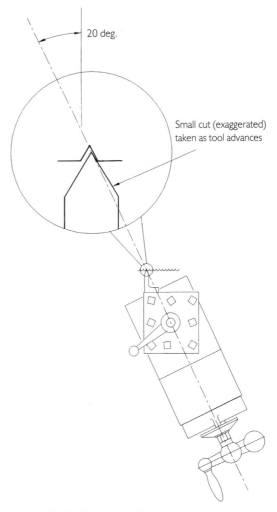

20 deg.

Small cut (exaggerated) taken as tool advances

Figure 16.43 Set-over topslide.

Many references to this technique suggest that the topslide should be angled over to exactly half the included angle of the thread, ensuring that the tool advances down the right-hand flank of the vee, and cuts only on the left. It is preferable, however, if the topslide is set over to a slightly smaller angle, say, 25 degrees. The tool then cuts a small shaving from the right-hand side of the vee and is permitted to perform its task as a form tool, in the way intended.

Since the topslide feed is being used to advance the cut, the tool is not advanced directly into the cut thread. The total in-feed needs to be increased by a small amount to compensate for the fact that the topslide is not pointing directly into the cut, but is approaching obliquely. The required increase is dependent upon the angle to which the topslide has been set, but for an offset of 25 degrees, the increase is very conveniently calculated by multiplying the required depth of the vee by 1.1. It is simply a matter of referring to Table 16.3 and calculating the necessary feed by adding 10 per cent to the listed figure for the thread pitch being cut. For example:

For 26 tpi:

Listed full vee form depth = .037in.

In-feed for 25-deg. off-set
of topslide = .037
+ 10% = .0037
———
= .0407in.

In which case, a total in-feed of .041in. can be used.

If adopting this method of working, it is convenient to set the topslide feed dial reading to a convenient point such as zero and then to advance the tool into the work to pinch a cigarette paper, using the cross-slide. The cross-slide dial can then be set to zero, if it is settable, or the reading noted, which means writing it down.

When cutting starts, the feed is by using the topslide feed. The cross-slide feed can be used for withdrawing the tool at the end of the cut, returning it to its initial, or zero, position when the tool is once more clear of the end of the work. The tool can then be advanced for subsequent cuts using the topslide feed, and the process is somewhat simpler than when using only the cross-slide to advance and retract the tool.

Female threads

Female threads are cut in exactly the same way as a male thread, except that the operation is made more difficult by the lack of visibility, hence the suggestion that it is frequently convenient to purchase a tap to cut the female thread, and then to cut the male thread on the lathe.

The lack of space inside the hole means that a weaker tool will almost certainly be used and there will, therefore, be more spring in the arrangement than is really desirable. Smaller cuts therefore need to be taken, and the task has to be performed a little more delicately. The hole to be threaded should be drilled or bored to the nominal diameter of the thread, **less** twice the full depth of thread, which can be read from Table 16.3. The procedure for cutting the thread is then similar to that for the male counterpart, except that the feed, when cutting, is towards the operator.

After completing the thread, truncation can be achieved by passing a suitable drill down the bore, after which the thread should be cleaned thoroughly and the male thread tried in the female. It is equally possible to enlarge the initial hole to take account of the truncation, and then cut the internal thread until the male thread fits, making the necessary mental adjustment in the depth of thread which needs to be cut, in order not to remove too much material.

The worst situation is when the female thread must be formed in a blind hole, in which case an undercut at the bottom of the hole is absolutely essential to attempt to prevent the tool touching the end. Even so, it is an extremely difficult operation to carry out. One way which is frequently recommended is to fit a handle to the mandrel and wind it round by hand, counting the turns as you go, and stopping when the tool has just cleared the end of the threaded portion. While this may seem laborious, it is an effective way to achieve the desired result, and that, after all is what counts.

The handle might also be used for cutting coarse external threads, since relatively few turns of the mandrel are required, for each inch of length.

Cutting metric threads on an imperial lathe

With an imperial lathe, it is obviously impossible to cut accurate metric screw threads since a 1mm pitch equates with 25.4 tpi and the lathe is not equipped to cut fractions of a tpi. One way to cut approximately correct pitches is to purchase either a 127-tooth or a 63-tooth gear, both of which are close to providing the '254' scale factor which is required. However, these gears only allow approximations to be cut, and it is possible to come quite close to many metric pitches without purchasing more gears.

Although it is somewhat cumbersome, three pairs of gears can be used to approximate a 1mm pitch, as follows:

$$\frac{\text{Driver}}{\text{Driven}} = \frac{65}{30} \times \frac{20}{50} \times \frac{20}{55}$$

With an 8 tpi leadscrew, this ratio cuts 25.38 tpi, or 0.9992mm pitch.

In practical terms, such a pitch is so close to 1mm that, for the normal lengths required, the error is insignificant. (An 0BA nut, is 5.4mm, or 0.2124in. long.) This is true even if the approximation is not so close as the one above, and it is well worthwhile to work out gear ratios which will cut pitches close to those required. As well as the regular metric threads, the much-used BA thread series is actually based on metric pitches, the 0BA, with 1mm pitch, being the largest size.

There is one vital point to bear in mind regarding metric pitches – the thread dial indicator on an imperial lathe is arranged to cater for odd, even and something-and-a-half pitches. If you are cutting 25.38 tpi, the half nuts must never be disengaged once they have been closed for the first cut, otherwise the chances of picking up the thread are a very large number to one against. The technique has to be varied, and the cut is brought to an end on each pass simply by withdrawing the tool and then rotating the mandrel backwards until the tool is once more clear of the work.

17

Buying a lathe

General

The first point for consideration when purchasing a lathe is its size. Within reason, the bigger the lathe, the better, so unless you have a specific need for a small, light-duty lathe, it pays to buy as large a machine as seems reasonable, which is within your capacity to accommodate and pay for. Naturally, compromises must be made, and there are other considerations which apply, but it is preferable to consider a larger, rather than smaller machine, given just a simple choice of two. This argument may be extended to allow comparison between a small, new machine and a larger, secondhand lathe. If the costs are comparable, and the secondhand item is in good condition, this may well be the better choice.

The benefit of a large-size lathe comes not only from the increased rigidity of the individual parts of the machine but also from the increased motor power which is likely to be provided. The rate at which material can be cut from the workpiece is directly dependent on the power which the motor can provide. This may not at first sight appear relevant, since it is

easy to suppose that maximum rate of metal removal is not of prime importance and if the machine won't 'take it' it is only necessary to reduce the cut. However, for various reasons, small cuts are not good practice, are difficult to sustain, and, if there is insufficient motor power, it may be impossible to keep the tool cutting as it should. The usual result is that a cut has to be attempted which is beyond the capacity of the motor to keep the machine turning, and the result is a stall, and turning therefore stops.

What power can be considered adequate? Details of two lathes are shown in Chapter 12. They are of similar size, being described as of 3⅜in. and 3½in. (86mm and 89mm) capacity but the smaller of the two is much more lightly constructed and is sufficiently long to accept work only 12in. (300mm) between centres. This machine is powered by a ¼hp motor. The larger machine, more robust and heavier in all its major parts, is designed to accept work up to 19in. (480mm) between centres. It has a significantly more substantial spindle and mounting for workholding devices, and is powered by a ⅓ hp motor. The more modern versions of this lathe have redesigned spindle bearings and an increased range of spindle speeds. A ½hp motor is recommended for the more recent versions and this, or the ⅓ hp alternative, should be considered as the preferred motor power for this size of machine. Since greater power is required for milling operations than for turning, the larger motor is certainly preferable if the lathe is to be adapted for milling.

Concerning motor power, it is worth noting that current continental practice is to specify motor power in watts rather than horse power. The correspondence between these units is such that 1 hp equates with 746 watts, making the power of the larger lathe described in Chapter 12 equal to 250 watts (or 375 watts for its developed version). It is also worth noting that it is important to distinguish between the power input to the motor and the output that it produces, as the input, electrical power is frequently stated by specifying the maximum current at starting, rather than the running, input power. In any event, the motor is not 100 per cent efficient, which means that it consumes more power than the output which it produces. From its labelling, a ⅓ hp motor might, therefore, appear to be equivalent to 600 watts (VA, strictly) which is misleading, to say the least, if you are not used to dealing with these matters.

Motor power and machine size naturally go together. It is bad practice to overload any machine habitually. Its life will be seriously shortened by overloading and it may well not be capable of producing good results when working at its maximum capacity, both in consideration of its physical capability and its motor power. It always pays to use a lathe (or any machine tool) within its capacity, which means that its likely use must be borne in mind when making a purchase. The general tendency for modellers to want a 'bigger and better' model 'next time' should also be considered, and allowance for this made at the outset.

Type of lathe

The lathes shown in Chapter 12 may be described as old-fashioned, belt-driven types. Indeed, the smaller of the two was originally driven by a flat, leather belt, rather than by vee belt. Both have a restricted range of speeds, provided by a 3-step pulley and the back gear. Due to the design of the spindle bearings, neither has a particularly high maximum speed and this may be considered as a slight disadvantage when considering the purchase of such a machine. In practice, the belt drive (and back gear) actually provides a simple and practical arrangement, its only real disadvantage being the time taken to change speed – a not very frequent operation – and the fact that one must get inside the belt guard to effect the change, where it is usually rather dirty. The lack of a fast top speed is a disadvantage and later versions of the larger lathe do have a considerably increased capability in this respect.

For the most part, current production lathes are designed with an enclosed, geared drive for the spindle, with speeds being selected by the use of external levers. This type of speed change selection is also sometimes implemented by utilising a multi-belt drive, one belt being engaged for the drive by positioning the external levers. In both cases, it is naturally unnecessary to gain access to the drive when making a speed change so this is more readily accomplished, and is cleaner. For these machines, the concept of a back gear is frequently not relevant as the lathe is considered to have 12 (or whatever) selectable speeds which are directly accessible and a 2-stage selection process may not always be required. Lathes of recent design also normally have the built-in capability to cut threads on the workpiece and are therefore described as screwcutting lathes.

A screwcutting lathe is to be preferred, but in seeking a machine on the secondhand market, it is possible that lathes without this capability will be encountered. In order to reduce costs, it was at one time customary for plain lathes to be produced, specifically not designed for screw-cutting. Some of these machines are not capable of adaptation for this purpose and might, therefore, be less valuable than an equivalent screwcutting type. Such a machine is less versatile since the screwcutting facility is useful also for operating the machine in self-acting mode in which a cut is continuously (and evenly) applied to the work and this can be valuable when working on long shafts or spindles.

The screwcutting operation requires the tool to be driven along the axis of the machine bed in a controlled manner. Recently, the capability to provide a power feed across the machine has been introduced into smaller lathes which are of interest to the amateur worker. This is described as power cross feed. This is a feature without which one can manage quite well, but it is extremely useful when working on the face of large-diameter work (facing) since the achievement of a fine and even manual feed can be somewhat difficult.

Adapting the lathe

The adaptation of the lathe to non-turning operations is also an important point to consider at the outset. With suitable attachments, the lathe may be adapted for milling, boring and dividing operations, and it is valuable if the manufacturer has allowed for this in the design by making a range of attachments available to suit the machine. A lathe from such a range commands a higher price on the secondhand market, due to its better potential, but easy adaptation, particularly for milling, is vital unless a separate milling machine is to be purchased.

Among more recently designed machines, are several which may be adapted for milling by

mounting a vertical column to the rear centre of the bed. The column carries a motor-driven spindle assembly, not unlike the lathe spindle itself, to which cutters may be mounted in order to perform milling operations on work which is clamped or held to the cross-slide. The milling spindle may be raised and lowered in a controlled manner, providing the facility to machine the work in three dimensions, using the lathe's longitudinal and cross feeds to supplement the elevating head.

The arrangement confers on the lathe the facilities normally provided by a vertical milling machine, usually at reduced cost, but with the possible disadvantage that the arrangement does not provide such a large table as is provided by a vertical mill of equivalent size. Some designs provide for an extension of the capability of the lathe-mounted vertical head by making available a separate compound table and base unit to which the column and head may be mounted, allowing conversion to a full vertical mill.

Alternative forms of this type of bolt-on milling head are available for some lathes, which mount on the end of the bed and are suitable for use if the lathe is not specifically adapted by the manufacturer to accept this type of attachment. Milling heads of these types are also normally capable of acting as a drill by the fitting of a drill chuck into the spindle. They can provide all of the facilities which are required, but with the disadvantage that a well-known law operates to ensure that the machine is never set up to perform the operation which you require.

Adaptation of the lathe for other than turning is described in Chapter 16, although the descriptions given are intended only to be introductory and are confined to operations in which cutters are mounted to the lathe spindle itself rather than by use of supplementary milling attachments.

The secondhand lathe

Naturally, a secondhand lathe costs less than the equivalent new machine and is also more likely to be 'complete'. That is to say, it will most likely be sold as a going concern, especially if purchased from an amateur source, and can be expected to have some essential accessories which allow it to be used without further major expense. The same may not be true of a new machine, at least at the advertised price, and the acquisition of two chucks to fit the lathe spindle, a drill chuck for use at the other end of the machine and some basic tools, can well add £200 to the basic price.

However, there are potential difficulties in buying secondhand and if you are unfamiliar with lathes, at least read Chapter 12 before embarking on a search for a machine, and collect some brochures and price lists so that you have some idea of new prices. Try also to become familiar with the sort of prices asked for secondhand machines of the general type and size which is of interest.

If you are a member of a model engineering society, or have joined a local evening class, it should be possible to arrange for someone to give you a conducted tour, or to demonstrate some of the basic processes on the lathe. It may even be possible for you to make some simple item for yourself. This should at least give you confidence that you understand the basic principles. If possible, when going to look at a secondhand lathe take a knowledgeable model engineer with you. Some experience of lathe operation can be very valuable when assessing the quality of a used machine, and if you do buy, you may need a hand to get it into the car, anyway!

One further point about the secondhand lathe is also worth bearing in mind. If the machine, though old, is the product of a manufacturer who is still in business, it is possible that spares are readily available should they be

required. If the lathe is still in production, albeit in modernised or updated form, it is also possible that accessories may still be obtainable which will fit directly on to the machine without modification, thus making it relatively easy to extend its capability, if this is required. It may be necessary to pay a little extra to purchase a machine which meets these requirements, but prices on the amateur market are so variable, depending as they do on the perceived value of the item to the seller, that there may indeed be no penalty in determining to seek such a machine for preference.

Assessing the condition of a secondhand lathe

General appraisal

Before going to see a secondhand lathe, try to contact the seller by telephone to determine whether it is in a usable condition and whether the seller has been in the habit of operating the machine. If the lathe has been idle for some time, it is likely to be dirty, and may even be rusty (especially its accessories) given the usual conditions in which such unused items are stored, and the owner may not even be sure what accessories are with the machine. A useful supplementary question is to enquire whether the machine can actually be run under power. If it cannot, this rather mitigates against making a proper judgment of its suitability, and probably indicates that the machine has not been in recent use. Make a judgment accordingly, being prepared for disappointment or determining that the cost must be sufficiently low for you to consider the purchase.

Once the lathe is available for inspection, the first step is naturally an all-over visual examination which should reveal if it is a going concern or not. If the sale is being arranged by someone with knowledge of these things, the machine should have been cleaned up – but beware the quick paint job, it is not necessary that the lathe should look good – and should be displayed with the accessories with which it is being sold.

If the seller is a reputable dealer, the above conditions will almost certainly be met, and the machine is likely to be in a usable condition, but no information is likely to be available concerning the use to which the machine has been put. If the sale is being arranged privately, the interests of the seller do provide a guide to the use of the machine and it is useful to know whether he has a current project on the go, and also the reason for the sale.

The bed

For the initial approach, ignore the accessories and instead make a careful visual examination of the bed. Does the machine appear to have been abused? Look first at the tailstock end of the bed. This part of the machine seldom gets much use during turning and the surfaces on which the carriage bears should be in the best condition at this end unless the same surfaces are also used to support the tailstock.

The working surfaces of the bed should not show signs of damage along the length, but at the headstock end, where most turning takes place, there will ordinarily be wear which may be discernible by eye. If the surfaces are badly scratched or scored, this may suggest that the machine has been habitually used in a dirty condition, or has not been adequately cleaned or lubricated during its life. In this case, there will also be scoring (wear) on the inside surfaces but this is not a major problem provided that the scoring is not excessive, since the working parts can be stripped and cleaned and once relubricated and adjusted will most likely be satisfactory.

Of more importance is evidence of damage at the headstock end of the bed. It is inevitable that one occasionally fails to realise that a large, awkwardly shaped item has been mounted to the lathe mandrel in such a manner that it cannot rotate completely without hitting the bed. It also happens that the chuck jaws are extended too far, with the same result. The careful operator will normally rotate the spindle by hand to check clearances before starting the drive motor, but it is amazing how many operators appear happy to switch on and await results! The result is that the lathe bed suffers.

Damage to the bed also occurs if the operator is in the habit of using a hacksaw to part work from a bar held in the chuck. This again seems to be a regular occurrence in some workshops, either to avoid the necessity to put the parting tool in the tool holder, or simply to avoid the difficulties which parting off sometimes causes. The result of sawing off in this fashion is frequently that the bed suffers and the evidence remains.

The front end of the carriage may also suffer in the same ways and the visual inspection should therefore continue with this item.

The chucks

A further indication of the general condition of the machine is provided by an examination of the accessories with which the lathe is being sold, in particular, the chucks. Their general condition will be apparent from a visual examination and the next check is to see that the jaws match the chuck body and have been inserted into the correct positions. If so, things look promising. The jaws should be withdrawn a little to allow the slideways to be seen and, if the chuck is of the type in which a scroll drives the jaws in and out, the scroll can be examined, and also the teeth on the jaws. These should be clean and undamaged. Wear on the scroll and the teeth of the jaws may give a guide to the likely age of the chuck, or the amount of work it has done.

The mandrel and its bearings

Supposing that the three-jaw chuck is not mounted on the lathe spindle, check that the thread and register are clean, also the thread in the chuck backplate or body, and screw the chuck fully onto the mandrel. The register should guide the chuck onto the mandrel and locate it in alignment with the axis of rotation, and an estimate of the fit between the chuck and the register can be obtained while screwing the chuck into position. Mount a stout bar in the chuck, with 6in. (150mm) or so protruding, and check for sloppiness in the mandrel bearings by pulling on the bar, both along and across the lathe axis. Longitudinal play is not in itself serious since it can usually be adjusted, but if it is discernible, it does indicate that the user has not been in the habit of keeping things as they should be, and this may influence your decision concerning the purchase.

Any play which is noticeable at right angles to the spindle axis is far more serious. This indicates that the spindle and/or its bearings are worn and major work may be required to restore them to a satisfactory working condition. Spindle bearing designs differ widely, so it is difficult to be specific, but there may not be provision for making any adjustments and poor bearings ought really to condemn the machine unless it appears that spares may be available or there are other significant reasons for its purchase.

At this point, the machine can be run for the first time, but with the three-jaw chuck removed from the spindle and replaced by the faceplate. The purpose of this test is merely to see whether the spindle nose and register run truly and this can be judged by eye if the faceplate is observed as the machine starts up and runs down. Both the outside diameter and the face should appear to run truly.

The faceplate is likely to have been less well used than the chuck and its mounting thread and register should be in much better condition. During its fitting and removal, its fit to the spindle thread and register will provide a guide to the condition of these items and since the accuracy of mounting is determined by the register, it is bad news if this seems to be in poor condition and is likely to mean that the spindle will have to be remade.

A check for wear in the chuck bodies can be made, once it appears that the spindle is running truly, by mounting the chucks on the spindle and checking that the body runs truly, again just observing by eye with the spindle rotating under power. The bodies should run reasonably truly if correctly mounted to the spindle. Chuck mounting may be by use of a backplate to which the chuck is bolted and the backplate should have been made true to the lathe axis by turning the register for the chuck body while the backplate was mounted on the mandrel. If the chuck is mounted to a backplate, but the body does not run truly, observe the backplate itself. It may require restoring to truth by turning, and, provided that there is sufficient material in the plate, this is not a difficult operation to perform. However, if you do find that this is the case, it once again shows that the machine has been in the hands of a careless operator and this may again influence your decision regarding the purchase.

If the chuck is directly mounted to the spindle (the usual term if the chuck body is itself threaded) then rectification is more difficult.

Mandrel and tailstock tapers

If the bearings and the fitting of the chucks and faceplate appear satisfactory, the tapers inside the mandrel and the tailstock sleeve should now be examined. Slow tapers such as the Morse series are designed to be self-holding and are ordinarily simply jammed into the socket, where they will hold in providing that excessive rotational force (torque) is not applied to them. The sockets do become worn and damaged, harm being done to them if a slip does occur, particularly if the taper and its socket are not clean, since dirt both prevents the tapers fitting together correctly and guarantees that scoring occurs if relative rotation is allowed to happen.

Clean both of the internal tapers carefully with a soft rag or paper tissue and check for ridges and burrs by inserting a finger as far as it will go into the taper. If all seems well, insert a clean (and burr-free) centre very gently into the mandrel and check that there is no discernible shake or looseness of fit between the two parts. Run the lathe under power and check that the taper appears to run truly. If the taper is not in good condition, remedial work will ultimately be required. If the mandrel is to accept standard tapers, this may mean that the end of the mandrel (or even the complete item) will have to be remade, or the spindle replaced. An alternative, which may be possible, is to bore the mandrel out very slightly, but this may mean that standard tapers will no longer be usable, and it may be necessary to turn up some specials to suit the modified spindle. This is not a difficult operation, but it is a nuisance if non-standard fittings are required.

The tailstock taper is normally used much more than the mandrel taper, and is also more likely to have suffered damage due to slippage of accessories. It must be examined with equal care therefore, but can be expected to show more signs of wear, and is another indication of the amount of use to which the machine has been subjected.

The tailstock

While at the tailstock end of the machine, it is as well to examine this assembly generally. The first point to note is whether the assembly is in two parts, allowing the upper part, containing the sleeve and handwheel, to be adjusted across the bed of the machine. This feature is always described as being of value for the turning of tapers, and so it is, but its real virtue lies in the fact that the tailstock centre can be brought into precise alignment with the spindle axis, an operation which is vital to the correct drilling of axial holes in the work, and indeed to the turning of parallel work mounted between centres. This feature is a very good plus for the machine.

With a centre pushed lightly into the taper, some discernible looseness is normal, certainly in the tailstock sleeve, in almost any second-hand lathe, depending of course on its age, and the amount of use it has had. This is certainly true of the larger of the two lathes described in Chapter 12, although it is dependent upon how firmly the centre is actually pressed into the sleeve. This lathe is some 50 years old, and was a one-owner model when purchased second-hand, having had about average amateur use, as far as can be judged.

What is also important in the tailstock is the fit of the movable sleeve in the tailstock bore. This should move freely in and out, but without significant slop or shake, under control from the handwheel. The sleeve is prevented from rotating by a peg or key which engages in a longitudinal groove in the sleeve, and the free play between the sides of the groove and the retainer should be tested, both with the sleeve retracted and with it extended toward the end of its travel.

The saddle

The saddle straddles the lathe bed and carries the cross-slide and topslide. A handle or handwheel is provided which operates through gearing to move the saddle along the bed. In use, wear occurs on the sliding faces on saddle and bed, predominantly at the headstock end of the machine, since that is where most use of the saddle takes place. Adjustments are provided to allow the clearances to be adjusted to their working values and these should be maintained by periodic adjustment in service. Increased wear at one end of the bed means, however, that movement may be stiff at the tailstock end when the clearances are correctly adjusted at the other, and a check should be made on the movement along the whole length of the bed.

As with all of the lathe's motions, movement should be smooth and with a nice, even feel throughout the traverse. Any adjustments should be set up to give a firm feel to the motion, that is, not sloppy but definitely not stiff. As you can gather by the descriptions used, judgment is somewhat subjective, and what is stiff to one person may be just right to another.

It is worth noting that on some small lathes the saddle motion is implemented very simply by fitting a rack to the bed and gearing a handle directly to the rack through a small spur gear. This produces a reversed motion to the handle, so that turning it anti-clockwise produces motion away from the headstock. In practice, it takes only a little time to get used to this reversal, but if you don't believe that you can live with it easily, then it becomes a feature which may influence your decision.

Having checked the freedom of motion of the saddle along the bed, the clearances between the two must be assessed. There is usually a clamping or locking arrangement for the cross-slide, normally two or more screws which can be tightened to lock the slide to the saddle. If these are provided, tighten them (gently!) to lock the cross-slide roughly at the mid point of its travel.

It is now possible to check the wear and clearances in the saddle by clamping a rectangular bar in the toolpost, using it to push and pull across the bed, and rotating both clockwise and anti-clockwise, to test if there is significant free play between the saddle and the bed. This test must be performed at both ends of the bed as a means of further checking for uneven wear. If there is just general sloppiness, it might be necessary to attempt to make an adjustment as a further means of discovering the extent of any differential wear, but the owner may not be too happy at this suggestion. In any event, it can be very time consuming and is not very satisfactorily accomplished if the machine is dirty and you are under pressure to make a decision.

If the cross-slide cannot readily be locked to the saddle, the saddle-to-bed clearances should be examined only after considering the cross-slide feedscrew, as described below.

Cross-slide and topslide

On most lathes intended for the amateur market, adaptation to processes other than turning is normally envisaged by the designer. Consequently, the cross-slide is provided with tee-shaped slots (tee slots) so that an accessory or a workpiece may be fixed down to it. Correctly, the fastenings provided for the tee slots should be specially machined to fit the slot snugly and to distribute the load imposed on the underside of the slot over as wide an area as possible. In practice, it is common for hexagon-headed bolts to be machined down in the head so that they will enter the slot and these inappropriate fastenings are used directly in the slot. The imposed load is, however, much more concentrated when such a makeshift fastening is used and the underside of the slot becomes damaged. Habitual use of these shortcuts is sure to have left its evidence and a careful examination of the slots should be made, particularly under the topslide, if its fixings make use of the tee slots.

The next test should be of the cross-slide traverse (don't forget to unlock it first!). The feed should produce a smooth motion of the cross-slide throughout its length, without any significant difference in the feel of the action during a complete traverse.

During this test, a check for free play, or backlash should be performed. This can be detected if the feed handle is rocked backwards and forwards, when any lost motion in the cross-slide becomes evident. It is worthwhile checking at both extremes of cross-slide travel since most turning is performed at small diameters and this may lead to differential wear once again.

Lost motion may be due to two causes – either wear between the feedscrew and its nut, or incorrect setting of the feedscrew endfloat in the cross-slide endplate. This is adjustable, and a careful operator will set the clearance periodically. If excessive play is discernible, it is worth enquiring how the endfloat adjustment is made, with a view to attempting to correct the problem. It is usually a simple procedure to follow and if the play cannot be reduced to an acceptable level, it may indicate that there is significant wear between the feedscrew and its nut.

During this check, it is appropriate to test for free play of the cross-slide on its guides. Clearances here are adjustable and should be maintained at reasonable values. Excess play is discernible by pushing and pulling on the cross-slide along the axis of the bed, but the feel of the traverse is usually the clue to whether this check is necessary. If no change in stiffness is detected during the complete traverse, and the motion is firm, but not stiff, without significant difference in the backlash at the feed handle,

things are probably OK. Sloppy motion, or change in discernible backlash at the handle, or indeed, excessive backlash, all indicate the desirability of checking the clearance between the cross-slide and its guides.

It is usually possible to examine the cross-slide feedscrew from below with the slide itself extended, and may give a guide to the wear which the machine has suffered. It is also significant if there is no lubricant in evidence.

The topslide feed arrangement must naturally also be examined to determine if the same faults which can afflict the cross-slide feed are evident here and the same basic procedure should be followed.

If it was not possible to lock the cross-slide to the saddle to assess wear between the saddle and the bed, a bar clamped to the toolpost can now be used to assess these clearances, bearing in mind the free play (or otherwise) which has been detected in the feedscrews, which may mask other problems. If the cross-slide feedscrew is unworn, and the endfloat well adjusted, the test should reveal any play between saddle and bed. If free play has been detected between the cross-slide feedscrew and its nut, or endfloat of the cross-slide on its feedscrew, observation will have to be relied on to reveal where the problem lies.

The leadscrew and its drive

The leadscrew is a vitally important part of the machine. Not only is it used to drive the saddle along when screwcutting, or using the lathe in its self-acting mode, but it is also used in many cases for locking the saddle to the bed. Its condition, and the condition of its bearings and supports, is therefore important in determining the quality of work the lathe is capable of performing.

Wear naturally occurs in the leadscrew during use and is most prominent at the headstock end where most turning is performed. The ends of the screw receive little use and can be examined for use as a comparison when assessing the wear, using a watchmaker's glass. Wipe a short section of the screw where it is unworn, say, at the tailstock end, to remove any surplus oil and remove any major debris which is evident, and then examine the screw under a glass. Repeat the procedure at the headstock end, in the region in which the saddle is likely to have been most used, and make an assessment of the wear.

Scratching is likely to be in evidence, caused by the occasional trapping of swarf between the half nuts and the screw. Some rounding (wear) of the crests of the thread will also be evident, but there should be no major loss of material from the thread flanks. If the thread is of an essentially square form, which is normal, the crests and troughs should have a 1:1 ratio, at least as discerned by eye, and this provides a guide to the general appearance to be expected. If there is significant wear on the flanks of the thread, it will most likely not extend to the full depth of thread and should be visible as a pronounced ridge on the flank.

If the leadscrew is provided with a handle or handwheel, rotate it several times without the saddle locked to it and the drive from the mandrel disengaged. Check for freedom of rotation and note whether the screw is bent or otherwise appears not to run correctly. If no handwheel is provided, it will be necessary to engage the drive in order to carry out this check and the seller should be able to arrange this, either through the changewheel cluster or the quick-change gearbox, if one is fitted.

If the general condition of the leadscrew appears to be satisfactory, the fit of the half nuts to it must next be tested. The half nuts are normally engaged by operating a lever mounted to the saddle apron. To engage the half nuts to the leadscrew, it is naturally necessary to position the saddle correctly to allow full engagement. Operate the half nut lever gently while rocking

the saddle backwards and forwards along the bed for a short distance until the half nuts engage the screw. Now try the saddle traverse to check for any free movement of the saddle. If free motion is detected, this may be due to incorrect adjustment of the leadscrew itself in its bearings at the ends of the bed, or may be caused by worn half nuts or leadscrew. End-float of the leadscrew should readily be discernible, if it is present, and is not usually a significant problem since it is easily adjusted in most cases. Since the leadscrew itself has already been examined for wear, it should be possible to determine whether there is likely to be a problem with the half nuts, although it is conceivable that they could be significantly worn without there being much evidence of wear in the screw.

Overall assessment

By the time you have carried out the sort of appraisal described, you may well feel that everything on the machine is worn and there is therefore no point in proceeding with the purchase. Do not be disheartened. If wear has been detected, as distinct from general lack of adjustment and care, it is probably wise not to proceed with the purchase. It is possible to improve a worn lathe, either by seeking a professional service, where this is available, or by employing traditional hand skills, such as filing and scraping, to remove the high spots and restore the surfaces to parallelism and flatness. However, professional services are expensive, and as a beginner you may not possess the necessary manual skills to attempt rectification unaided.

The above comments naturally apply to the bed and to the sliding surfaces which guide and steady the various slides on the machine. These are the most difficult to deal with from the point of view of rectification, should wear be present, but other worn components are not that difficult to rectify. Some of these have been referred to above. They include the fitting of new backplates to chucks, which does require careful and accurate work, and also needs the use of the screwcutting facility of the machine, but the making of a new chuck backplate is in reality not all that difficult to carry out. Castings are available especially for use as chuck backplates, and provided that they can be mounted somehow to the machine, they can be turned on the back, bored and screwcut to fit the mandrel and its register, and then mounted to the mandrel and turned truly to suit the chuck.

Once screwcutting has been mastered, it is surprisingly straightforward to make a new mandrel for a lathe. The basic turning of the outside of the spindle can be carried out between centres, including the cutting of the thread which is usually required at the extreme left-hand end to take the screwed collar (or whatever) that allows the endfloat of the mandrel to be adjusted. It can then be mounted into the machine for drilling through and for screwcutting and boring the front end, or nose.

Leadscrews and half nuts are more difficult to rectify, although the nuts should not be impossible if two blocks of material can be dowelled together and held in the four-jaw for the screwcutting operation. But threads which have a square form are that bit more difficult to screwcut.

The leadscrew is difficult only because its length prohibits its manufacture on the lathe for which it is intended, but if access to a larger machine is possible, even this item can be remade. By the same argument, feedscrews and their nuts might also be replaced, but the cutting of coarse-pitch threads on thin shafts is relatively difficult, and this should be borne in mind. However, a start has to be made somewhere, and it is arguable that this might as well be in refurbishing a machine as anything else, especially if you are reducing the cost of enter-ing the hobby. In any event, there is great satisfaction in using a machine which you have rebuilt to a usable condition, but do bear in mind that there are limits, especially for a beginner.

Other repair work on a secondhand machine is likely to be straightforward and well within the capability of a mechanically minded enthusiast. The machine will require stripping, cleaning and lubricating and subsequent reassembly. During this process, the internal condition of the parts will become evident and any rectification can then be put in hand, or at least planned for the future. The adjustments of the slides can be carried out once the machine is clean and properly lubricated and the best compromises adopted for the clearances, according to the wear which is present.

After this stripping, cleaning and adjustment, the lathe should be in a perfectly usable condition, even though it is elderly, provided that it has not been abused and has been kept clean, and well lubricated. It naturally helps if the machine has not been used in an intensive manner and some knowledge of its history is a guide to its likely condition. If it has been in amateur hands all, or most, of its life, and has been in recent use by a modeller whose work you can see, and who appears to have kept the machine up to scratch and used it carefully, then its age probably does not matter.

Essential accessories

General

At one time, it was common practice to price new lathes in as basic a form as possible, in order to reduce the advertised selling price. The argument was, and perhaps still is in some instances, that the items without which the lathe was priced were subject to a certain choice on the part of the purchaser and might therefore vary to suit the individual. However, items not priced frequently included the drive motor and the chucks, without which the lathe is useless (certainly in the case of the drive motor) so the process was self-defeating and has now more or less been dropped.

What accessories are useful or essential? The lathe, whether new or secondhand should at least be supplied with the following:

- Drive motor (!)
- Faceplate
- Catchplate
- Centres for the mandrel and tailstock bores
- Cross-slide
- Topslide
- Toolpost (or clamping arrangement for tools)
- A set of changewheels (if the leadscrew drive is by this means) and the relevant quadrants and studs.

It is possible that a secondhand machine will not be provided with all of these items, due to them becoming damaged or lost. Provided that the major parts of the machine are present, the faceplate and catchplate can be regarded as unimportant. Unmachined castings are available for faceplates, which can be bored and screwcut to suit the mandrel nose and afterwards faced off, although holding the large casting for the initial operations without having something similar available may tax your ingenuity. A catchplate is certainly not essential in itself since a faceplate may be used in its stead, but in any event, a commercial casting intended for use as a chuck backplate can readily be adapted once screwcutting has been mastered, as noted above. If a catchplate is available, a faceplate casting might readily be secured to it for the initial turning and screwcutting.

Naturally, a catchplate (or chuck backplate) casting, being smaller than a faceplate can probably be held in a chuck for its basic turning, and will afterwards allow a faceplate casting to be held for its initial machining. For a secondhand lathe which lacks faceplate and catchplate, some sort of chuck would appear to be essential. If this is in poor condition, it matters not (provided that the price of the machine is right) since it is required only for the first-stage machining of chuck backplates etc. and can ultimately be replaced.

Workholding chucks

The essential chuck is the four-jaw. Without it, turning operations on rectangular workpieces cannot be performed. The four-jaw is available as a self-centring chuck but this type is not very common and references to this chuck may be taken to refer to the independent type in which the jaws are individually adjustable.

If only a four-jaw chuck is available it is a nuisance to be forced to centralise circular work, but at least the turning of circular bars can be accomplished, whereas a three-jaw chuck cannot hold rectangular or irregularly shaped work.

An accurate three-jaw is naturally a great asset, but, as noted in Chapter 13, if concentricity is required, either all diameters must be turned at one setting i.e. without removing the work from the chuck, or the four-jaw must be used to set up one diameter on the work to run accurately to allow the production of further diameters which are concentric with it. Three-jaw chucks are not guaranteed to hold circular work without run out, so cannot be used in instances where concentricity with an existing diameter is required.

The three-jaw is valuable as a ready means to hold circular bar reasonably accurately and without the need to carry out any setting-up before commencing turning. It is, therefore, the first choice when circular bars are to be machined and is essential for holding hexagonal bars. However, if you have to buy chucks, and can only afford one immediately, consider the four-jaw first unless you see an immediate need for much use of hexagonal bar.

Tailstock chuck

With chucks and faceplate available, the lathe may appear to be fully equipped, but before much work can be done, a drill chuck fitted with an arbor to suit the tailstock bore is required. Like the work-holding chucks for the mandrel, its size is dependent on the size of the machine. Although a large chuck may appear to be desirable, the machine should not be overstretched, and it is likely that the motor will limit the size of drill which can be used. Bear in mind, also, that a large chuck may not be capable of holding very small drills, which will also be required from time to time, and two chucks may be needed in due course.

Since the stock of available drills will also be dictated by the capacity of the drilling machine chuck, there is a good case for utilising roughly the same size of chuck for both machines. A drilling machine intended for the amateur market might have a ½in. (12.5mm) capacity chuck, and this size, or perhaps ⅝in. (16mm), is appropriate to the typical lathe described in Chapter 12.

Desirable accessories

Equipped with the mandrel-mounted work-holding devices, centres, tailstock drill chuck and some suitable tools, the lathe is equipped to carry out all of the essential turning operations. As experience is gained, additional accessories may appear to be desirable, according to the general type of work which is undertaken. Some of these are suggested in the description of the lathe, and its use, in Chapters 12 and 15.

The first practical problem which becomes apparent during turning is the fairly frequent need to change tools. It does depend on the workpiece, of course, but if there is much material to remove, roughing and finishing tools may well be used, even on a straightforward job, to maximise metal removal and yet achieve the required finish. If the work is a repetitive task, such as the production of boiler stays, for example, several tools may be required. Without the benefit of an expensive, quick-change tool holder, which allows both precise setting to height and rapid replacement of a tool and its holder, individual tools will be set to height with loose packing. Changing tools becomes an extremely time-consuming task and the pleasure derived from making things soon palls when more than 100 stays are required!

The alternative to the quick-change tool holder is a multi-tool turret for the toolpost. Commercial versions of these are usually square in plan and designed to mount four tools but designs have been published for types with a triangular planform, which naturally take three tools. An indexing, four-tool turret is shown in Chapter 15. This is machined from the solid, but it is perfectly feasible to adopt a built-up construction in which top and bottom plates are secured by screws to a central block. This form of construction avoids the need to carry out heavy milling operations and the turret can be made relatively straightforwardly.

Like many accessories, a multi-tool turret may not be necessary much of the time, but when it is, the only substitute is a time-consuming tool change for each operation which is required.

Threading of turned work is a frequent requirement and this is most readily carried out from the tailstock. A tailstock-mounted die holder is illustrated in Chapter 15, together with an adaptation of it for holding taps. Both accessories are extremely useful, and mounting taps and dies into the tailstock will be found to be far superior to handheld methods.

If the lathe's cross-slide is sufficiently long to accept a second toolpost, at the rear, tools may be mounted in it, but in the inverted condition. This is a particularly beneficial way in which to use parting-off tools since it provides more support for the tool and helps to avoid problems associated with this operation. The real benefit of the arrangement derives from the permanent availability of the parting-off tool, but this can only be arranged if the cross-slide is sufficiently long to allow the rear tool to be permanently mounted without unduly restricting the gap between the tools. Thus, it is the longer cross-slide which is desirable since it is this which allows both toolposts to be mounted at the same time.

Non-essential accessories

The quick-change, screwcutting gearbox is perhaps the one non-essential accessory which most readily comes to mind. The price differential between new lathes with and without this facility is perhaps 25 per cent of the cost of the basic lathe not fitted with the gearbox. This represents a significant number of other items and/or materials. In the amateur's workshop, most turning is performed by manual manipulation of the feedscrews. In one way, this is part of the fun. The method of working also suits the one-offs which are the modeller's normal stock-in-trade and unless much very diverse screwcutting is envisaged, the quick-change facility can well be dispensed with.

This is not to say that the self act is not useful. It is very helpful in providing a steady feed during normal turning and is particularly useful when long shafts need to be turned (or for any surfacing of long work) since the known steady feed allows the tool to be flattened at the point, as described in Chapter 14, so that successive cuts overlap. Thus, automatic fine feed can be useful, but even without a quick-change gearbox, the changewheels may be set up in a suitable condition, when not screwcutting, perhaps to give .010in. (0.25mm) per rev or 100 tpi.

Other accessories which 'come in' for only occasional use are the steadies. These are arranged as two basic types – the fixed steady and the travelling steady. Both are designed to provide support for the workpiece, but in two distinctly different ways. A fixed steady attaches to the bed of the machine and is designed to support a long bar which has been set up in the chuck so that the outer end (towards the tailstock) runs truly. The fixed steady is useful when long material which is too large to pass through the mandrel bore needs to be turned.

The above method of using the fixed steady to provide support, supposes that the bar is reasonably substantial. If it is long and thin, turning its unsupported centre may not be possible due to its inability to resist the cutting force and a cut cannot be taken unless the centre is supported. Such support is provided by a travelling steady. This attaches to the cross-slide and has two fingers which are adjusted to just touch the work, one above and one behind the workpiece, immediately opposite the tool. Properly lubricated, the fingers will run on the work and provide support just at the point of cutting, travelling along the workpiece with the tool as the saddle and cross-slide are traversed along the bed.

Work on long, thin shafts has not featured in my workshop, the only work on small-diameter bars having been confined to very short lengths. Support may be desirable, even on shorter work, but it is possible to turn small diameters by dealing with not more than ¼in. (6mm) or so at a time, drawing the bar progressively out of the chuck or collet and turning short lengths successively.

It is possible to support thin work by hand, provided that care is taken, simply by pressing on the bar with a piece of wood. It is, of course, vital that the tool is kept sharp and set correctly at centre height, especially for smaller work, which otherwise just bends up out of the way if the cut cannot proceed cleanly.

Setting up the lathe

General

Unless the lathe is properly installed and set up it cannot produce the level of accurate work for which it is designed. In simple terms, this means that the lathe needs to be held securely to whatever bench or stand is provided for its accommodation. This does not mean, however, that a few bolts passing through the mounting feet into any old bench will be satisfactory.

When two items are screwed together, their mating faces do not normally match perfectly, and there is consequently some distortion of at least one member of the assembly. Although the lathe bed may appear to be very substantial, it is usually long in relation to its width and can easily be distorted if bolted to an 'un-level' or out-of-square bench, the usual problem being that the bed becomes twisted, thus giving to the saddle a distorted motion as it moves along the bed. This naturally means that the tool does not move along a line parallel to an extension of the mandrel axis. Parallel work cannot be produced, and this suggests one way in which the setting-up may be checked.

Whatever the mounting platform for the lathe, its installation must start with a firm floor. In this respect, it should be remembered that a suspended wooden floor must be expected to deflect under load, and may be expected to settle under the loads imposed by a heavy lathe and its benching or mounting cabinet. This means that those with first-floor workshops should periodically check the instal-

lation. If working on a concrete floor, this will provide the necessary rigidity, but unless it is smoothly screeded, it may present problems due to its unevenness.

The immediate mounting for the lathe may be a wooden or metal bench which incorporates some inherent flexibility that allows it to stand firmly, even on an un-level and uneven floor. Thus, the top surface of the bench may well be twisted and if the lathe is bolted firmly to the distorted bench, distortion of the bed almost inevitably results. The lathe should be mounted to the bench with levelling (jacking) screws at each corner. If the bench has a wooden top, the screws must be provided with large washers to spread the load, and should be individually adjustable and lockable. Effectively there should be four studs fitted into the bench top, with large washers to spread the load, and provided with nuts and washers both above and below each mounting hole on the lathe bed.

If the lathe is mounted on an all-welded, metal cabinet stand, this may well be sufficiently rigid to rock when stood on an uneven floor, when in its unloaded condition i.e. with the lathe not mounted on it. In these cases, all rocking due to the uneven floor should be removed by placing metal shims beneath the appropriate feet until the cabinet stands firmly on all four corners.

Provided that jacking screws are fitted, it should not matter if the bench or cabinet is somewhat distorted since the lathe can be set up independently at each corner and be brought to the correct alignment. An excellent starting point for setting up the installation is to level the bed, both longitudinally and laterally. Indeed, if a sufficiently accurate spirit level is available, say, reading to 1 or 2 minutes of arc, this can be used alone to complete the alignment, basically by confirming that tightening of the holding-down bolts does not distort the bed and thus affect the level.

Alignment by levelling

A sufficiently accurate level is required for this method to be successful, and to achieve the 1- or 2-minute capability a good quality engineer's level is necessary. This will most probably have its accuracy specified in terms of its capability to indicate an out-of-level condition in terms of 1/1000in. per foot (or ⅒mm per 300mm). As a guide, a good level, capable of indicating a tilt of 0.12mm in 300mm, by movement of the bubble of one division on its scale (a typical commercial specification) is measuring a displacement of approximately 1.3 minutes of arc. This equates with .048in. per foot (.047mm per 300mm).

If the level of the bed is to be adjusted, the lathe must be provided with jacking or levelling screws at each corner of the bed. Initially, the lower nuts which support the bed must be adjusted so that the bed is level longitudinally (headstock-to-tailstock) and is free from rock. The upper, or locking, nuts should be loose at this stage.

The next adjustment is to establish a level condition at the headstock end, across the bed, adjusting at the tailstock jacks to maintain the 'not rocking' condition and then adjusting at the tailstock end also for a front-to-back level. Once this condition is achieved, the locking nuts (above the lathe feet) should be gently tightened, monitoring the bed with the level as this is done.

The front-to-back (across the bed) levelling must then be repeated at the headstock and tailstock ends, and the locking nuts tightened a little further, the process afterwards being repeated, gradually bringing the bed to the level condition with both the jacking and locking nuts fully tightened.

Alignment using a DTI

The alignment of the path of the tool with the longitudinal axis of the work can be judged by using a dial test indicator (DTI), mounted to the toolpost, in conjunction with a stout bar some 6in. or 8in. (150mm or 200mm) long. The size does not matter, but you must be sure the bar is straight.

Before proceeding with the test, the lathe should be levelled both longitudinally and laterally. This is just to provide a starting point and a really accurate level is not required for this since the DTI, in conjunction with the test bar, is used to judge the alignment. The bed must initially be set up on its jacks, however, and the locking nuts tightened moderately, before starting the final test.

For this, the (straight) test bar must be held in the three-jaw chuck with some 5in. or 6in. (125mm or 150mm) protruding towards the tailstock end. The DTI must be mounted in the toolpost with the saddle positioned so that it can monitor the test bar close to the chuck. Since the three-jaw will not hold the test bar absolutely truly, it is first necessary to rotate chuck and bar, by hand, to determine the high and low points of rotation and then to rotate the chuck to the centre position as shown by the DTI.

Without moving the chuck, the saddle is traversed 4in. (100mm) along the bed to check that the DTI reading is the same. If it is not, there may be a need for adjustment of the lathe mounting. If the work leans only a little towards the operator, say, .0003in. to .0005in. in 4in. (.0075mm to .0125mm in 100mm) there is no need to adjust the lathe mounting since this amount of lean is allowable to counter any tendency for the work to deflect away from the tool when the cut is taken.

When making the final adjustments to the jacks, it is best to make these only at the tailstock end, assuming that the lathe has already been levelled across the bed. The headstock end, being more robust is less likely to distort in practice, but in any event, the heavier of the two ends ought to be used as the reference against which to judge the position of the other.

The 'no strain' method

A variation of the above method may be employed by using the DTI to detect distortion of the bed directly. If the lathe is first brought to a level condition, along the bed and transversely at the headstock end, and a stout bar held in the chuck as before, a DTI clamped to the toolpost and monitoring the bar 4in. (100mm) away from the chuck can measure any distortion which occurs in the bed as the lock nuts or jacking nuts are tightened.

Starting with the lathe level across the bed at the headstock end, and the locking nuts tight at this end, the locking nuts at the tailstock end should be tightened and the effect of this monitored on the DTI. If tightening a lock nut causes distortion, this indicates that adjustment is required in the associated jacking screw which should be repositioned as required and the lock nut afterwards tightened. Adjustment should be performed only at the tailstock end since the headstock has already been levelled and is best used as the reference, as noted above.

Turning a test piece

In Chapter 15, a description of the method of setting up the tailstock for the parallel turning of between-centres work was given. This involves the use of a test bar, having its centre reduced in diameter to leave two short, large-diameter portions, one at each end. The bar is mounted between centres and the larger diameters are turned, using a sharp tool, to confirm that the tailstock is adjusted correctly for parallel turning, this being indicated by equality of the two diameters.

A similar method may be used for checking lathe alignment, except that the support of the tailstock is not used since the test checks the ability of the lathe to turn work parallel when it is mounted to the mandrel. The largest diameter bar that can be inserted inside the chuck body is therefore mounted into the three-jaw with about four times its diameter protruding. This is roughly turned away at its centre, and close to the chuck, to leave two large-diameter portions, as described, and these are then turned away sufficiently to remove any run-out and produce true surfaces. A really sharp tool is then put in the toolpost and a cut of .005in. (0.125mm) is taken over both diameters, using a fine, automatic feed (if available). Measurement of the diameters then indicates whether the bed is twisted, a smaller diameter at the tailstock end indicating that the rear of the tailstock end needs to be jacked up.

A slight increase in size towards the tailstock end is again allowable, to compensate for the tendency of work which is not supported by the tailstock centre to deflect away from the tool when the cut is taken. This should not exceed .0003in. to .0005in. in 4in., as described above.

Index